McGraw-Hill Ryerson

Mathematics of Data Management 12

Study Guide

AUTHOR

David Petro
B.Sc. (Hons.), B.Ed., M.Sc.
Windsor Essex Catholic
District School Board

REVIEWERS

Dr. Steven J. Desjardins
Department of Mathematics and Statistics
University of Ottawa

Carol Miron
Toronto District School board

Larry Romano
Toronto Catholic
District School Board

Toronto Montréal Boston Burr Ridge, IL Dubuque, IA Madison, WI New York
San Francisco St. Louis Bangkok Bogotá Caracas Kuala Lumpur Lisbon London
Madrid Mexico City Milan New Delhi Santiago Seoul Singapore Sydney Taipei

McGraw-Hill Ryerson
Mathematics of Data Management 12 Study Guide

ISBN-13: 978-0-07-090880-2
ISBN-10: 0-07-090880-X

http://www.mcgrawhill.ca

1 2 3 4 5 6 7 8 9 0 MP 1 9 8 7 6 5 4 3 2 1 0

Printed and bound in Canada

PUBLISHER: Kristi Clark
PROJECT MANAGER: Maggie Cheverie
DEVELOPMENTAL EDITOR: Paul McNulty
COPY EDITOR: John Green
ANSWER CHECKER: Carol Miron
MANAGER, EDITORIAL SERVICES: Crystal Shortt
SUPERVISING EDITORS: Shannon Martin, Alexandra Savage-Ferr
EDITORIAL ASSISTANT: Erin Hartley
MANAGER, PRODUCTION SERVICES: Yolanda Pigden
PRODUCTION COORDINATOR: Jennifer Hall
COVER DESIGN: Vaild Design/Michelle Losier
ELECTRONIC PAGE MAKE-UP: Laserwords
COVER IMAGE: John Warden/Getty Images/Stone

Contents

Note: Titles shown in grey are sections of the Mathematics of Data Management student edition that cover topics that were omitted when the Ontario MDM4U curriculum was revised in 2007. They are not now required for this course.

Overview

Data management plays an important role in many activities. Collection and analysis of data are fundamental in business and economics, and vital in the social, medical, and physical sciences. *McGraw-Hill Ryerson Mathematics of Data Management 12 Study Guide* is designed for students planning to qualify for college or university. The study guide is designed to complement the *McGraw-Hill Ryerson Mathematics of Data Management* student edition and to supplement that text book to help students with the small changes made in the 2007 Ontario MDM4U curriculum revision. This study guide may also be used as a review of the course.

Study Guide Organization

- This study guide mirrors the student edition, providing supplementary material for the sections that are required for the MDM4U course for each of chapters 1 to 8.
- The Mathematics of Data Management course (MDM4U) requires a culminating project. Chapter 9 of the student edition is designed to help students prepare their project. In this study guide, some case studies are presented to give the reader an opportunity to consider what might be expected in their project.

Study Guide Features

- Each section begins with Key Concepts. This feature provides a point-form summary of the concepts covered in the section of the student edition. In some sections, particularly those where the revised curriculum has a somewhat different emphasis, worked examples supplement those provided in the student edition.
- Exercises are organized into three sections: A (practise), B (apply, solve, communicate), and C (extend and challenge).
- Selected questions in each section are marked by a star that indicates that full worked solutions are provided at the back of this book. Answers to all other questions are also provided.
- A practice exam at the end of the study guide gives you the opportunity to determine whether you are ready for the final examination.

Technology

- Computers and calculators make handling large amounts of data easier. Many sections of this resource include examples and exercises involving data files that can be downloaded from *www.mcgrawhill.ca/MDM12*. Most files are posted in both Excel® and Fathom™ format. Fathom™ is dynamic statistical software that is licensed for school use in Ontario. For those who do not have this software or Internet access, all sections do include some exercises that stand alone.
- Use of a graphing calculator is strongly recommended, particularly for the later chapters, chapters 4 to 8.

Formulas

Statistics of One Variable

	Population	Sample	Weighted Mean
Mean:	$\mu = \frac{\sum x}{N}$	$\overline{x} = \frac{\sum x}{n}$	$\overline{x}_w = \frac{\sum w_i x_i}{\sum w_i}$
Variance:	$\sigma^2 = \frac{\sum (x-\mu)^2}{N}$	$s^2 = \frac{\sum (x-\overline{x})^2}{n-1}$	
Standard Deviation:	$\sigma = \sqrt{\frac{\sum (x-\mu)^2}{N}}$	$s = \sqrt{\frac{\sum (x-\overline{x})^2}{n-1}}$ or $\sqrt{\frac{n\sum x^2 - (\sum x^2)}{n(n-1)}}$	
Z-score:	$z = \frac{x-\mu}{\sigma}$	$z = \frac{x-\overline{x}}{s}$	
Grouped Data:	$\mu \doteq \frac{\sum f_i m_i}{\sum f_i}$	$\overline{x} \doteq \frac{\sum f_i m_i}{\sum f_i}$ where m_i is midpoint of ith interval	
	$\sigma \doteq \sqrt{\frac{\sum f_i (m_i-\mu)^2}{N}}$	$s \doteq \sqrt{\frac{\sum f_i (m_i-\overline{x})^2}{n-1}}$	

Statistics of Two Variables

Correlation Coefficient

$$r = \frac{s_{XY}}{s_X s_Y}$$

$$= \frac{n\sum xy - \sum x \sum y}{\sqrt{n\sum x^2 - \left(\sum x\right)^2}\ \sqrt{n\sum y^2 - \left(\sum y\right)^2}}$$

Least Squares Line of Best Fit

$$y = ax + b, \text{ where } a = \frac{n\sum xy - \sum x \sum y}{n\sum x^2 - \left(\sum x\right)^2}$$

$$\text{and } b = \overline{y} - a\overline{x}$$

Permutations and Organized Counting

Factorial: $n! = n(n-1)(n-2)(n-3) \ldots (3)(2)(1)$

Permutations

r objects from n different objects: ${}_nP_r = \frac{n!}{(n-r)!}$

n objects with some alike: $\frac{n!}{a!b!c!\ldots}$

Combinations and the Binomial Theorem

Combinations

r objects chosen from n different objects: ${}_nC_r = \dfrac{n!}{(n-r)!\,r!}$

at least one item chosen from n distinct tems: $2^n - 1$

at least one item chosen from several different sets of identical items: $(p + 1)(q + 1)(r + 1) \ldots - 1$

Pascal's Formula: ${}_nC_r = {}_{n-1}C_{r-1} + {}_{n-1}C_r$

Binomial Theorem: $(a + b)^n = \sum_{r=0}^{n} {}_nC_r a^{n-r} b^r$

Introduction to Probability

Equally Likely Outcomes: $P(A) = \dfrac{n(A)}{n(S)}$

Complement of A: $P(A') = 1 - P(A)$

Conditional Probability: $P(A \mid B) = \dfrac{P(A \text{ and } B)}{P(B)}$

Independent Events: $P(A \text{ and } B) = P(A) \times P(B)$

Dependent Events: $P(A \text{ and } B) = P(A) \times P(B \mid A)$

Mutually Exclusive Events: $P(A \text{ or } B) = P(A) + P(B)$

Non-Mutually Exclusive Events: $P(A \text{ or } B) = P(A) + P(B) - P(A \text{ and } B)$

Discrete Probability Distributions

Expectation: $E(x) = \sum_{i=1}^{n} x_i P(x_i)$

Discrete Uniform Distribution: $P(x) = \dfrac{1}{n}$

Binomial Distribution: $P(x) = {}_nC_x p^x q^{n-x}$ $\quad E(x) = np$

Geometric Distribution: $P(x) = q^x p$ $\quad E(x) = \dfrac{q}{p}$

Hypergeometric Distribution: $P(x) = \dfrac{{}_aC_x \times {}_{N-a}C_{r-x}}{{}_NC_r}$ $\quad E(x) = \dfrac{ra}{N}$

Continuous Probability Distributions

Exponential Distribution: $y = ke^{-kx}$, where $k = \dfrac{1}{\mu}$

Normal Distribution: $y = \dfrac{1}{\sigma\sqrt{2\pi}} e^{-\frac{1}{2}\left(\frac{x-\mu}{\sigma}\right)^2}$

Normal Approximation to Binomial Distribution: $\mu = np$ and $\sigma = \sqrt{npq}$ if $np > 5$ and $nq > 5$

Normal Approximation to Hypergeometric Distribution: $\mu = np$ and $\sigma = \sqrt{npq\left(\dfrac{N-n}{N-1}\right)}$

Distribution of Sample Means: $\mu_{\bar{x}} = \mu$ and $\sigma_{\bar{x}} = \dfrac{\sigma}{\sqrt{n}}$

Confidence Intervals:

For normally distributed data:

$$\bar{x} - z_{\frac{\alpha}{2}}\left(\frac{\sigma}{\sqrt{n}}\right) < \mu < \bar{x} + z_{\frac{\alpha}{2}}\left(\frac{\sigma}{\sqrt{n}}\right)$$

For a sample, where $\hat{p}$ is the proportion found in the sample:

$$\hat{p} - z_{\frac{\alpha}{2}}\left(\sqrt{\frac{\hat{p}\hat{q}}{n}}\right) < p < \hat{p} + z_{\frac{\alpha}{2}}\left(\sqrt{\frac{\hat{p}\hat{q}}{n}}\right)$$

Margin of Error: $E = z_{\frac{\alpha}{2}}\left(\dfrac{\sigma}{\sqrt{n}}\right)$

1.2 Data Management Software

KEY CONCEPTS

- Thousands of computer programs are available for managing data. These programs range from general-purpose software, such as word processors and spreadsheets, to highly specialized applications for specific types of data.

- A spreadsheet is a software application that is used to enter, display, and manipulate data in rows and columns. Spreadsheet formulas perform calculations based on values or formulas contained in other cells.

- Spreadsheets normally use relative cell referencing, which automatically adjusts cell references whenever cells are copied, moved, or sorted. Absolute cell referencing keeps formula references exactly as written. This is ideal for “what if?” scenarios.

	A	B	C	D	E
1	Month	Temp. (°F)	Deviation From Mean	Absolute Deviation	Squared Deviation
2	Jan	39	-20.25	20.25	410.06
3	Feb	42	-17.25	17.25	297.56
4	Mar	50	-9.25	9.25	85.56
5	Apr	[illegible]	.25	0.25	0.06
6	May	[illegible]	.75	7.75	60.06
7	Jun	[illegible]	.75	14.75	217.56
8	Jul	[illegible]	.75	18.75	351.56
9	Aug	[illegible]	.75	17.75	315.06
10	Sep	71	11.75	11.75	6
11	Oct	60	=B11-B15	0.75	6
12	Nov	51	-8.25	8.25	6
13	Dec	43	-16.25	16.25	264.06
14	sum	711			=SUM(E2:E13)
15	mean	59.25			200.75
16				SD Using Table	14.17
17				SS Using Formula	14.17

An example with relative cell referencing (B11) and absolute cell referencing (B15). The $ is an indication of absolute referencing.

An example with relative cell referencing only.

- Spreadsheets can produce a wide variety of charts and perform sophisticated sorts and searches. You can add worksheets to a file and reference these sheets in cells in another sheet.

- Non-numerical data can be sorted by attribute easily with software such as Fathom™. For example, the relative sizes of various groups can be shown graphically using a breakdown plot or a bar chart.

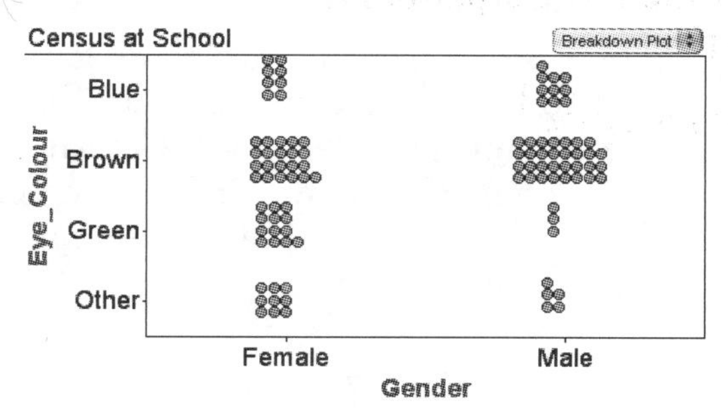

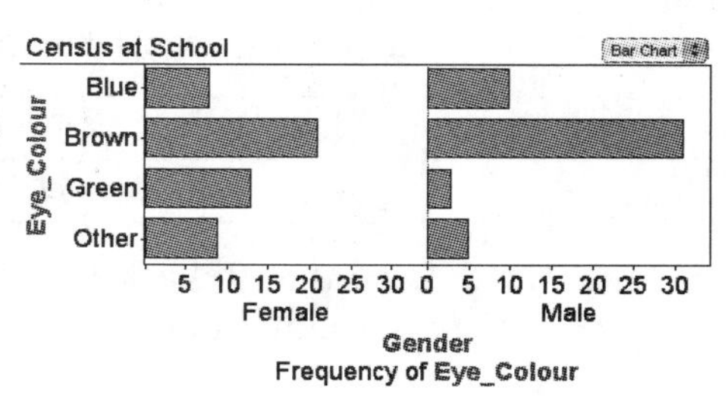

- Several histograms can be created simultaneously on the same graph.

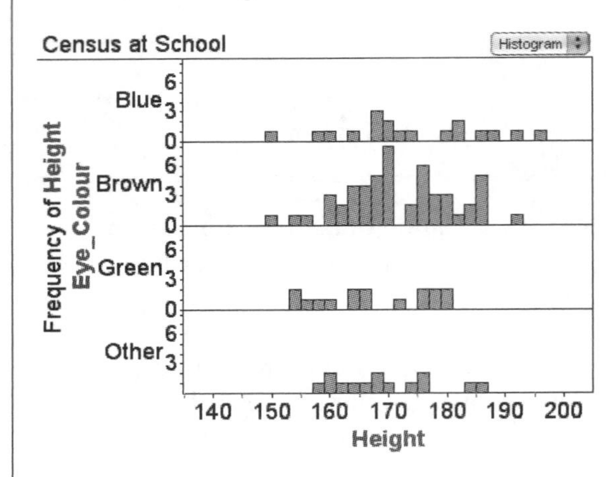

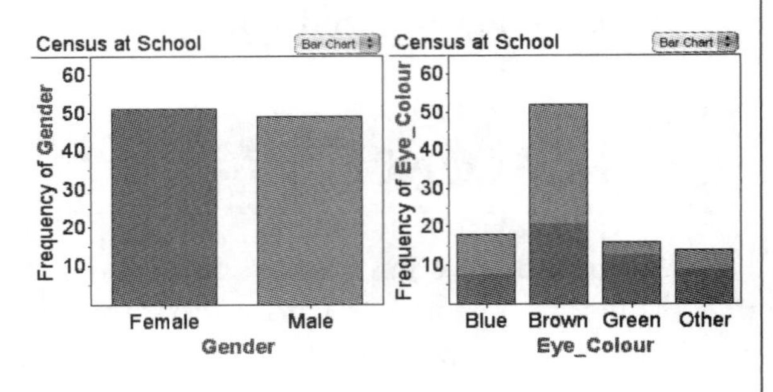

- Several graphs can be linked dynamically by clicking in any bar in any graph. In the gender graph, the bar for females is selected. In the eye-colour graph, the frequency among females is highlighted at the bottom of each bar.

- Software can be used to analyse data found on the Internet. Data in tables can often be imported directly into software applications. For example, from the sample web page shown, agricultural data can be imported into a spreadsheet by copying and pasting the table, or it can be imported into Fathom™ by dragging the URL.

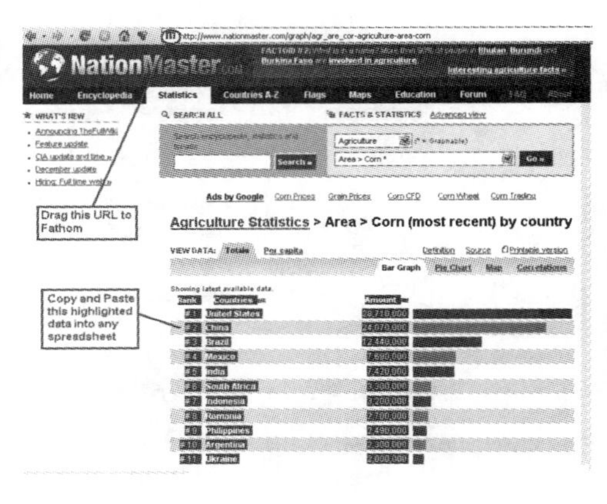

A

1. **Create and Format a Spreadsheet** The table shows sales, in dollars, at five hotdog stands during the first hour of a ball game. Enter the data into a spreadsheet.

Stand #	1	2	3	4	5
Sales ($)	37.5	18	31.5	46.5	54

a) Format the data so that dollars are expressed to two decimal places.

b) Use a function to determine the mean of the data.

c) Use a function to determine the median of the data.

d) Convert the data from row data to column data by using the **Copy** command and the **Paste Special** command with the **Transpose** check box selected. Sort the data from greatest to least.

2. **Work With Multiple Data Sets** A group of students measure and record their heights and arm spans. Enter the data shown in the table into a spreadsheet.

Name	Height (cm)	Arm Span (cm)
Sunita	160	156
Joan	165	163.5
Manuel	165	168
Ty	170	164
Clair Marie	171	177
Charles	163	165
Brittany	175	179.5

a) Find the mean of each numerical column.

b) Format each numerical column to have one decimal place.

c) Sort the data by

i) name

ii) height

iii) arm span

☆3. **Use Fathom™**

a) Repeat question 2 using Fathom™. To get help when working with Fathom™, from the **Help** menu choose **Fathom Movies**. You may also review the introduction to Fathom™ on pages 24 to 27 of *Mathematics of Data Management*, Student Edition.

b) Show the units for the numerical columns by right-clicking in the table, choosing **Show Units**, and typing "cm" into the space provided.

B

4. **Use Multiple Worksheets** Go to *www.mcgrawhill.ca/links/MDM12* and select **Study Guide**. Follow the links to obtain a copy of the file **Autosales.xls**.

a) Determine the totals for each column in the division sheets using the **Sum** function.

b) Reference the **Totals** from each division on the **Totals** sheet.

c) On the **Totals** sheet, determine the grand total for each column.

d) Create a chart to display the Totals data.

5. **Relative and Absolute Cell Referencing** Enter the following data into a spreadsheet.

◇	A	B	C	D	E
1	Commission Rate:		15%		
2	Item	Price	Quantity	Total Value	Commission
3	TV	$899	3		
4	Camera	$250	6		
5	Laptop	$1,200	2		
6			Totals		

a) Use relative cell referencing to calculate the **Total Value** of the television sales in cell D3.

b) Use absolute cell referencing to calculate the **Commission** for television sales in cell E3.

c) Use **Copy** and **Paste** or **Fill** to calculate the **Total Value** and **Commission** for cameras and laptops.

d) Determine the **Totals** for each column. Which type of referencing did you use?

6. Go to *www.mcgrawhill.ca/links/MDM12* and select **Study Guide**. Follow the links to obtain a copy of the file **Magazines.xls**. In this data set are the price, total number of pages, number of ad pages, and type of magazine for several magazines. Use a filtered search to

a) show only business/news magazines

b) determine how many magazines have more than 100 pages

c) show only magazines that cost between $4 and $5, inclusive

★7. **Use Fathom™**

a) Repeat question 6 using the file **Magazines.ftm**.

b) Create a **Summary Table** and determine

i) the total number of health/fashion magazines

ii) the average cost of science/ technology magazines

8. Go to *www.mcgrawhill.ca/links/MDM12* and select **Study Guide**. Follow the links to the NationMaster web site. Under **Facts and Statistics**, from the **Select Category** list choose **Media**. Then, from the **Select Media Stat** list, choose **Cinemas**. Click **Go**.

a) Copy these data into a spreadsheet.

b) Drag the URL into Fathom™.

c) In each program, calculate the average number of cinemas per country.

9. **Use Fathom™**
Go to *www.mcgrawhill.ca/links/MDM12* and select **Study Guide**. Follow the links to obtain a copy of the file **OntarioYouths.ftm**. This is data gathered from the National Longitudinal Survey of Children and Youth.

a) Drag a blank graph down to the workspace. Double-click the **Ontario Youths** collection and drag the **Sex** attribute to the horizontal axis of the graph. How many males and how many females took this survey? Hint: Hold your mouse over each bar and look at the bottom left corner of the Fathom™ window.

b) Drag a new blank graph down to the Fathom™ workspace. Drag the **Like_Math** attribute to the horizontal axis and drag the **Like_English** attribute to the vertical axis. Drag the items on the horizontal axis so that "I like it a lot" is nearest to the origin and "I hate it" is farthest from the origin. Which is the largest group?

c) Click on the **M** bar on the **Sex** graph. What seems to be true about youths who like math a lot but hate English?

10. Using a spreadsheet, create a table of values that calculates the week-by-week total cost of renting a guitar for one year if there is a one-time fixed fee of \$20 and a rental charge of \$7.50 per week. If the cost of a new guitar is \$150, when does the rental option no longer make sense?

11. Use a spreadsheet to create a cell phone cost table similar to the table shown. When you have created the table, change any of **Monthly Service Charge, Per Minute Rate**, or **Data Cost** to recalculate the entire spreadsheet.

	A	B	C	D	E
1		Monthly Service Charge:		20	dollars/month
2			Per Minute Rate:	15	cents/minute
3			Data Cost:	5	cents/kb
4					
5	Minutes\Data	0	100	500	1000
6	0	\$20.00	\$25.00	\$45.00	\$70.00
7	20	\$23.00	\$28.00	\$48.00	\$73.00
8	40	\$26.00	\$31.00	\$51.00	\$76.00
9	60	\$29.00	\$34.00	\$54.00	\$79.00
10	80	\$32.00	\$37.00	\$57.00	\$82.00
11	100	\$35.00	\$40.00	\$60.00	\$85.00
12	120	\$38.00	\$43.00	\$63.00	\$88.00

C

12. The Fibonacci numbers follow the pattern 1, 1, 2, 3, 5, 8, 13, … , where any number is the sum of the previous two numbers. Create a spreadsheet that generates the first 100 Fibonacci numbers and then calculates the ratio of successive pairs of values.

KEY CONCEPTS

- A database is an organized collection of related records. A well-organized database can be easily accessed through searches on multiple fields that are cross-referenced.
- Although most databases are computerized, many are available in print form.
- Metadata is data about data. Most computer files have metadata associated with them, such as information about file sizes and dates when files were created or modified.
- Global Positioning System (GPS) units and geographic information systems (GISs) are databases of geographical information.

A

1. Go to *www.mcgrawhill.ca/links/MDM12* and select **Study Guide**. Follow the links to obtain a copy of the file **OntarioYouths.ftm**.

a) Describe what the database represents.

b) How many people is there information about in this database? In Fathom™, this is called the number of cases.

c) For each case, how many different pieces of information are there? In Fathom™, these pieces of information are called attributes.

d) What do the attributes of the database provide information about?

2. Do you know that each MP3 audio file is part of a database? Open up a program that plays music on a computer. What information in addition to song name and artist is stored in each MP3 file?

3. List three sources of metadata.

4. Open up any word-processing document that you have. From the **File** menu choose **Properties**. What type of metadata exists in your document?

B

5. Go to *www.mcgrawhill.ca/links/MDM12* and select **Study Guide**. Follow the links to obtain a copy of the file **Library.ftm**. This is an example of an MP3 music library.

a) Double-click **Collection**. List the attributes for each music file.

b) Drag an empty graph onto the workspace. Drag the **Genre** attribute from the case table to the horizontal axis of the graph. Arrange the values on the horizontal axis from greatest to least by dragging the **Genre** names. Which type of music is most prevalent in this library?

c) Based on these data, which artist do you think is the favourite of the person who created the library?

6. Explain how the Internet is a database.

7. Go to *www.statcan.gc.ca/estat/licence-eng.htm* and click **Accept and enter** to access the E-STAT site.

a) Click **Table of contents - Data Tables**.

i) Under the **Economy** topic, click **Information and communications technology**.

ii) Under the **CANSIM** theme, click **Telecommunications industries**.

iii) From the list of tables, click **354-0006**, the table for Internet service providers, to go to the **Subset selection** page.

iv) In the **Summary Statistics** list, select both **Operating revenue** and **Operating expenses**. Hint: To select more than one item in a list, hold down the **Ctrl** key and click all required items.

v) In the **Reference period** box, change the **From** year to 1997 and click **Retrieve as individual time series** to reach the **Output specification** page.

vi) Click **Retrieve now** at the bottom of the page. Based on the resulting graph, when did the telecommunications industry make a profit?

b) Go back to the **Output specification** page. What other type of graph can you choose in order to convey the same information as in step vi) of part a)?

c) On the **Output specification** page, under **SCREEN OUTPUT: Table**, select **Plain text: Table, time as rows**. Click **Retrieve now**. Copy and paste the resulting table into a spreadsheet. Create a double bar graph similar to the one produced if you select **Horizontal bar chart** on the **Output specification** page.

d) How is having the tabular data more informative than having only the graph? How is having the graph more informative than having only the table?

e) Go back to the **Subset selection** page. What other selection can you make in the **Summary statistics** box to give you the same information as in step vi) of part a)?

f) What type of metadata exists for each of these graphs?

g) On the **Subset selection** page, click **latest article from *The Daily***. *The Daily* is a report produced by Statistics Canada each day that focuses on recent data. How can this report help to describe why Internet service providers have done so well lately?

8. Go to the **Welcome to E-STAT!** page.

a) On the **E-STAT** menu on the left, click **Search CANSIM in E-STAT**.

i) **Type cancer incidence** in the search field and click **Search**.

ii) In the resulting list of tables, click table **103-0204** (July 2007 CCR file).

iii) Select the following options:

Geography: Canada
Sex: Males
Selected sites of cancer: Lung cancer
Characteristics: Cancer incidence
Reference period: 1976–2004

iv) Click **Retrieve as individual time series** at the bottom of the page. On the **Output specification** page, click **Retrieve Now** to create the graph. Describe how the incidence of lung cancer changed during the years reported.

b) As accurately as possible, determine the maximum incidence rate and when it occurred. Hint: hold the mouse pointer over any part of the line where there would be a point.

c) Go back to the **Output specification** page. Under **SCREEN OUTPUT: Table**, select **Plain text: Table, time as rows** and click **Retrieve now**. Copy and paste these data into Fathom™. Note: You will not be able to drag the URL onto the Fathom™ workspace. Instead, while in Fathom™, create a new collection, right-click it, and paste the data there. Create a scatter plot of these data by dragging the time attribute to the horizontal axis and the incidence rate attribute to the vertical axis.

9. In addition to E-STAT, Statistics Canada has other useful database features on its public web site. One such feature is the **Maps and geography** section, where recent census data is dynamically integrated with geographic maps of Canada.

a) Go to the Statistics Canada web site at *http://www.statcan.gc.ca/start-debut-eng.html.*

i) In the **Browse by** area, click the **Key resource** tab. Under **Maps and other geographical tools**, click **Maps and geography**.

ii) On the **Maps** tab, under **Interactive maps**, click **Example of interactive map**.

iii) Zoom into southern Ontario by clicking in southern Ontario in the map. On the **Layers(1)** tab, select **Census Metropolitan Areas / Census Agglomerations**. Hint: you may have to zoom several times before the **Census Metropolitan Areas / Census Agglomerations** option becomes available.

iv) Under **Map tools** on the left, select **Identify**. On the map, click in the **London** area on the map. What is London's population and how many private dwellings does it have?

b) Click the **Thematic maps** tab under the map. Under **Choose profile**, select **% of allophones** and click **Create thematic map**. What percent of Londoners have a mother tongue other than English or French? Name a nearby community that has a lower percent of allophones than London has. Scroll around the map and identify an area that has a higher percent of allophones than London has.

c) What sorts of people might be interested in the data available on this site? Explain.

☆ 10. A geographic information system (GIS) incorporates numerical data and metadata into interactive maps. Many communities have a GIS department that is available online for their residents. One such instance is found on the Town of Essex web site.

a) Go to *www.mcgrawhill.ca/links/MDM12* and select **Study Guide**. Follow the links to the Town of Essex web site.

i) Hover the mouse pointer over the **Residents** button. From the submenu, choose **GIS**.

ii) Under **Town of Essex Web Based Mapping**, click the link that says **clicking here**. On the next page, click **Launch Map Viewer (Public Site)**.

iii) A new window should appear. On it, click **I Accept**.

iv) You should now have a map showing the boundaries of the Town of Essex. On the right is the active **Map Layers** menu. To see how this works, open the **Water** folder and select **Hydrant**. How can this display tell you where the population centres are?

b) Zoom into the Town of Harrow. Its location is shown on the map below.

With the **Map Layers** tools, use the **Landmarks & Public Services** folder to determine the number of **Municipal Parks** in Harrow.

c) Click the **Identify** icon ❶ next to **Schools** to make this layer active. Find the name of each of the schools shown by clicking the school symbols on the map. Hint: your Web browser options must be set to allow pop-up windows.

d) Open the **Airphoto** folder and select the **2001 Airphoto** map layer. Just to the northwest of Harrow you will see three square water reservoirs. Use the airphotos from different years to determine approximately when a fourth reservoir was created.

e) Use the **Measure Area** tool found in the toolbar at the top of the page to determine the area of one of the reservoirs.

f) Use examples from the layers seen to explain how this tool can be useful for area residents. How can the tool help a local public-works department?

C

11. Search your own community's web site and determine whether your community has a GIS system. What information can you discover about your own neighbourhood that you did not know before?

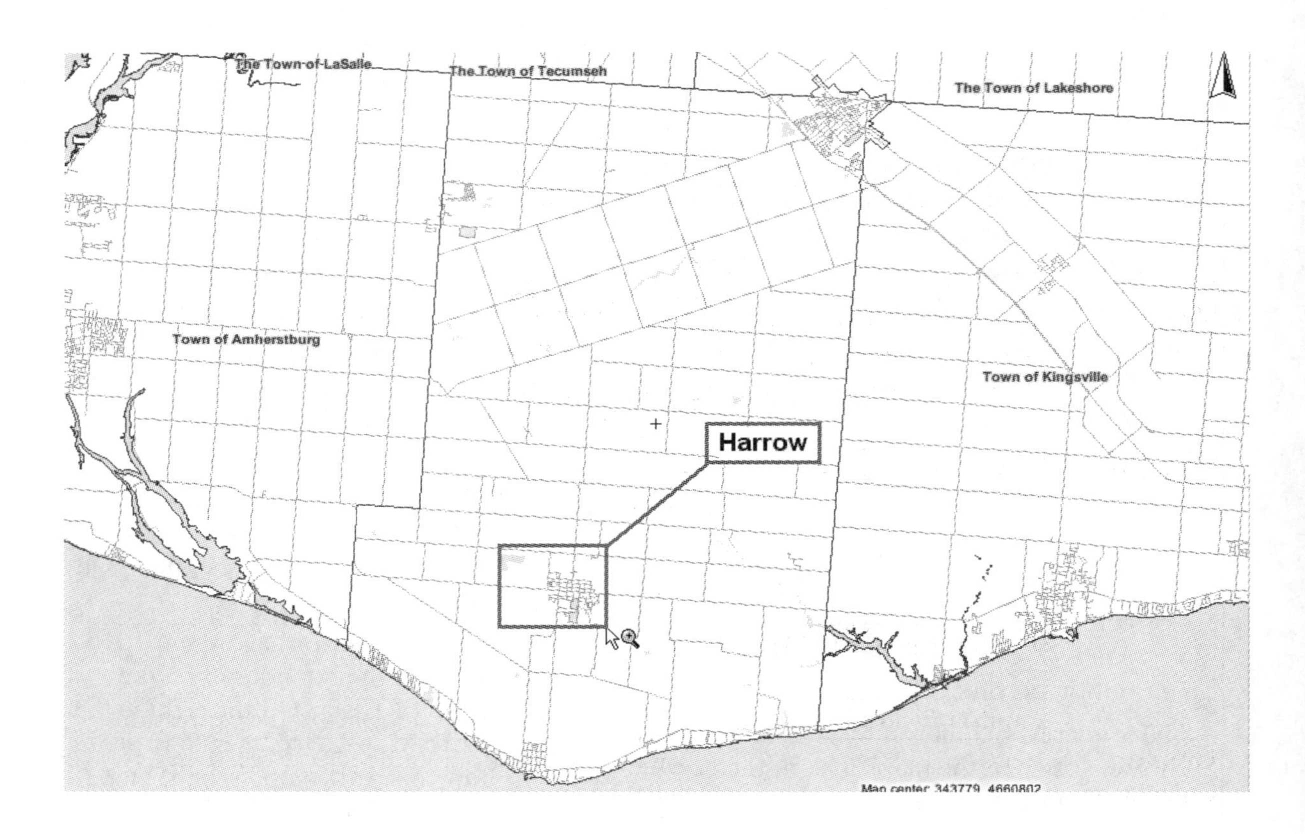

KEY CONCEPTS

- Simulations can be useful tools for estimating quantities that are difficult to calculate and for verifying theoretical calculations.
- A variety of simulation methods are available, ranging from simple manual models to advanced technology that makes large-scale simulations feasible.
- Coins, dice, and cards are all common tools that can be used to construct simulations of various situations.
- When simulations are done, in order for their results to be valid, a large number of trials must be completed.
- When generating random numbers, the following functions may be useful.

Microsoft® Excel	RAND()	Returns a random number greater than or equal to 0 but less than 1. To generate a random number between 0 and k, multiply RAND() by k.
	RAND()*(b–a)+a	Returns a random number greater than or equal to a but less than b.
	INT(number)	Round any number down to the nearest integer.
	COUNTIF(range, criteria)	Counts the number of occurrences of the criteria in a given range of values.
Corel® Quattro Pro®	@RAND	Returns a real number between 0 and 1.
	@RANDBETWEEN(N,M)	Returns a random number between N and M.
	@INT(number)	Removes the decimals from any number.
	@COUNTIF(range, criteria)	Counts the number of occurrences of the criteria in a given range of values.
Fathom™	random(N,M)	Returns a random number between N and M. If only one argument is supplied, it returns a random number between 0 and that number. If no arguments are supplied, it returns a random number between 0 and 1.
	randomInteger(N,M)	Returns a random integer between and including N and M.
	randomPick(a,b,c …)	Returns one of the elements in the expression.
Graphing Calculator	rand	Returns a random number between 0 and 1. To generate a random number between 0 and k, multiply rand by k.
	randInt(a,b,c)	Returns a random integer between a and b, inclusive, for c trials.

A

1. **Use a Graphing Calculator**
   a) Generate a random number between 0 and 26.
   b) Generate 20 random integers between 25 and 50.

2. **Use a Spreadsheet**
   a) Generate a set of 10 random numbers between 0 and 20.
   b) Generate a set of 100 random integers between 10 and 20.
   c) For the set generated in part b), create a graph to show how many there are of each value. What would you expect this graph to look like?

☆ 3. **Use Fathom™**
   a) Generate 30 random numbers between –10 and 10.
   b) Generate 30 random integers between –10 and 10.
   c) Use the randomPick function to generate 100 random picks of the colours red, green, and blue.
   d) Create a graph to show how many times each colour was selected in part c).

B

4. Driving simulators and flight simulators are often included as part of computer games. Some games have options to turn off the simulation features. Explain why someone might want to do that.

5. Describe how you could develop a non-computer-based simulation that would randomly select
   a) a day of the week
   b) a day of the month
   c) a day of the year
   d) a random date of birth, during the years 1950 to 2010

6. Repeat part d) of question 5 using
   a) a spreadsheet **b)** Fathom™

7. Describe a situation that you could simulate using a bag containing three black marbles and seven red marbles.

8. A student forgot to study for a 20-question true-or-false test and so answers questions by flipping a coin—tails for true and heads for false. The teacher watches the student from beginning to end and is surprised when the student starts the test over and begins to flip the coin again. When asked why, the student proudly states, "I'm checking my answers." What other method could the student have used to choose the answers?

9. CAPTCHA is a type of test often used to verify that it is a human filling in a form on a web page and not a computer. When making a purchase on a secure web site, you might be asked to type letters or phrases that are hard to read due to distortion. In most cases, a human can interpret these letters while a computer cannot.

   kbpsh 3m573 Vzpk2Z

   Create a simulation to generate a five-letter word that could be used as a CAPTCHA.

10. The infinite monkey theorem states that if you had an infinite number of monkeys randomly typing at an infinite number of keyboards, eventually one of them would type out the entire works of Shakespeare. Create a bag of letters that spell out the phrase *to be or not to be*. Draw letters one at a time from the bag and write them down in order. Each time you draw a letter, return it to the bag before drawing again.
    a) How many times did you have to draw a letter before you spelled a recognizable three-letter word? How many times would you have to do this experiment to get a reasonably accurate average number of draws?

b) How many times would you have to draw to spell the word *not*? Would you expect this average number to be greater than or less than the answer to part a)? Explain your reasoning.

c) How would you expect the average number of draws to compare if the three-letter word is *too* instead of *not*?

d) What do you expect happens to the number of times you have to draw if you are looking to find a four-letter word instead of a three-letter word? Explain.

To view an applet that simulates this, go to *www.mcgrawhill.ca/links/MDM12,* select **Study Guide**, and follow the links.

11. Create a simulation to represent a baseball player at bat if the player normally gets a hit 20% of the time. Run the simulation 20 times. How do the results of the simulation match the actual hitting percent? Explain why this might happen.

☆ **12.** A classic prize simulation is the cereal box problem. Each time you buy a box of cereal, you have a chance to get one of six different prizes.

a) Create a simulation to represent buying a box and having a one-in-six chance of winning any particular prize. Determine how many boxes you have to buy to get all six prizes. Complete this simulation 20 times. On average, how many boxes do you need to buy to collect all six prizes?

b) Use the Fathom™ file **Cereal.ftm** to verify your results.

c) Compare the methods in parts a) and b).

13. Go to *www.mcgrawhill.ca/links/MDM12* and select **Study Guide**. Follow the links to obtain a copy of the file **Coin.ftm**. Use the following steps to simulate flipping a coin 100 times.

a) Drag the corner of the collection open to see the two pieces of data.

b) Right-click the collection and choose **Sample Cases** from the menu.

c) A new collection appears called **Sample of Coins**. To represent flipping a coin 100 times, change the number of cases to 100. Drag the collection window open enough to see the **Sample More Cases** button and click it. (You may wish to clear the **Animation On** check box to speed up this process).

d) Select the **Sample of Coins** collection and drag a new table to the workspace. Now drag a new graph to the workspace. Drag the **Face** attribute to the horizontal axis of the graph. Click the **Sample More Cases** button (if you do not see this, drag the corner of the **Sample of Coins** collection open). As you continue to click this button, comment on how often the number of tails equals the number of heads.

e) To keep track of the results of several trials of flipping a coin 100 times, double-click the **Sample of Coins** collection to open the **Collection Inspector**. Click the **Measures** tab. Where it says **<new>**, type **FlipResult**. Double-click the **Formula** box, type the following formula, and click **OK**:

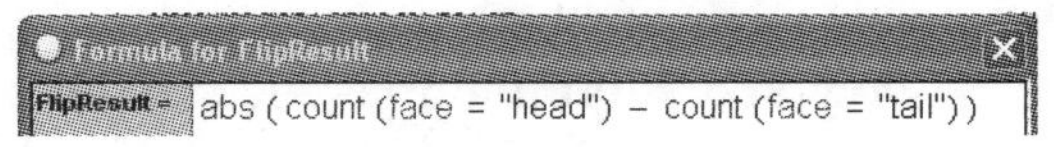

f) Right-click the **Sample of Coins** collection and choose **Collect Measures**. A new collection called **Measures from Sample of Coins** is created. Select it and create a new table. This table should have the results of 5 trials of flipping a coin 100 times. To change this to 20 trials, double-click **Measures from Sample of Coins** to open the inspector (if it is not open already) and under the **Collect Measures** tab, change 5 to 20. Select **Replace Existing Cases** and click **Collect More Measures**.

g) Create a new graph and drag the **FlipResult** attribute to the horizontal axis. What does each dot represent? How often are the numbers of heads and tails equal? What is the greatest difference in values? Click the **Collect More Measures** button to get several results.

☆ **14.** Go to *www.mcgrawhill.ca/links/MDM12* and select **Study Guide**. Follow the links to obtain a copy of the file **Dice.ftm**.

a) Follow the procedure in question 13 to simulate rolling a single die 100 times and displaying the number of times each face comes up.

b) To create a simulation to represent rolling two dice, complete the following:

i) Right-click the **Single Die** collection and choose **Sample Cases**. Under the **Sample** tab, in the **Sample of Single Die** inspector, change the number of samples to 2. This represents rolling two dice. Click the **Sample More Cases** button. Drag the corner of the **Sample of Single Die** collection to see the result of the "roll."

ii) To get the value of the total of the two dice, click the **Measures** tab and where it says **<new>**, type **Total**. Double-click in the **Formula** box and type **sum(face)**.

iii) Right-click the **Sample of Single Die** collection and choose **Collect Measures**. This will create a collection that will hold the result of 5 rolls of two dice. Change this to roll the dice 20 times instead. In this case, clear the **Replace existing cases** check box so you can continue to collect data.

iv) Select the **Measures from Sample of Single Die** collection and drag a new table to the workspace. Drag a graph to the workspace, and from the table, drag the **Total** attribute to the horizontal axis of the graph. What does each dot represent? Do all possible sums come up? What happens if you collect the **Measures** until there are more than 100 dots?

c) A furniture company advertises a promotion. How do the results of part b) explain why the company assigned rolls to particular dollar amounts?

Leo's Furniture
Roll the dice and save! With every purchase you get a chance to save by rolling a pair of dice. Here's what you can save.
* Roll 5, 6, 7, 8, or 9 and save $50 *
* Roll 3, 4, 10, or 11 and save $100 *
* Roll 2 or 12 and save $200 *

15. The local weather forecast predicts a 70% chance of rain tomorrow. Create a simulation to model this situation. Describe any limitations your simulation method has.

C

16. Investigate what pseudo-random numbers are and how they relate to computers.

☆ **17.** Go to *www.mcgrawhill.ca/links/MDM12* and select **Study Guide**. Follow the links to obtain a copy of the file **Deck.ftm**. Use the file to create a simulation that draws two cards from a deck to see how often Black Jack (i.e., a total of 21) comes up when you play 100 games.

18. Use the probability simulator in the **Apps** section of a graphing calculator to

a) flip a coin 1000 times

b) represent the probability that a dart thrower will win if he needs to throw a 4

Chapter 2 Statistics of One Variable

2.1 Data Analysis With Graphs

KEY CONCEPTS

- Data exist in many forms. These forms can be organized in the following way:

  Numerical (quantitative) data can be summarized with mean and median.

Discrete data can have only certain values, with gaps in between. The number of members of a family is an example of discrete data. A bar graph can be used to represent discrete data.	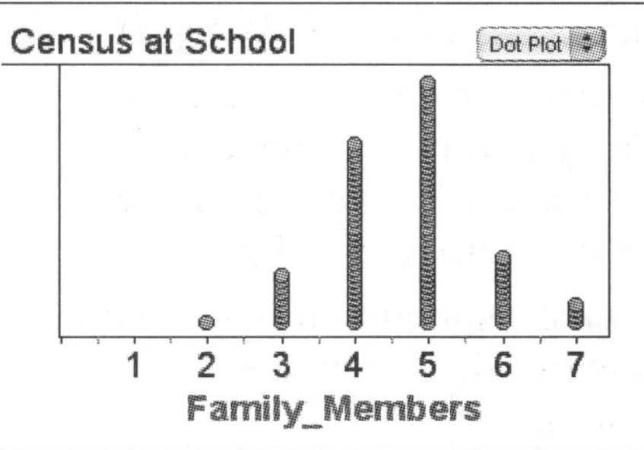
Continuous data can have any value within a range. The height of a person is an example of continuous data. A histogram can be used to graphically represent continuous data.	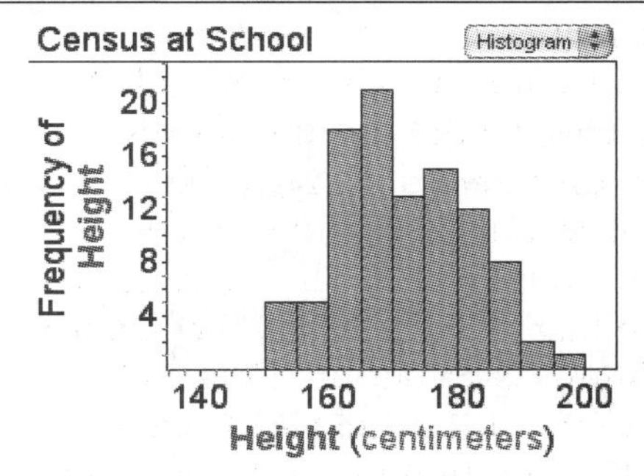

Categorical (qualitative) data can be put into distinct groups and summarized as totals or percents.

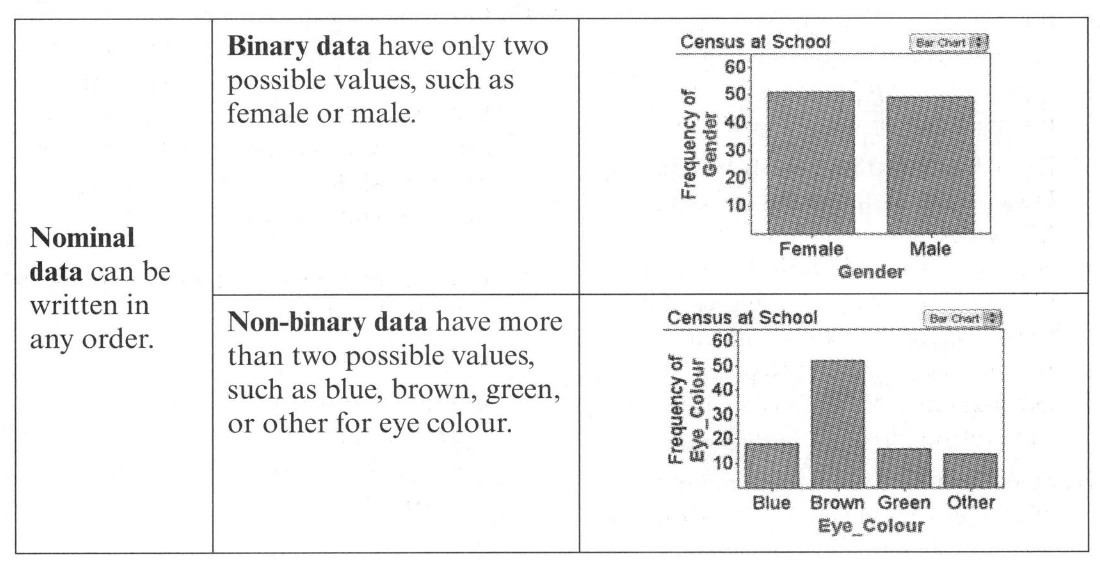

Nominal data can be written in any order.	**Binary data** have only two possible values, such as female or male.	
	Non-binary data have more than two possible values, such as blue, brown, green, or other for eye colour.	

Ordinal data make sense being ordered in a specific way.	**Non-binary data** have more than two possible values, such as a lot, some, very little, or none.	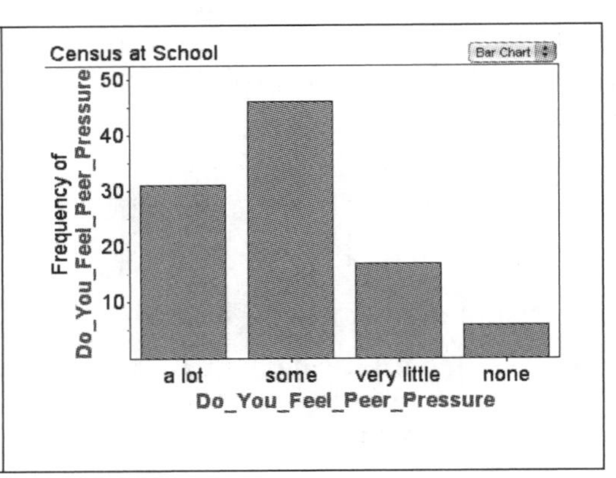

- Continuous numerical data are often displayed using frequency tables. Data are usually grouped in classes or intervals when the number of measured values is large. The size of these classes or intervals is called the bin width. Generally, it is convenient to use from 5 to 20 equal intervals that cover the entire range of the variable.
- A cumulative-frequency diagram shows the running total of frequencies from the least interval up.

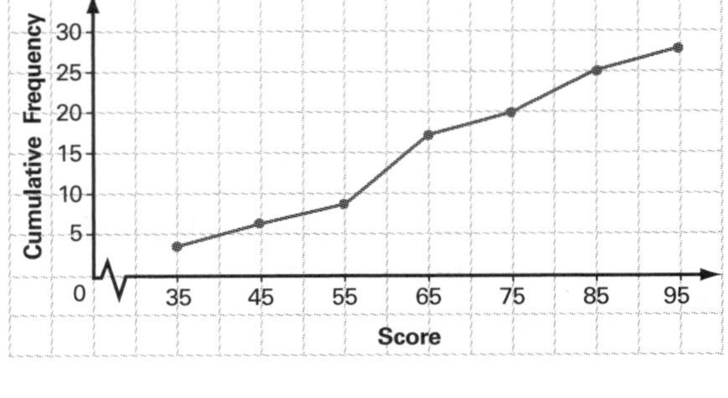

- Relative-frequency diagrams show the frequency of each interval as a proportion of the entire data set.
- Categorical data can be presented in various forms, including bar graphs, circle graphs (pie charts), and pictographs.

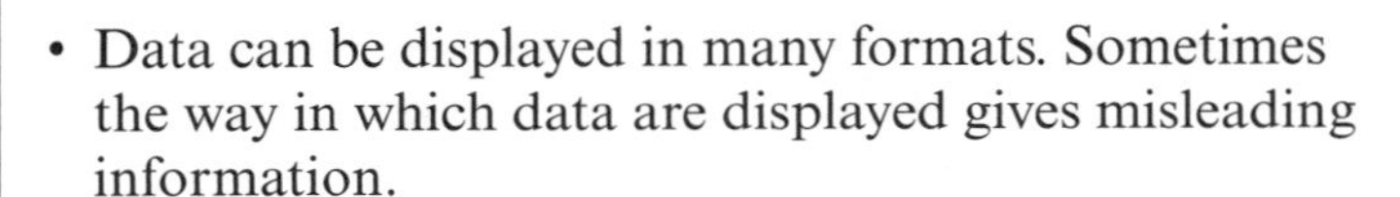

- Data can be displayed in many formats. Sometimes the way in which data are displayed gives misleading information.

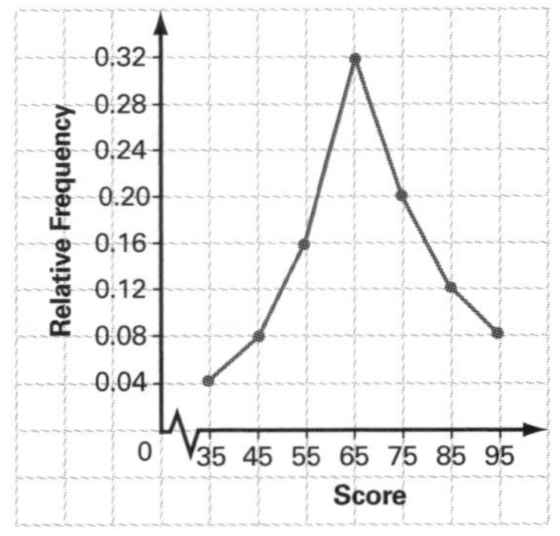

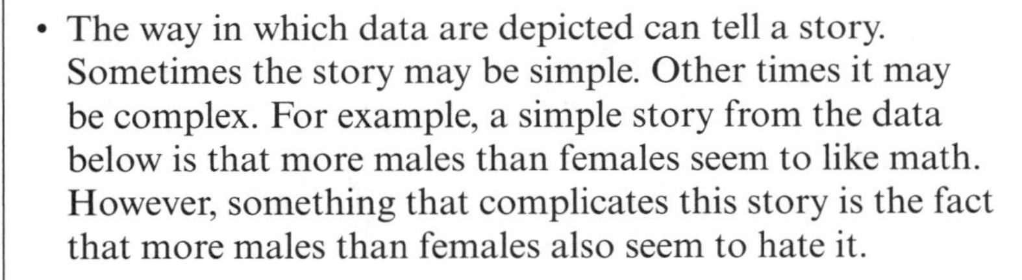

- The way in which data are depicted can tell a story. Sometimes the story may be simple. Other times it may be complex. For example, a simple story from the data below is that more males than females seem to like math. However, something that complicates this story is the fact that more males than females also seem to hate it.

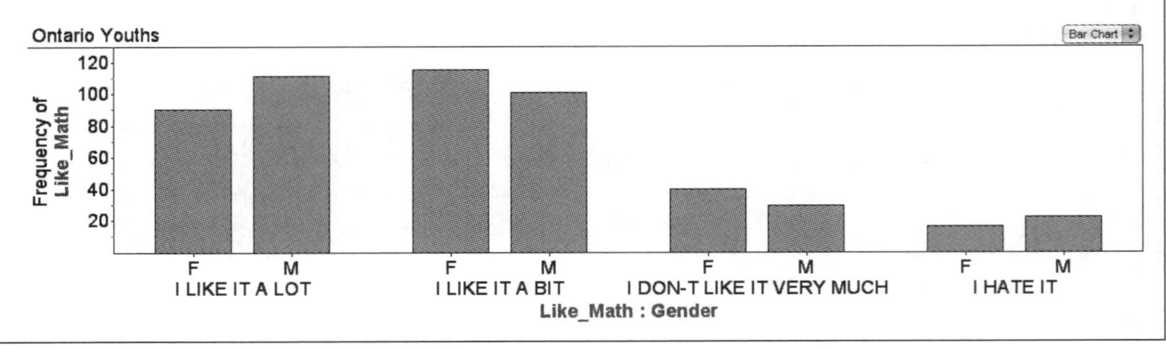

A

1. In each case, identify the type of data using the descriptors in the Key Concepts chart. Be as specific as possible.

   a) 

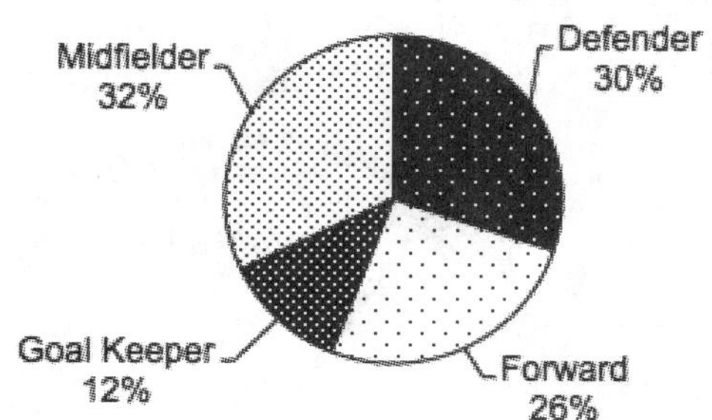

   b) 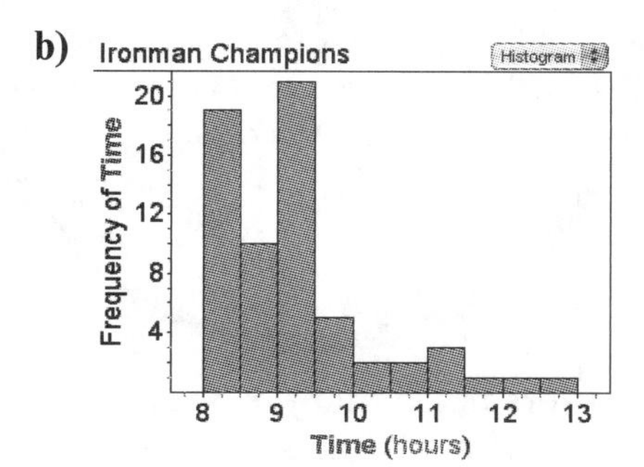

   c) 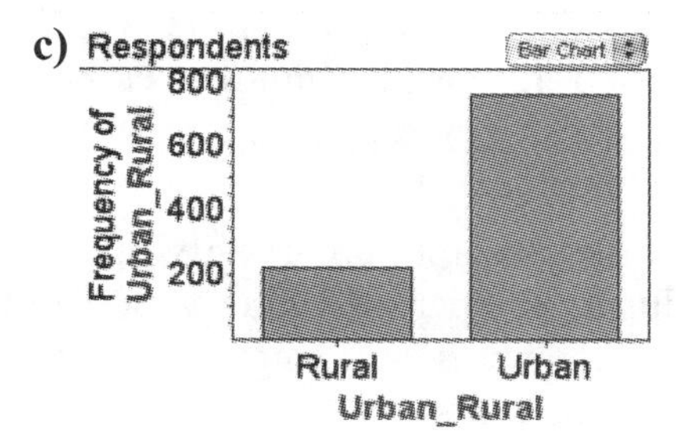

   d) 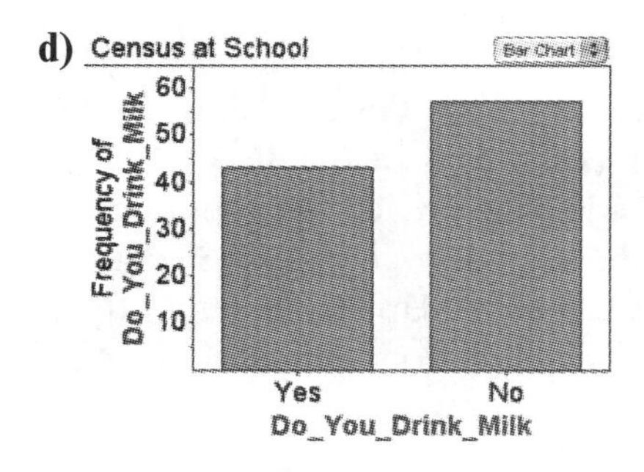

2. For each graph in question 1, what is a simple story that the data tell?

3. Use examples to describe the difference between a histogram and a bar graph.

4. This figure is an example of a dot plot. Describe what a dot plot represents.

5. This figure is an example of a histogram.

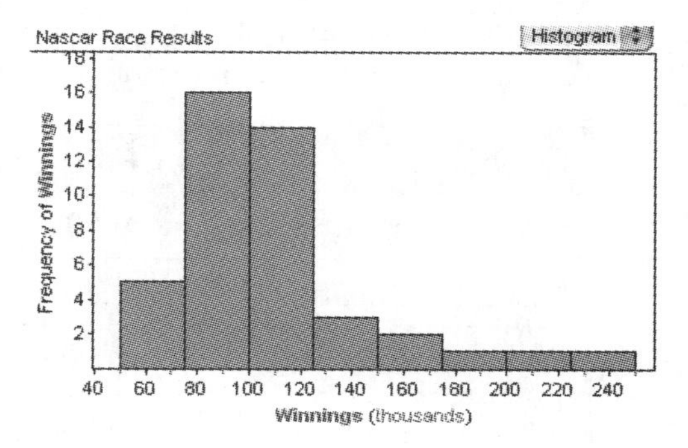

   a) What do the data appear to be about?

   b) What information does each bar tell you?

   c) Which interval has the greatest number of data values in it?

   d) How could you use this graph to approximate the total prize money for this competition?

   e) It is said that a picture is worth a thousand words. What story does this "mathematical picture" tell?

6. For each range of data, identify an appropriate bin width and state the first few intervals.

   a) 24 − 115

   b) 115 − 524

   c) 21 304 − 28 950

   d) 1.3 − 2.5

7. A frequency table has intervals 0 − 10, 10–20, 20–30, etc. In which intervals would you put the following data? Explain.

   a) 26 **b)** 30 **c)** 39.9

8. Pictographs are often used in the media to show data.

   a) Why do you think media outlets might use a pictograph rather than a more standard graphing method?

   b) Create a pictograph of the following data about the cost of a movie.

Location	Cost ($)
Super Cinema	9.50
Discount Theatres	6.50
Red Carpet Rentals	16.00
Star Time Video	15.00
Netwatch	8.00
iRent	4.50

   c) Explain why a pie graph cannot be used for these data.

9. Describe a set of data to fit each type of data.

   a) nominal, binary

   b) nominal, non-binary

   c) ordinal, binary

   d) ordinal, non-binary

   e) numerical, continuous

   f) numerical, discrete

B

10. Describe what this graph says about Canada's provinces and territories. What do you think the size of each circle represents?

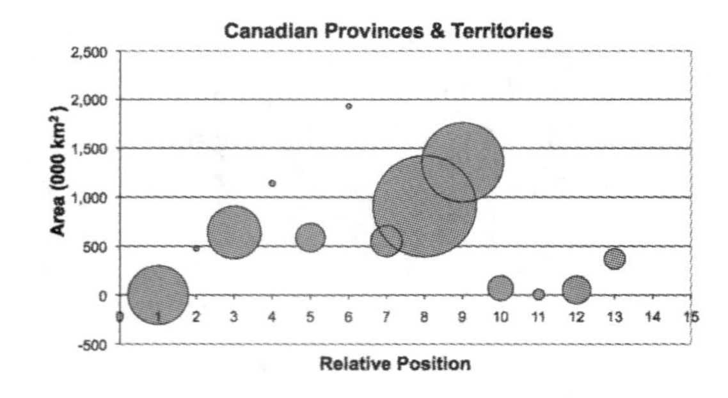

11. The graph shows obesity rates among Americans for certain years.

    a) How are these data depicted in a way that relates to the topic?

    b) In 2006 the rate of obesity among Americans was 30.6%. Where on the graph would that value be placed?

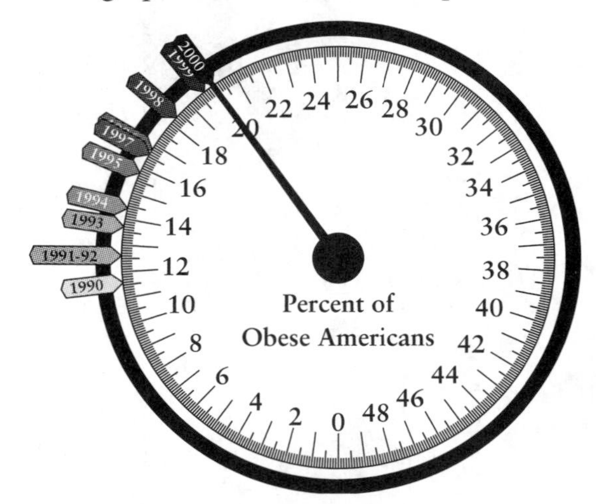

    c) It is predicted that the obesity rate in the United States in 2018 will be 43%. Plot this value on the graph. Based on this information, is the rate of obesity predicted to increase at a quicker rate? Explain.

    d) Look up Canadian obesity rates during the period 1990 to 2000 and summarize them using a different kind of graph.

12. Go to *www.mcgrawhill.ca/links/MDM12* and select **Study Guide**. Follow the links to obtain a copy of the file **DrinkingAge.csv**. This file contains data on the legal drinking age in 83 countries. Create a pictograph to represent the data. What kind of story could these data tell?

13. An article titled "Why You're Still Getting Gouged" appeared in the August 18, 2008, issue of *Maclean's* magazine. Some of the data reported in that article are shown.

Item	Cost ($ CAN)	Cost ($ US)
Toy microphone	14.99	10.99
Online movie purchase	19.99	14.99
Razor blade	10.99	8.99
Hair dryer	51.65	33.23
Perfume	80.00	63.00

a) Create a double bar graph of this data.

b) Based on the data, what do you think the *Maclean's* magazine article was about?

☆ **14.** The graph is from a pamphlet published in 2000 about health care spending in Ontario.

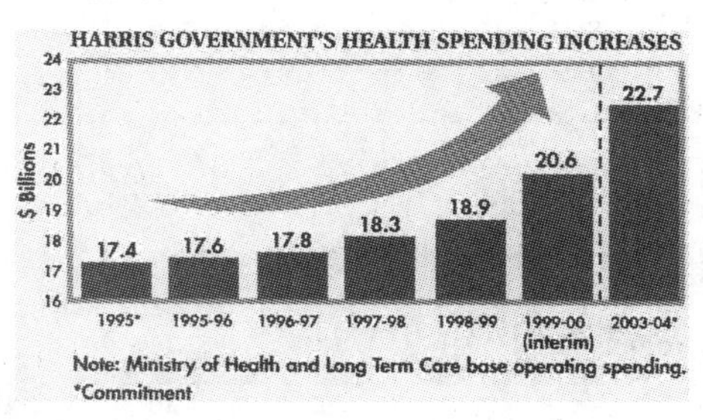

a) What story do you think the graph is trying to tell?

b) Describe ways in which the graph may be giving a false impression of the data.

c) What would be a more realistic representation of the data?

d) Does a more realistic representation tell a different story?

15. The magnitudes of all earthquakes that occurred during October 2009 and that had a magnitude of at least 6.0 are listed.

6.6 6.1 6.0 6.1 6.6 6.8 7.7 7.8
6.6 6.8 6.0 6.0 6.0 6.2 6.2 6.5
6.4 6.3 6.0 6.0 6.1 6.0 6.2 6.0
6.0 6.0 6.2 6.8 6.9 6.0 7.4

a) Create a frequency table with bin widths of 0.2. Include relative frequency.

b) Create a histogram of the data.

c) Which interval has the greatest percent of data values?

d) An earthquake with a magnitude greater than 7.0 is classed as severe. What percent of the earthquakes reported in the data were severe?

e) There are three severe earthquakes reported in the data. Each occurred on October 7, 2009. Use the Internet to determine where they were located.

f) Go to *www.mcgrawhill.ca/links/MDM12* and select **Study Guide.** Follow the links to obtain a copy of the file **Earthquake.ftm**. Create the same histogram as in part b) using Fathom™.

g) What other information about these earthquakes is included in the Fathom™ file?

h) Create a new graph. Drag the **Lon** attribute from the collection to the horizontal axis of the graph. Drag the **Lat** attribute to the vertical axis and drag the **Mag** attribute to the centre of the graph. What does this graph represent?

16. The length, in seconds, of each of the 50 songs on Lydia's MP3 player is shown.

193 285 424 342 418 210 224 388
250 301 265 510 240 342 279 265
352 204 313 383 176 304 262 301
303 442 254 143 504 293 202 344
270 302 158 290 291 297 167 199
339 376 229 269 288 504 270 170
319 152

a) Keeping in mind that these figures represent time, what is an appropriate interval, or bin width?

b) Create a frequency table with these data. Include relative frequency.

c) Create a histogram from the frequency and create a frequency polygon from the relative frequency.

d) Go to *www.mcgrawhill.ca/links/MDM12* and select **Study Guide**. Follow the links to obtain a copy of the file **LibrarySample.ftm**. Recreate your histogram using that file.

e) Go to *www.mcgrawhill.ca/links/MDM12* and select **Study Guide**. Follow the links to obtain a copy of the file **LibrarySample.csv**. Recreate your frequency polygon in a spreadsheet using that file.

f) What percent of the songs are more than 3 min in length?

17. Go to *www.mcgrawhill.ca/links/MDM12* and select **Study Guide**. Follow the links to obtain a copy of the file **Canada.ftm.**

a) Using the census data, create one graph for each of the data types described in the Key Concepts chart.

b) What story is told in each of the graphs you made in part a)?

c) Create a new histogram with **Total Income** on the horizontal axis and **Gender** on the vertical axis. What story does this graph tell about wages and gender?

C

★ **18.** Go to *www.mcgrawhill.ca/links/MDM12* and select **Study Guide**. Follow the links to obtain a copy of the file **Pennies.ftm**. This file represents the mass, year, and country of 200 pennies.

a) Create a dot plot of these data. What story do these data tell?

b) Drag the **Country** attribute to the vertical axis. How does this representation add to the story?

c) Create a second graph by making a histogram of the **Year** attribute. Use your mouse to highlight parts of the data in your first graph. How does this add to the story?

d) Drag the **Year** attribute to the centre of the first graph. How does this affect the graph? Search on the Internet for "minting Canadian pennies" to find reasons these data appear as they do.

19. The graph below represents the years of working experience of employees of a company that has more than 600 workers. What story could these data possibly tell?

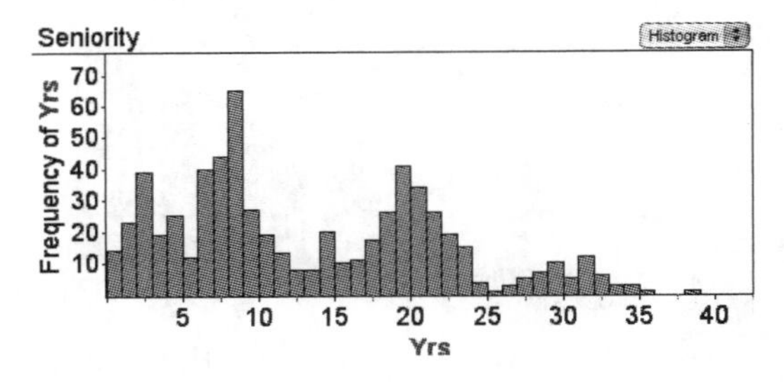

KEY CONCEPTS

- An index can summarize a set of data. Indices usually compare the values of a variable or group of variables to a base value. Indices have a wide variety of applications in business, economics, science, and other fields.

- The actual value of the index is often arbitrary. The index values, however, are often compared to a base value. Often the base value is set to be 100 for easy percent comparison.

- A time-series graph is a line graph that shows how a variable changes over time.

  This time-series graph shows changes in a consumer price index over time.

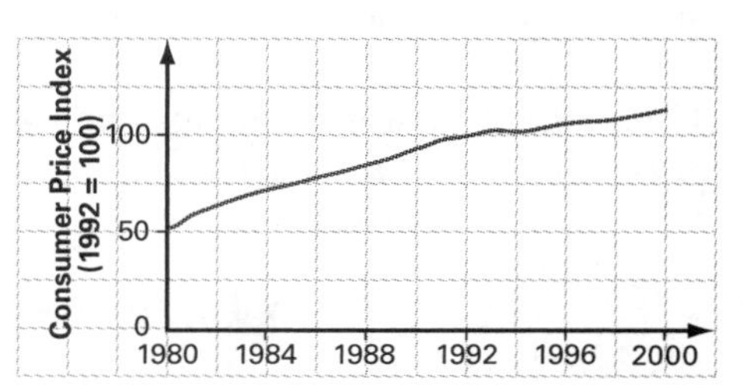

- Inflation is the general measure of the increase in prices over time. In a period of inflation, goods that cost \$100 today will likely cost more than \$100 in the future. For example, if you buy something today for \$50 and annual inflation is 3%, then the same item will cost \$51.50 (3% more) one year from now.
  One measure of inflation is the consumer price index (CPI). Economists calculate Canada's CPI by looking at the prices of more than 600 items representing typical household expenditures, such as food, shelter, clothing, furniture, transportation, and recreation. The cost of this "basket" of goods is tallied each month and prices are tracked by Statistics Canada.

A

1. Consider this graph of the CPI.

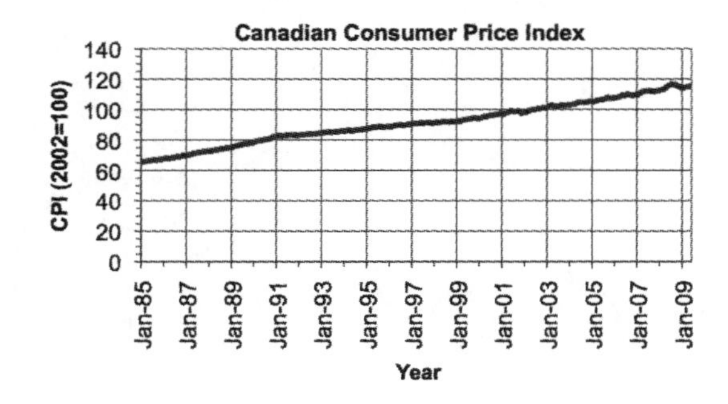

a) In which year is the base value?

b) By approximately how many points does the index rise in the period reported on the graph?

c) If a hair dryer cost \$24.50 in 2002, how much did it cost in 2009?

d) A sweater cost \$45.00 in 2009. How much would you expect the same sweater to have cost in 2002?

2. Name three indices commonly used in the media.

3. Determine the change in index points and the percent change for each case.

	Initial Value	New Value	Index Change	Percent Change
a)	100	130		
b)	100	67		
c)	245	300		
d)	150	100		

4. The graph shows the index for banana prices for the last 60 years.

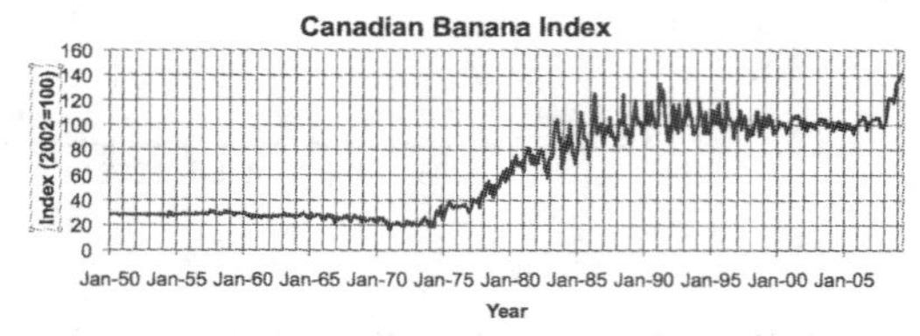

a) Describe the trend of banana prices during the time span of this graph

b) What happened to the price of bananas in the last two years that are reported?

c) Currently bananas cost about $1.30/kg. Based on the index, what would have been their approximate price at their lowest cost? When did bananas have this price?

B

5. The costs to attend home games of various hockey teams are shown in the table below.

a) Create an index to represent the cost for one adult to park, attend a game, and have a hot dog and soft drink.

	Cost Per Item ($)				
Team	Adult Ticket	Child Ticket	Parking	Soft Drink	Hot Dog
Bears	15	5	5	1.00	1.50
Tigers	18	2	0	1.75	3.00
Flash	19	0	2	1.25	1.50
Riders	9	2	10	2.00	3.50

b) Create an index to represent the cost for a family of four (two adults, two children) to attend a game, including the cost of parking and the cost of a hot dog and soft drink for each person.

c) What do parts a) and b) suggest about considering only ticket prices when comparing the cost of attending hockey games?

6. The body mass index (BMI) is a simple way to determine whether you are overweight. To calculate your BMI, divide your mass, in kilograms, by the square of your height, in metres. Using this index, you are considered overweight if your BMI is greater than 25. You are considered obese if your BMI is greater than 30. Normal weight is in the range from 18.5 to 24.9. Being underweight is associated with values less than 18.5.

a) Calculate the BMI and identify the applicable weight category for each person.

	Mass (kg)	Height (m)	BMI Category
i)	85	1.53	
ii)	51	1.78	
iii)	68	1.64	

b) What problems, if any, do you see with this index?

7. The table gives CPI information for Canada, the United States, and Japan.

Year	Canada	U.S.	Japan
1986	62.0	56.4	88.4
1987	64.0	58.2	88.7
1988	66.6	60.6	89.3
1989	69.9	63.5	91.4
1990	73.2	66.9	94.2
1991	77.4	69.7	97.2
1992	78.6	71.8	98.9
1993	80.0	74.0	100.1
1994	80.1	75.9	100.8
1995	81.8	78.0	100.7
1996	83.1	80.4	100.8
1997	84.5	82.2	102.6
1998	85.3	83.5	103.3
1999	86.8	85.6	103.0
2000	89.2	88.2	102.2
2001	91.4	90.7	101.4
2002	93.5	92.1	100.5
2003	96.0	94.2	100.3
2004	97.8	96.7	100.3
2005	100.0	100.0	100.0
2006	102.0	103.2	100.2
2007	104.2	106.2	100.3
2008	106.6	110.3	102.4

a) What is the base year for these data?

b) Calculate the average rate of change between 1998 and 2008 for each country's CPI.

c) Create a graph to compare the rates of inflation among the three countries. What story does this graph tell?

☆ **8.** The yearly average price for one litre of gasoline is shown in the table.

Year	Average Price (¢/L)	CPI
1990	55.7	80.2
1991	53.8	82.9
1992	52.4	83.8
1993	51.1	85.2
1994	49.8	86.4
1995	52.4	88.3
1996	56.1	89.6
1997	56.1	91.0
1998	51.6	91.8
1999	57.5	93.4
2000	70.8	95.9
2001	67.8	98.2
2002	67.3	100.0
2003	70.9	102.6
2004	76.6	104.4
2005	89.0	106.7
2006	93.4	109.1
2007	97.4	111.7
2008	110.2	115.0

a) What is the percent increase in the CPI from 1990 to 2008?

b) One thing that the CPI can tell is how much more, or how much less, could be purchased each year with the same amount of money. Use your answer from part a) to determine the price, in 2008 dollars, of gasoline in 1990.

c) Go to *www.mcgrawhill.ca/links/MDM12* and select **Study Guide**. Follow the links to obtain a copy of the file **GasPrices.csv**. For each year, calculate the price of gasoline in 2008 dollars. In what year was the price of gasoline actually the least?

d) What story do these data seem to tell?

9. In 2009, the New York Yankees moved to a new baseball stadium that cost \$1.6 billion to build. In contrast, Wrigley Field in Chicago cost \$250 000 to build in 1914. Without adjusting for inflation, it is hard to compare these costs. Go to *www.mcgrawhill.ca/links/MDM12* and select Study Guide. Follow the links to obtain a copy of the file **BaseballStadium.csv.** The file contains information about the cost of building major league baseball stadiums and the U.S. CPI. Use the data to determine which stadium was the most expensive to build and which was the least expensive.

10. The Dow Jones Industrial Average (DJIA) is a common index used to describe the state of financial markets.

a) Describe what the DJIA actually represents.

b) In 2008, the global economy plunged into a well-publicized financial crisis. Below are the historical data for the DJIA from 1930 to the present. How could this graph have been used to predict the crisis?

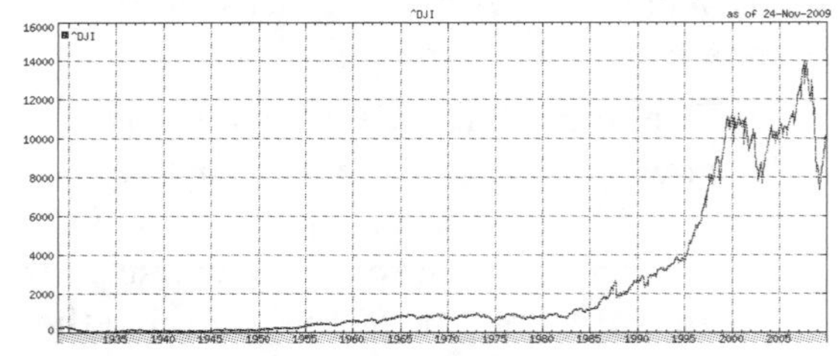

c) Evaluate the reasonableness of the method used.

C

11. A student makes the following graph based on the data from question 7 and states, "In 2005, Canada, the United States, and Japan had the same rate of inflation."

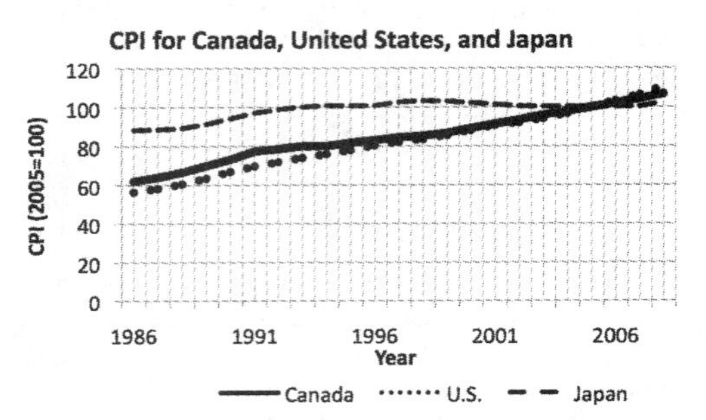

How can the following graph help explain why the student's statement is incorrect?

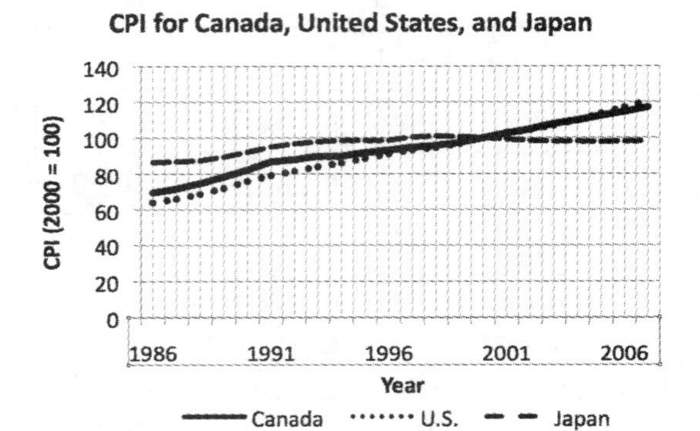

12. *This Is Spinal Tap* is a mockumentary about a fictional heavy metal band. In one scene, a guitarist is showing off the band's custom amplifiers. He is bragging because rather than having controls that go up to 10, their amplifiers have controls that go to 11. "That's one louder, isn't it?" he says. Do you think that the amplifiers are actually louder? How could knowledge of indices explain any misconception that the guitarist has?

KEY CONCEPTS

- Data that you collect yourself for a particular purpose are called primary data. Data that are collected by others and reused by you for a different purpose are called secondary data.

- Data about an individual person, household, or business are often referred to as microdata. These data are usually taken directly from an experiment or survey and often are reported collectively as aggregate data. Aggregate data are sometimes called summary data and are often given in grouped format.

- Experimental data are typically collected when the groups involved are assigned criteria, and the effects of those criteria are recorded. For example, a medical study to test a new headache medicine organizes participants into three groups. One group gets the drug, one is given a placebo (a dummy drug with no active ingredients), and the remaining group receives nothing. Researchers then look to see if one group has more headaches or fewer headaches than the others.

- Observational data are typically collected when the groups are determined by their characteristics. For example, if a medical study wants to investigate whether men live longer than women, the groups are organized by the gender of the participants.

- When studying a group of people or information, the entire group is referred to as the population. However, when collecting data, it is often unreasonable to collect information from everyone or everything involved. For this reason, a carefully selected group is chosen instead. This is called a sample. Sampling is used to collect some data in the census of Canada.

- In order for any sample to accurately reflect the group, the sampling method must be such that every person or thing in the population has an equal chance of being selected for the sample. Otherwise, a bias can be introduced.

- There are several types of random sampling methods that give each person or thing the same opportunity to be selected. Choosing the appropriate sampling method depends on a number of factors, such as the nature of the population, cost, convenience, and reliability.

A

1. Classify the types of data in each case. Use data descriptors such as primary, secondary, micro, and aggregate. Be as specific as possible.

a)

Name	Eye Colour	Hair Colour
Joan	Green	Blonde
Sanjay	Brown	Brown
Ray	Blue	Brown

b)

Age Group	Percent of Adults Who Smoke
20–30	28
30–40	32
40–50	33

c)

Trial Number	Length of Frog Leg (cm)
1	8.3
2	7.8
3	9.5

2. Use examples to describe the difference between a population and a sample.

3. Identify the sampling method in each case, using descriptors such as simple random, systematic, stratified, multi-stage, voluntary-response, cluster, and convenience.

a) Your favourite web site asks whether you are in favour of banning cell phones in schools.

b) You write the names of all the students in your class on slips of paper, put the slips into a hat, and draw five of them.

c) You randomly select three schools. From each school, you select one class from each grade. Then, you randomly select ten students from each class.

d) The population of your town is 45% female and 55% male, so you randomly select 45 women and 55 men to survey.

e) The government randomly selects 20 towns and then surveys everyone in each town.

f) You walk into a mall and someone asks you to fill out a questionnaire.

g) Your principal gets the alphabetical list of students and then chooses every tenth student.

☆4. In your school you have the following demographics:

Grade	Gender	Number of Students
9	Male	180
	Female	120
10	Male	125
	Female	125
11	Male	50
	Female	100
12	Male	70
	Female	30

Determine how you could obtain the following samples of 100 students.

a) a systematic sample

b) a stratified sample by gender

c) a stratified sample by grade

5. Explain why someone would conduct research and collect data that are

a) experimental

b) observational

B

6. Put 15 Canadian pennies and 5 U.S. pennies into a bag.

a) For $n = 4$, 8, and 16, conduct the following experiment ten times for each value of n and record your results:

Randomly select n pennies, one at a time, replacing each penny after selecting it and recording each selection as either a U.S. penny or a Canadian penny.

b) Based on your results, on average, how many pennies do you need to select before your sample has proportions similar to the population in the bag?

7. Go to *www.mcgrawhill.ca/links/MDM12* and select **Study Guide**. Follow the links to obtain a copy of the file **500Students.ftm.** This is a selection of information about 500 students who filled out a Census at School survey. There are three complete graphs and three empty graphs in the file.

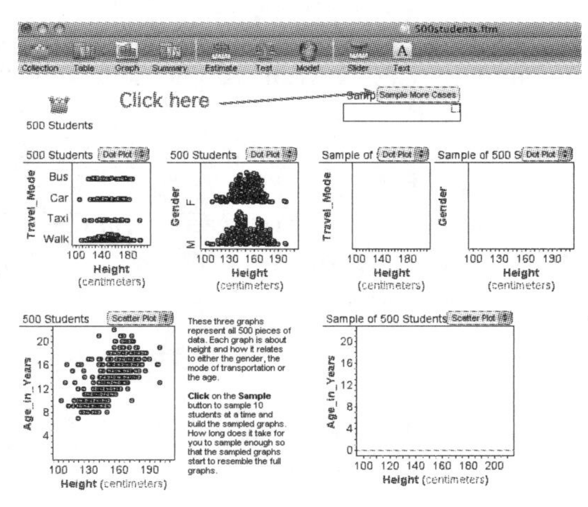

The purpose of this question is to determine how many clicks (i.e. how much data) are needed for the sample to reasonably represent the entire population. When you click the **Sample More Cases** button, Fathom™ randomly selects ten cases from the original set of data, places them into the empty collection, and creates a graph. Continue to sample cases and record the sample size when you reach the point where you have sampled enough that you are confident that your sample represents the original data.

a) Approximately how many cases need to be sampled so that the graph of height versus gender is a reasonable sample of the original set of data?

b) Approximately how many cases need to be sampled so that the graph of height versus travel mode is a reasonable sample of the original set of data?

c) Approximately how many cases need to be sampled so that the graph of height versus age is a reasonable sample of the original set of data?

d) In general, what do your answers to parts a), b), and c) say about how large a sample needs to be to accurately represent the original set of data?

8. A farmer has 1285 bushels of soybeans to sell. The beans were removed from their pods using an automated combine. However, the process is not perfect and occasionally pods get mixed with the soybeans. The bushels are sold by weight, but before the load is weighed, several samples are taken from random places in the load to determine the actual proportion of the load that consists of beans.

a) Why do you think that random samples are taken from several places rather than just taking one bushel of soybeans all at once? Use mathematical reasoning.

b) If a total of one bushel is sampled and 9.9 kg of pods are found, what total weight of pods is expected to be found if the entire load is 47 720 kg?

☆**9.** A sample of 2000 responses is taken from the 2001 Canadian census. Here are the numbers of people from each province or territory who are in that sample.

Area	Number in Sample	Population (000s)
Canada	2 000	29 639.0
NL	33	506.2
PE	5	132.3
NS	48	899.4
NB	51	720.8
QC	463	7 125.6
ON	773	11 285.5
MB	73	1 104.1
SK	75	944.4
AB	199	2 959.5
BC	273	3 868.9
YK, NT, NU	7	92.3

a) Create a bar graph that shows the proportion of the sample that is from each province or territory.

b) Because this is a sample of original census data, the proportions may not match the actual demographics of the country. Create a graph that shows the actual proportion of Canadians living in each province and territory.

c) Transfer your information from parts a) and b) to a spreadsheet and create a double bar graph. What does this tell you about the effectiveness of the sample size?

10. In 2004, the *Lancet* medical journal published a controversial study of the number of deaths in a war. In the study, data were collected by first randomly selecting 33 populated locations. Surveyors went to each location and drove around that city or town mapping it with a GPS unit. Then, they randomly selected one neighbourhood in the area and surveyed the 30 nearest households. Describe the type of sampling method and discuss its validity.

C

11. Read the following statement from a privacy officer discussing a drawback to collecting personal information online: "We have a suspicious number of 98-year-old Albanian millionaire grandmothers with Hotmail addresses who own 200-plus computers, head up technology firms, and who are interested in reading our newspaper."
Describe what the privacy officer means by this statement.

12. The Nielsen Ratings are a rating system for television and radio programs. The system uses a sample of people to report what they watch or listen to in order to get a sense of what everyone is watching or listening to.
Research the Nielson Ratings on the Internet. Describe how the Nielsen rating system works.

2.4 Bias in Surveys

KEY CONCEPTS

- Sampling, measurement, response, and non-response bias can all invalidate the results of a survey.
- Intentional bias can be used to manipulate statistics in favour of a certain point of view.
- Unintentional bias can be introduced if the sampling and data-collection methods are not chosen carefully.
- Leading and loaded questions contain language that can influence respondents' answers.
- When developing your own survey questions, to obtain valid results you must consider how you are asking the questions, to whom you are asking them, and how you are analysing the results.

A

1. Classify the bias in each situation.
   a) Your teacher asks you to raise your hand if you did your homework last night.
   b) The student council conducts a survey to see if people think they should fundraise for a new scoreboard and ask only people at a basketball game.
   c) A radio talk-show host asks callers to agree or disagree with a new handicapped-parking law.
   d) A survey asks whether your favourite type of music is country or western.
   e) A blog about comic books asks who the better superhero is: one who can only shoot silly webs from his hands or one who can shoot cool lasers from his eyes.
   f) A teacher asks members of the Math Club whether the school should continue to sponsor math contests.
   g) You are asked who the best rapper is and are given only the choices Half$, Smar-T, or MC Outlier.

2. You are conducting a survey on music interest. Describe a question or method that you could use that has
   a) sampling bias
   b) measurement bias
   c) non-response bias
   d) response bias

B

3. A patient taking medication sometimes gets better simply because he or she thinks that the medication is working. This is called the placebo effect. To compensate for this effect, trials of new medicines often include three participant groups. One group is given the actual medication, the second group is given a placebo (medication that has no active ingredient), while the third group is given nothing. Describe the purpose of each group.

4. What is a double-blind study?

5. A student hands out the following survey.

> Circle only one answer for each question.
>
> **1.** Gender: Female Male
>
> **2.** Age: 13 14 15 16 17
>
> **3.** Grade: 9 10 11 12
>
> **4.** Estimate the percent of healthy food that you eat.
>
> 0%–20% 20%–40% 40%–60% 60%–80% 80%–100%
>
> **5.** People should get into the habit of eating healthier foods, since it will reduce long-term health problems and chances of obesity.
>
> Strongly disagree Disagree Strongly agree Agree
>
> **6.** Would you like to see items in vending machines that promote healthy eating?
>
> Yes No

a) What information do you think the student hopes to obtain from the survey?

b) Are there any biased questions? If so, what could be done to remove any bias?

6. A company is proposing to erect 20 large wind turbines in your community. To alleviate any fears residents may have, the company holds an information meeting and shows their statistics about the safety of wind turbines.

a) Why might you be skeptical about any data presented?

b) What could the company do to help make the data as unbiased as possible?

7. An election is a sort of survey, with results determining who is voted into power. How elections are conducted could lead to some bias. Explain how the following have some type of bias associated with them.

a) The location where voting takes place is accessible only by car.

b) An opinion question on the ballot reads, "Knowing that global warming could damage Earth, would you support the approval of an increase in funding for the recycling program?"

c) Polling stations are open only from 1:00 p.m. to 3:00 p.m.

8. When companies advertise and give statistics, they are required by law not to make claims that are materially false or misleading. Superbubble Chewing Gum Corporation has the following line on all of its advertising: "Four out of five dentists recommend chewing Superbubble." Explain how there may be some biased methodology in getting this result.

9. A survey is being taken at a high school reunion. One question asks how much each respondent earns.

a) What type of bias is likely to occur?

b) Do you think the results of the salary question will underestimate or overestimate the average salary of people attending the reunion? Explain.

C

10. Polio was one of the most feared childhood diseases in the twentieth century. In the early 1950s, American medical researcher Jonas Salk developed a vaccine to fight the disease. When clinical trials of Salk's vaccine were started, statistical problems occurred.

a) Research the history of the polio vaccine and describe some of the statistical problems that affected the clinical trials.

b) Explain Canada's role in the development of the vaccine.

KEY CONCEPTS

- The three principal measures of central tendency are the mean, median, and mode. The values of these measures for a sample can differ from those for the whole population, especially if the sample is small.
- The average, or central tendency, is a single number that is supposed to represent a "normal" value for any particular measurement.
- The word *average* is often used to replace the word *mean*. However, in mathematics, *average* means a measure of central tendency, of which the mean is one type.
- The mean is the sum of all the values in a set of data divided by the number of values in the set.
- The median is the middle value when the values are ranked in order. If there are two middle values, then the median is the mean of these two middle values.
- The mode is the most frequently occurring value. For grouped data, the interval with the tallest bar is called the modal class interval.
- A single measure of central tendency often will have little or no meaning until more information is known about the sample.
- When the mean, median, and mode are very different from each other, none of them is a good measure of the centre.
- When the mean and median are near each other in value, the data usually have a symmetrical shape. However, when the mean and median are not close in value, the data are often skewed.

Skewed Left

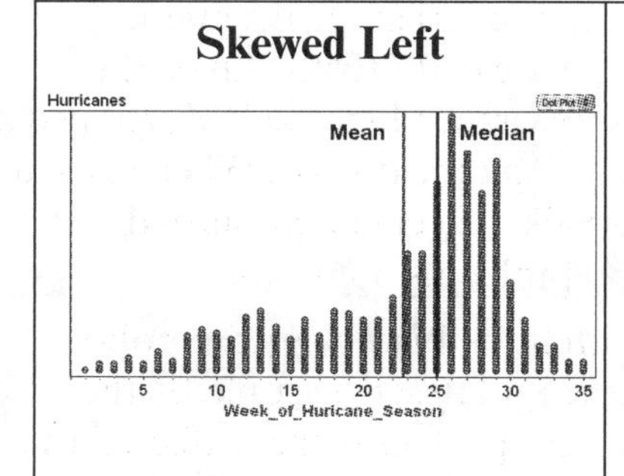

The mean is to the left of the median and data appear to trail off to the left.

Symmetrical

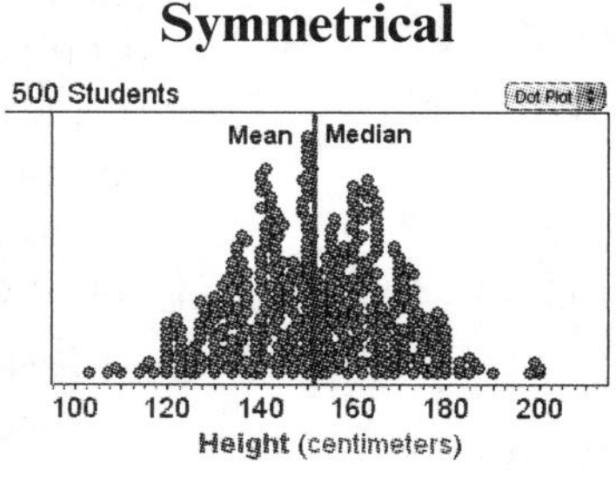

The mean and median are near each other and data appear to have a symmetrical shape.

Skewed Right

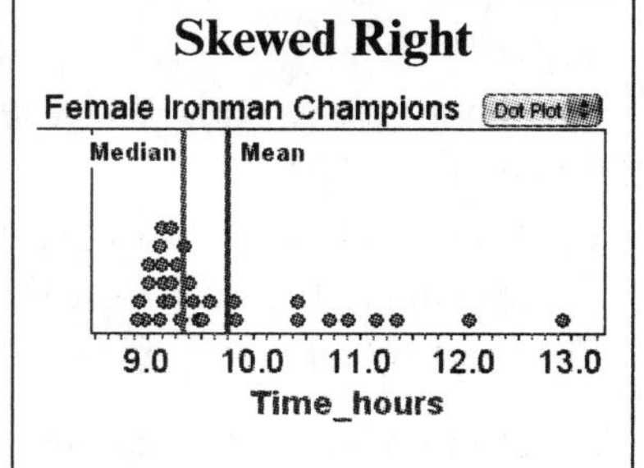

The mean is to the right of the median and data appear to trail off to the right.

- Outliers are data values that are very different from the rest of the data. They can have a dramatic effect on the mean more than any other type of average, especially if the data set is small. For example, the mean of the prices of five used cars is likely dramatically affected if four of the vehicles are sub-compacts and one is a rare antique Rolls-Royce.
- A weighted mean can be a useful measure when all the data are not of equal significance.
- For data grouped into intervals, the mean and the median can be estimated using the midpoints and the frequencies of the intervals.

A

1. For each set of data, determine the mean, median, and mode.
   a) 42, 44, 32, 57, 47, 54, 53, 37, 53, 57, 60, 61, 63, 52, 49, 53, 52, 59, 50
   b) 663, 679, 678, 681, 655, 654, 661, 659, 658, 665, 684, 673, 674, 665, 682, 669, 665, 655, 672, 683
   c) 1.27, 1.28, 1.21, 1.28, 1.28, 1.30, 1.24, 1.28, 1.25, 1.28, 1.20, 1.28, 1.29, 1.23, 1.29

2. Verify your answers for mean and median in question 1 using
   a) a graphing calculator
   b) a spreadsheet
   c) Fathom™

3. Identify whether each graph is skewed left, skewed right, or symmetrical. Where possible, identify the line that represents the mean and the line that represents the median.
   a)

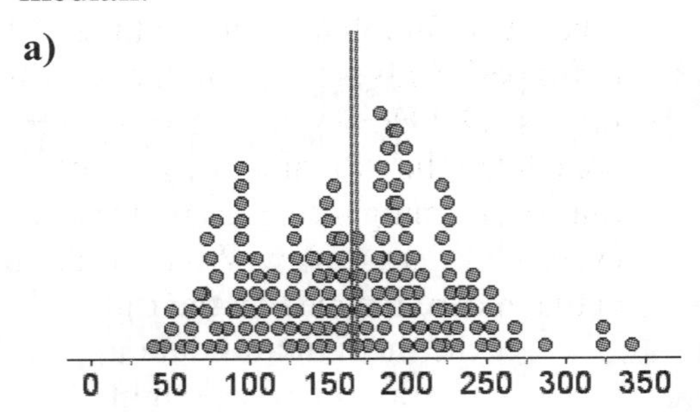

b)

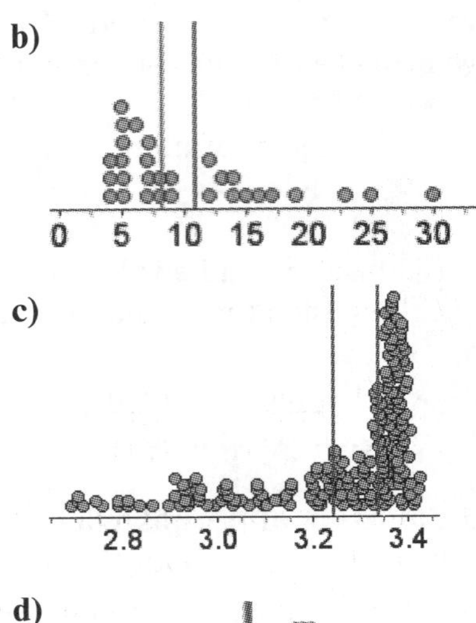

d)

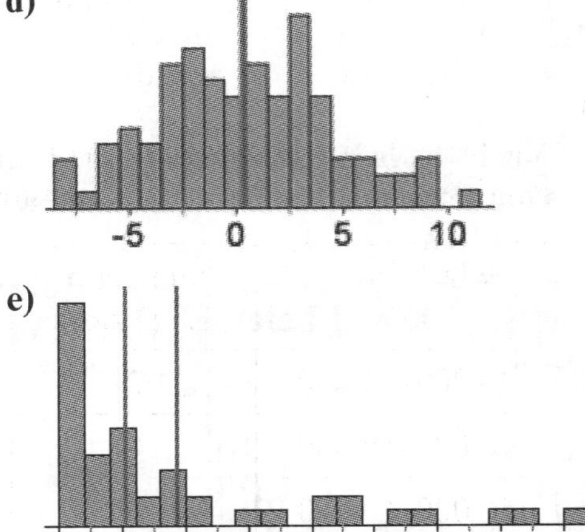

4. Consider the set of numbers 1, 2, 2, 2, 3, 9.
   a) Determine the mean, median, and mode.
   b) Why is the mean not an appropriate representation of average in this case?

5. Use a weighted mean to calculate the average mass of the following packages:

Package Mass (kg)	Number of Packages
5.0	2
7.5	8
8.0	3
9.25	8
11.5	5

6. The set of data represented in the graph has 67 elements.

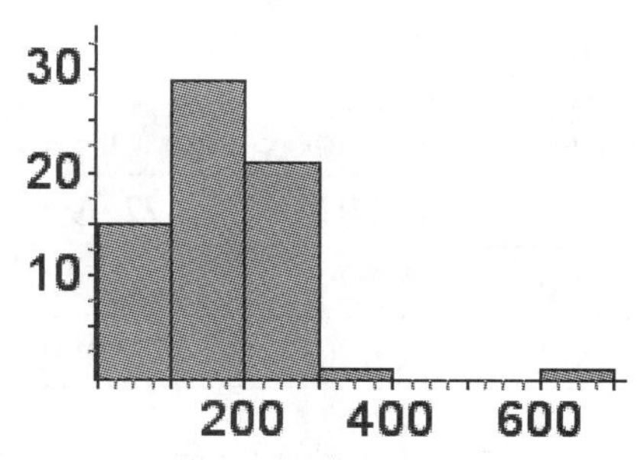

a) Determine the approximate median of the data.

b) Determine the modal class interval.

B

7. The table shows the distribution of salaries of employees of a small business.

Salary Range ($)	Number of Employees
30 000–40 000	2
40 000–50 000	8
50 000–60 000	3

a) What is the approximate mean of the salaries?

b) What is the least possible value of the mean? What is the greatest possible value?

c) Compare your answers from parts a) and b).

☆8. The table describes the distribution of salaries that a company pays its employees.

Salary ($)	Number of Employees
175 000	1
150 000	1
100 000	2
57 000	1
50 000	3
37 000	4
30 000	1
20 000	12

a) Determine the mean, median, and mode of the salaries.

b) What do the values found in part a) tell you about the average salary?

c) How could a person or group use one or more of the averages to misrepresent the salaries of employees?

d) How are the measures of central tendency affected if the owner's salary is increased from $175 000 to $450 000?

e) What does this tell you about how each of the averages is affected by an outlier?

9. You are conducting experiments and measure how much time each experiment takes. You find that the mean time of the ten experiments is 5.7 s and the median time is 5.9 s. However, after you have concluded the experiments, you notice that your timing mechanism measured everything 2 s too slow. What would the actual mean and median be?

10. Stuart receives the marks shown.

Category	Mark (%)	Category Weight (%)	Course Weight (%)
Knowledge	83	40	70
Application	70	30	
Communication	71	15	
Thinking	52	15	
Exam	75		30

a) Stuart determines his grade by adding up the marks and dividing by 5. Why is he wrong to use this method?

b) Determine Stuart's correct final mark.

c) If Stuart also receives a 5% bonus that he can apply to any category, to which category should he apply it to maximize his final grade? Why?

11. Stuart has just written a test that covers knowledge, application, and communication. His teacher gives Stuart his marks on knowledge (35/40) and application (13/20), but has yet to give him his communication mark. Based on his marks so far and the weightings from question 10, what grade does Stuart need on communication to have an average of 80% on the test?

12. A teacher lists the following measurements on the board and asks students to determine the median arm span. A student quickly states that the median is 163 cm. Is the student correct? Explain.

Student	Height (cm)	Arm Span (cm)
Sarah	154	158
Perdita	166	172
Luis	168	163
Frank	176	171
Shaun	179	180

13. The Cylindrix Company manufactures small pistons. To test the quality of production equipment, the company samples ten pistons each hour to see if they are within accepted size tolerances. A machine is considered to be operating incorrectly if the size of at least one of the pistons sampled is not within 12 μm (thousandths of a millimetre) of the mean size or if the mean and median are both not less than 3 μm from the desired diameter of 73 mm. Which of the three machines are operating correctly? Go to *www.mcgrawhill.ca/links/MDM12* and select **Study Guide**. Follow the links to obtain a copy of the file **Machine.csv** if you wish to use a spreadsheet.

Machine A	Machine B	Machine C
72.987	72.995	72.990
73.006	72.997	72.998
72.982	73.007	72.989
73.009	73.011	72.993
72.987	72.998	72.995
72.985	73.001	72.997
72.986	73.008	73.010
73.007	72.987	73.004
73.008	72.991	73.009
73.009	72.997	72.989

14. Explain why each statement is not valid.

a) At a craft store, the owner notes that the mean length of craft ribbon sold is 4 m. She concludes that half of the ribbons sold must be longer than 4 m.

b) A golfer determines the median score of his last ten games to be 83. He concludes that the total number of strokes was 830.

c) During a contract negotiation, a company complains that it must be a lie for a union to state that 70% of all the workers earn less than the mean salary.

15. Go to *www.mcgrawhill.ca/links/MDM12* and select **Study Guide**. Follow the links to obtain a copy of the file **Earthquake1900** (you have a choice of a Fathom™ file or a spreadsheet). The file represents the number of earthquakes of magnitude 7.0 and greater since 1900.

   a) How many earthquakes were of magnitude 7.0 and greater?

   b) For the **Mag** attribute, determine the mean, median, and mode.

   c) What type of skewing, if any, do these data have?

   d) Which measure of central tendency do you think best represents the data?

   e) Determine the strongest earthquake and use the other information given to describe the event.

☆ 16. Go to *www.mcgrawhill.ca/links/MDM12* and select **Study Guide**. Follow the links to obtain a copy of the file **Library.ftm.** This is a sample of songs from an MP3 collection.

   a) Create a single-variable graph to show the mean and median of the **Size** attribute. Explain your choice for the type of graph that you use.

   b) Change your graph to show a separate graph for each genre of music.

   c) If you want to store only one genre of music, which genre allows you to fit the greatest number of songs into a fixed amount of storage space? Justify your answer.

17. Go to *www.mcgrawhill.ca/links/MDM12* **Study Guide**. Open the files **Soccer2006** and **Soccer2009** (you have a choice of Fathom™ files or spreadsheets). These files have information about the base salaries for more than 300 players in the North American Major League Soccer league.

   a) For each year, compare the mean and median salaries.

   b) Determine the least salary and greatest salary paid in each year.

   c) Fathom™ only: For each year, create a histogram. What do these histograms suggest about the salaries of the soccer players?

   d) In each year, which player salaries would you consider to be outliers? Justify your reasoning.

   e) What is a data story that this information tells?

C

18. "This semester everyone's grade will be above the class average." Explain why this statement cannot be true.

19. There is controversy at a university because of claims of bias in admissions practices. Opponents of the admissions policy argue that male applicants are accepted at a greater rate than female applicants. This statement agrees with admissions data when overall totals are presented. However, when admissions are listed by department, the opposite appears to be true. How is this possible?

	Number of Applicants	Percent Admitted
Males	1735	57
Females	733	49

	Males		Females	
Dept.	Number of Applicants	Percent Admitted	Number of Applicants	Percent Admitted
A	925	61	98	88
B	495	65	32	69
C	315	34	603	42

☆ 20. The median of five numbers is 21, the mode is 7, and the mean is 18. What are the five numbers?

KEY CONCEPTS

- The variance and standard deviation are measures of how closely a set of data clusters around its mean. The variance and standard deviation of a sample may differ from those of the population the sample is drawn from.

	Population	Sample
Variance	$\sigma^2 = \frac{\Sigma(x-\mu)^2}{N}$	$s^2 = \frac{\Sigma(x-\bar{x})^2}{n-1}$
Standard Deviation	$\sigma = \sqrt{\frac{\Sigma(x-\mu)^2}{N}}$	$s = \sqrt{\frac{\Sigma(x-\bar{x})^2}{n-1}}$
***z*-score**	$z = \frac{x-\mu}{\sigma}$	$z = \frac{x-\bar{x}}{s}$
	Where x are the individual data values, μ is the mean of the population, and N is the total number of data values in the population.	Where x are the individual data values, $\bar{x}$ is the mean of the sample, and n is the number of data values in the sample.

- The standard deviation can be used as a measure of how meaningful any particular average is. A large standard deviation means that the average is not reliable as a representation of what a normal piece of data is like.
- The z-score of a datum measures how many standard deviations the value is away from the mean.
- Quartiles are values that divide a set of ordered data into four intervals with equal numbers of values in each interval. Quartiles are often represented with a box-and-whisker plot (also called a box plot) or modified box-and-whisker plot. In this example, the "whiskers" indicate the first and fourth quartiles, while the box represents the interquartile range (IQR). The box is split by the median into the second and third quartiles.

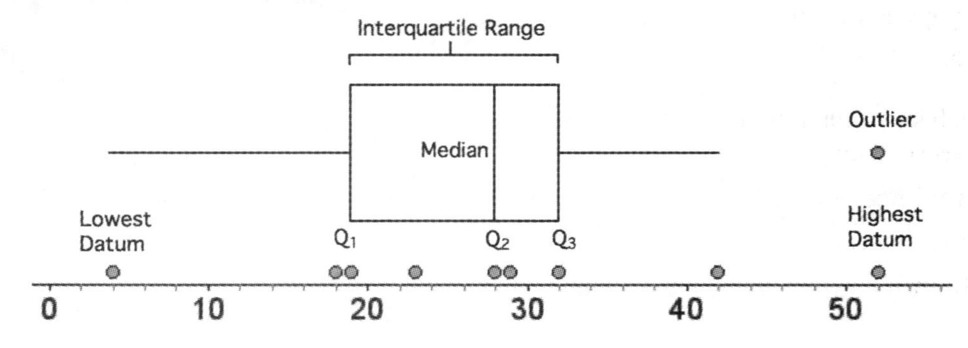

- An outlier is defined as any data point that is 1.5 times the interquartile range away from either side of the box.
- Percentiles are values that divide a set of ordered data into 100 intervals with equal amounts of data in each interval.
- The interquartile range and semi-interquartile range are measures of how closely a set of data clusters around its median.

A

1. For each of these sets of data, determine the sample mean, variance, and standard deviation.

a) 44, 67, 94, 36, 33, 68

b) 1.6, 1.5, 1.3, 1.7, 1.7, 1.0, 1.1, 1.8, 1.4, 1.2

c) 5.12, 5.92, 5.21, 5.29, 5.64, 5.92, 5.48, 5.58, 5.95, 5.94

☆**2.** For each set of data in question 1, verify your answer using

a) a graphing calculator

b) a spreadsheet

c) Fathom™

3. Determine the lowest datum, Q_1, Q_3, median, interquartile range, and highest datum for each set of data.

a) 56, 83, 24, 65, 91

b) 140, 127, 180, 177, 142, 166, 143, 145, 174

c) 13.3, 12.9, 13.3, 12.8, 13.4, 12.4, 12.6, 12.5, 13.0, 13.1

4. For each set of data in question 3, verify your answer using

a) a graphing calculator

b) a spreadsheet

c) Fathom™

5. The scores on a test worth 40 marks are shown:
17 20 21 23 23 24 25 26 29 30 30 31
31 33 34 34 36 37 39 40

a) If a student scores at the 80th percentile, what is the student's raw score?

b) In which percentile are the students who scored 30 out of 40?

c) If a student scores in the 70th percentile, what percent did the student get on the test?

d) In which percentile is a student who gets 60% on the test?

6. Draw a box-and-whisker plot for each set of data and indicate the approximate values of the lowest datum, Q_1, median, Q_3, and highest datum.

a)

b)

7. Determine the z-scores for each set of information.

	x	s	$\bar{x}$
a)	45	8.7	38
b)	126	14.5	150
c)	3.4	0.35	3.8
d)	567	78	452

B

8. A newborn baby is found to be in the 5th percentile for weight and the 95th percentile for height. What does this say about the characteristics of the baby?

☆ **9.** The Cylindrix Company is producing pistons but potentially has some quality control issues. The mean and median of the diameter of each piston from Machine A are 104 mm and 106 mm, respectively. For pistons made on Machine B, the mean and median are 103 mm and 104 mm, respectively.

a) Discuss any problems there might be with measuring only the mean and median for these machines if you are interested in how accurately they are producing parts.

b) One production problem is that pistons are "out of round," meaning that the units are not cylindrical but oblong. To test for this problem, random pistons are removed and the diameter of each is measured at ten different places. Here are results for a sample unit produced on each machine:

Machine A (mm)	Machine B (mm)
105.9	103.4
106.8	102.3
104.4	102.0
100.3	103.7
105.8	102.8
102.8	104.0
107.4	102.4
101.6	104.9
100.1	105.0
105.2	103.8

Which machine is more likely to produce more cylindrical pistons? Justify your answer.

c) When you take multiple measurements of an object, it is customary to report the standard error. The standard error is found in the following way:

$$SE = \frac{s}{\sqrt{n}}$$

Determine the mean measurement with the standard error for each machine (stated as $\bar{x} \pm SE$).

☆ **10.** Go to *www.mcgrawhill.ca/links/MDM12* and select **Study Guide**. Follow the links to obtain a copy of the Fathom™ file **Magazines.ftm.** This file has data about hundreds of magazines, including total number of pages, number of advertising pages, and type of magazine.

a) Create box-and-whisker plots of the **Total Pages** and the **Ad Pages** attributes. Use the same scale. Which set of data seems to be more consistent? Why?

b) On each box plot, drag the **Type** attribute to the vertical axis. Right-click each graph and choose **Plot Value** from the menu. In the equation box, type **mean( )**. What does this show for each graph?

c) Identify all of the outliers for each magazine type. Are there some magazines that are outliers in more than one graph?

d) Based on each of the box plots, which magazine

i) has the greatest mean?

ii) has the most consistent mean?

e) To get specific values for the mean of each type, drag a summary table to the Fathom™ workspace. Drag each of the numerical attributes to the arrow pointing down (the arrow will appear once you drag the attribute over it). Drag the **Type** attribute to the arrow pointing to the right.

What is the value of the highest mean of each attribute?

f) To determine the standard deviation for each attribute, right-click the summary table and choose **Add Formula**. In the equation box type **stddev( )**. How do these values confirm your answers about the most consistent mean?

g) What story could this data tell?

11. The average monthly temperatures for three cities are recorded.

Mathland	4, 6, 14, 19, 25, 26, 23, 22, 16, 11, 10, 4
Geometown	9, 14, 14, 17, 17, 18, 19, 14, 13, 18, 16, 11
Algeville	−7, −6, −5, 33, 34, 37, 31, 36, 34, −1, −4, −2

a) Determine the mean and median temperatures for each city.

b) If you saw only the answers from part a), what conclusion would you make about the climates of these cities?

c) Determine the standard deviation for each city. What do these values tell you about the reliability of the averages?

d) Create a dot plot for each city using the same scale. How do the dot plots support your answer from part c)?

12. The Cabot Trail Relay Race is a running race that circumnavigates the Cabot Trail on Cape Breton Island. Each team must run 17 legs for a total distance of 277 km. The top three teams in 2009 had the following times for the race.

Team A	16 h 59 m 48 s
Team B	17 h 5 m 17 s
Team C	18 h 6 m 41 s

a) Calculate the mean speed for each team.

b) The table gives data about the speeds over each leg. Use these values to determine the mean speed for each team. Compare these answers to those in part a). Should they be the same? (This data can also be found online: Go to *www.mcgrawhill.ca/links/MDM12* and select **Study Guide**. Follow the links to obtain a copy of the file **CabotTrail.csv** or **CabotTrail.ftm**)

Leg	Distance (km)	Team A Speed (km/h)	Team B Speed (km/h)	Team C Speed (km/h)
1	17.00	18.20	17.37	15.58
2	17.92	18.75	18.47	15.64
3	13.46	13.37	14.15	14.37
4	20.01	15.92	17.17	14.52
5	17.50	15.86	15.76	15.19
6	17.50	16.84	16.91	16.96
7	13.10	14.79	15.78	14.09
8	12.36	15.13	14.82	13.70
9	17.84	15.17	16.23	15.39
10	14.70	14.78	14.89	14.47
11	14.00	16.60	16.01	15.12
12	15.78	16.57	16.21	16.42
13	15.88	15.40	14.80	15.44
14	19.81	16.94	17.31	16.59
15	15.42	17.81	14.29	14.03
16	15.35	16.54	16.89	15.68
17	18.70	18.03	17.44	15.84

c) Calculate the standard deviation for each team. Which team had the most consistent runners? Explain.

d) Create a box-and-whisker plot for each team. How do the box plots support your choice for the most consistent team?

13. Go to *www.mcgrawhill.ca/links/MDM12* and select **Study Guide**. Follow the links to obtain a copy of the file **BaseballTeams** (you have a choice of a Fathom™ file or a spreadsheet). This file represents the total payroll of each Major League Baseball team in 2009.

a) Determine the mean, median, standard deviation, Q_1, and Q_3 for this set of data.

b) Determine the z-score for each team. Which team is an outlier when it comes to salaries? How does this team's payroll compare to that of other lower-salaried teams? How does this team's z-score compare to the other teams'?

c) Look back at the history of the last ten World Series. How has the outlier done? Do you think it has been worth it to have such a high payroll?

14. Go to *www.mcgrawhill.ca/links/MDM12* and select **Study Guide**. Follow the links to obtain a copy of the file **Subway.ftm.** This is data regarding the nutritional information for a sample of Subway® submarine sandwiches.

a) What type of sandwich is an outlier in terms of sugar content? Why is that not surprising?

b) Determine which of the following attributes has the most reliable mean: **Total Fat**, **Sugars**, **Protein**, or **Saturated Fat**. Justify your answer.

☆ **15.** Go to *www.mcgrawhill.ca/links/MDM12* and select **Study Guide**. Follow the links to obtain a copy of the file **Daytona5002005.ftm.** These are the qualifying lap times and average speeds for all of the drivers in the 2005 Daytona 500 NASCAR race.

a) Create dot plots of the **Speed** and **Time** attributes. What do you notice about the two graphs? What explanation do you have for this situation?

b) What story might this data tell?

C

16. Go to *www.mcgrawhill.ca/links/MDM12* and select **Study Guide**. Follow the links to obtain a copy of the file **Movies2009** (you have a choice of a Fathom™ file or a spreadsheet). This represents the one-day movie statistics for November 30, 2009. Looking at all of the data, do you see any stories that these data could tell?

17. The breakdown of wages in Canada in 2003 is given in the following table:

Annual Income ($000)	Frequency
0–5	2 375 060
5–10	2 575 760
10–15	2 748 630
15–20	2 444 450
20–25	1 884 500
25–35	3 335 080
35–50	3 420 850
50–75	2 696 580
75–100	884 910
100–150	437 020
150–200	115 620
200–250	51 860
250 and over	99 880

a) Based on these data, determine the approximate average wage in Canada in 2003.

b) What other stories do these data tell?

Chapter 3 Statistics of Two Variables

3.1 Scatter Plots and Linear Correlation

KEY CONCEPTS

- Single-variable analysis of data is the comparing of a certain variable to itself. For example, if you take the height of each student in your class, you are repeatedly taking the same type of measurement. Comparative measures of a single variable (such as height in this example) include central tendency and spread.
- Multi-variable analysis of data is the comparing of two or more variables to each other. For example, if you measure the height and arm span of each student in your class, you are repeatedly taking a set of measurements, one for each person. Once you take those measurements, you can see if the variables are connected to each other: as one changes, does the other?
- Two sets of data are correlated if changes in one set cause predictable changes in the other set.
- Statistical studies often find correlations between two variables. In most cases, one variable is called the independent (explanatory) variable and the other is called the dependent (response) variable. For a relationship that has a real correlation, the independent variable causes the dependent variable to change.
- A scatter plot can often reveal the relationship between two variables. The independent variable is usually plotted on the horizontal axis and the dependent variable on the vertical axis.
- Two variables have a linear correlation if changes in one variable tend to be proportional to changes in the other. Linear correlations can be positive or negative. When data points are plotted, they will appear to (at least approximately) lie along a straight line.

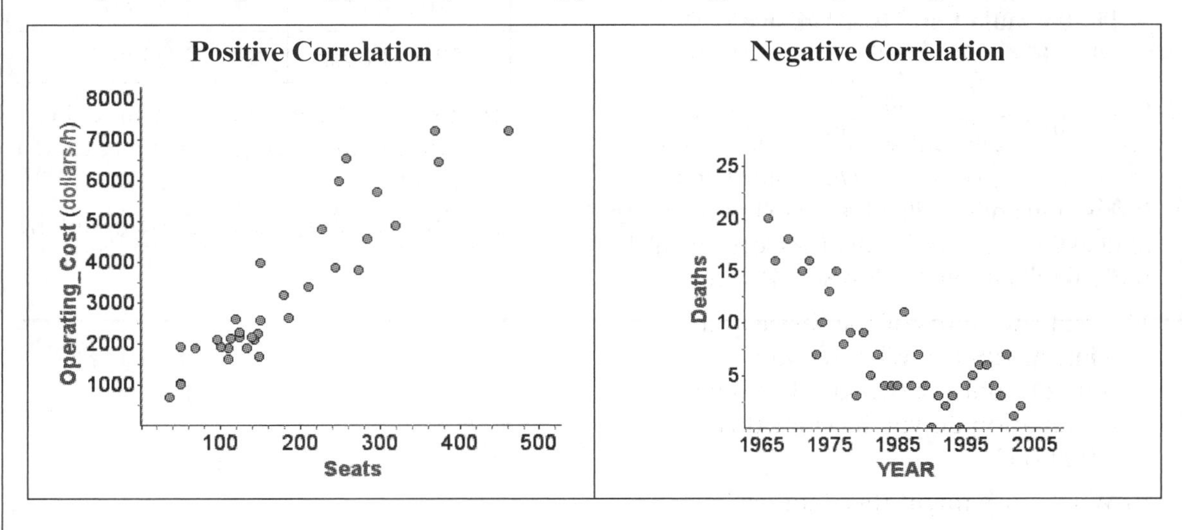

- The correlation coefficient, r, is a quantitative measure of the correlation between two variables. The correlation coefficient measures the fit of the data to a particular function model. The function model is linear in the case of linear correlations. Some correlations follow a non-linear pattern.

- Negative values for the correlation coefficient indicate negative correlations, while positive values indicate positive correlations. The correlation coefficient ranges from -1 (perfect negative correlation) to 0 (no correlation) to $+1$ (perfect positive correlation).

- The strength of a linear correlation is classified as being in one of three ranges: strong, moderate, or weak.

Negative Linear Correlation | Positive Linear Correlation

Perfect, Strong, Moderate, Weak | Weak, Moderate, Strong, Perfect

-1 $\quad -0.67$ $\quad -0.33$ $\quad 0$ $\quad 0.33$ $\quad 0.67$ $\quad 1$

Correlation Coefficient, r

- Sometimes the correlation is reported via the coefficient of determination, r^2, which is the square of the correlation coefficient and ranges from 0 to 1. Often it is difficult to guess the correlation visually, as illustrated by the data sets below.

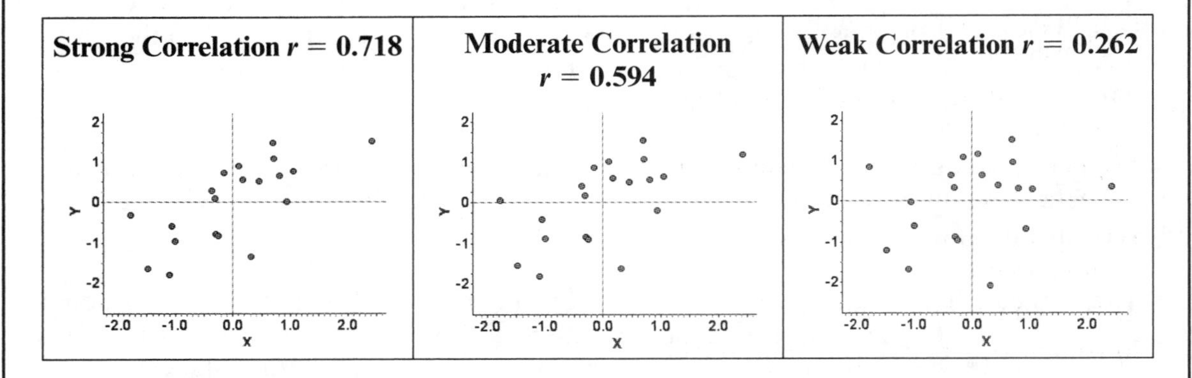

- The formula for calculating r is

$$r = \frac{n\Sigma xy - (\Sigma x)(\Sigma y)}{\sqrt{[n\Sigma x^2 - (\Sigma x)^2][n\Sigma y^2 - (\Sigma y)^2]}}$$

- Manual calculations of correlation coefficients can be quite tedious, but a variety of technology tools, such as spreadsheets, Fathom™, and graphing calculators, are available for such calculations.

A

1. Classify the strength (strong, moderate, or weak) and type (positive or negative) of the linear correlation that you would expect between the following pairs of variables.
   a) speed of a car, stopping distance
   b) shoe size, intelligence
   c) hours of practice, golf score
   d) a person's height, the person's arm span
   e) amount of time playing a video game, the number of points earned
   f) year, fastest 100-m sprint time
   g) year, earnings of the highest-grossing movie
   h) body mass index, resting heart rate

2. Identify the independent variable in a correlational study of
   a) smoking and cancer rates
   b) balance and hours of yoga practice
   c) dental cavities and candy consumption
   d) number of rock concerts attended and hearing loss
   e) a person's salary and the number of years of school completed
   f) temperature and the number of cases of sunburn
   g) stopping distance and speed
   h) volume of a balloon and air pressure

3. **a)** Use a spreadsheet to complete the table.

x	y	x^2	y^2	xy
1.1	2.5			
2.8	8.2			
1.3	7.3			
3.5	13.4			
5.4	9.5			
5.8	21.5			
7.0	19.5			
7.1	29.1			
Σx	Σy	Σx^2	Σy^2	Σxy

   b) Use the values in the last row to calculate the correlation coefficient, r, and classify your correlation.
   c) Use the correlation function in your spreadsheet to verify your answer in part b).
   d) Create a scatter plot in your spreadsheet.
   e) Repeat parts c) and d) using Fathom™.
   f) Repeat parts c) and d) using a graphing calculator.

4. Describe the correlation in each graph.

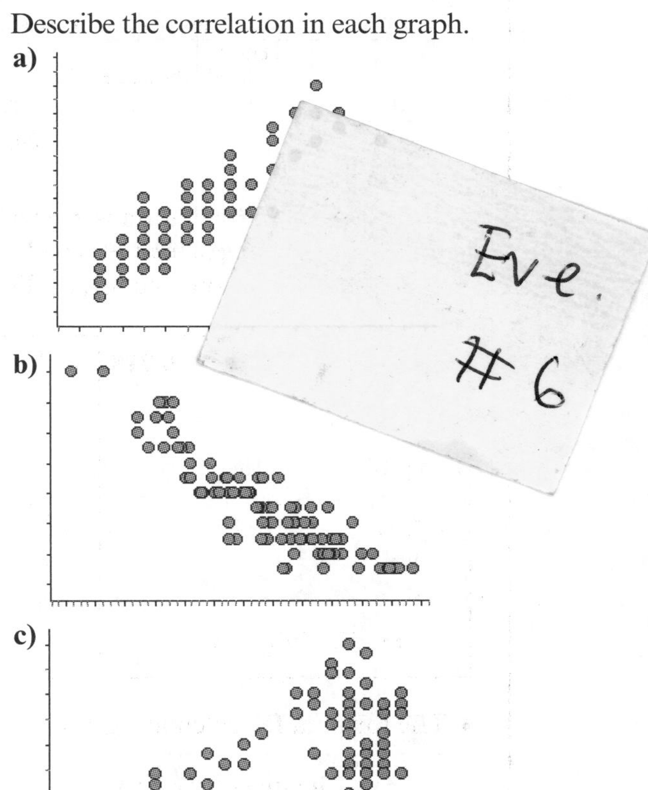

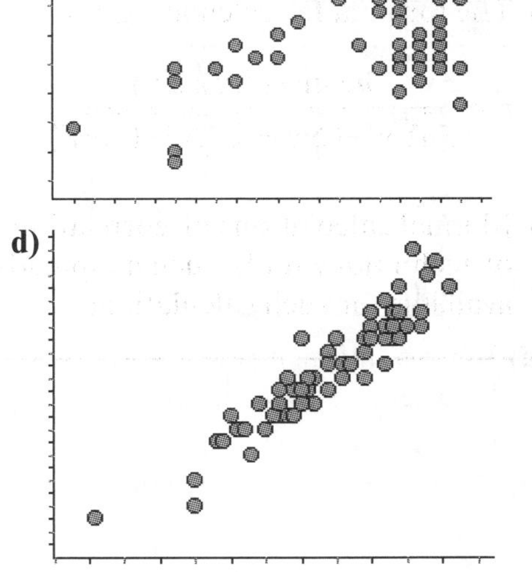

5. For each situation, indicate whether single- or two-variable analysis should be used. Justify your answers.

   a) The populations of 100 countries are listed.

   b) The temperature of a chemical mixture is taken every 30 s.

   c) The number of hits in each game for a baseball player are recorded.

   d)

Sandwich	Calories	Fat (g)	Sugar (g)
Salami	580	15	37
Turkey	360	8	14
Ham	450	18	21
Club	380	6	15

B

6. What range of values does the coefficient of determination have for a correlation that is

   a) weak?

   b) moderate?

   c) strong?

7. The table shows the average price of a movie ticket in each year of a ten-year period.

Year	Ticket Price ($)
2009	7.80
2008	7.80
2007	7.57
2006	7.27
2005	7.09
2004	6.85
2003	6.63
2002	6.41
2001	6.20
2000	5.93

   a) What relationship seems to exist between the year and the average ticket price?

   b) Determine and classify the correlation.

   c) Could these data be used to predict the average price of movie tickets in 2010? Explain.

☆8. Go to *www.mcgrawhill.ca/links/MDM12* and select **Study Guide**. Follow the links to obtain a copy of the file **BaseballHomers.csv**. This file contains data for the New York Yankees, who won the World Series in 2009, and for the Baltimore Orioles, who placed last in regular-season play that year. For the players on each team, there is information about each player's number of times at bat and number of home runs.

   a) Which of at bats and home runs would be the independent variable?

   b) Determine which team has the higher correlation between the variables.

   c) What do the correlation values tell you about how the number of home runs is related to the number of at bats?

   d) Create a scatter plot for each team. Use the same scales on each graph. What does this tell you about how the number of home runs may be related to a winning season?

9. In the Monopoly® board game, each player tries to buy all the properties in order to win the game. Go to *www.mcgrawhill.ca/links/MDM12* and select **Study Guide**. Follow the links to obtain a copy of the file **Monopoly.csv**. The file contains data about the position, cost, and rent of each property.

   a) How is the position of each property related to the cost?

   b) How is the rent related to the cost?

   c) Which three properties, when removed, allow the relationship in part b) to have a perfect correlation?

   d) Using the remaining properties from part c), determine the rent of a new property that costs $500. Justify your answer.

10. Aaron says, "I heard that a person's height is the same as the person's arm span." Chandra says, "I heard that the length of a person's forearm is the same as the length of the person's foot." Go to *www.mcgrawhill.ca/links/MDM12* and select **Study Guide**. Follow the links to obtain a copy of the file **Measurements.csv**. In this file are the height, arm span, forearm length, and foot length for 93 students. Use this file to determine which of Aaron's claim or Chandra's claim is more likely correct.

☆ 11. Go to *www.mcgrawhill.ca/links/MDM12* and select **Study Guide**. Follow the links to obtain a copy of the file **Hockeytop20.csv**. The file contains data about the total career salaries of the 20 highest-paid NHL players since 1989.

  a) How do you think the number of games played will relate to the number of years active? Justify your answer.

  b) A player gets one point for each goal and each assist. How is the number of points related to the number of games played? Justify your answer.

  c) Why should the correlation between the number of points and the number of goals be strong and positive?

  d) What relationship would you expect there to be between the number of years active and the total salary? Do the data back this up? Explain why or why not.

  e) What relationship would you expect to see between the salary and the number of points? How do the data support your claim?

  f) Based on the data, which hockey player is the most cost-effective? Which player is the least cost-effective?

☆ 12. Go to *www.mcgrawhill.ca/links/MDM12* and select **Study Guide**. Follow the links to obtain a copy of the file **NASCAR.csv**. This file has the results for one of the last NASCAR races of 2009.

  a) What relationship would you expect to see between the finish position and the amount of money won? Is this relationship true for each driver? Justify your answer with a graph.

  b) In NASCAR, a driver must first qualify for a particular race. A driver's start position is based on how fast the driver completes one lap in qualifying. The two drivers with the fastest qualifying times start in the first row. The two drivers with the slowest qualifying times are positioned in the last row. How important do you think having a good start position is to having a good finish position? How do the data support your answer?

C

13. Go to *www.mcgrawhill.ca/links/MDM12* and select **Study Guide**. Follow the links to obtain a copy of the file **Subway.ftm**. The file contains information about several varieties of Subway® sandwiches.

  a) What relationship would you expect to see between the **Total Fat** and **Calories** attributes? Determine the correlation to support your statement.

  b) Create a scatter plot of the **Total Fat** and **Calories** attributes. Does this support your answer to part a)? Explain.

  c) Drag the **Type** attribute to the middle of the graph. How does this change the perspective?

  d) There is one sandwich that is not identified as "healthy" or "regular." What should it be labelled? Justify your answer.

14. Go to *www.statcan.gc.ca/estat/licence-eng.htm* and click **Accept and enter** to access the E-STAT site. Locate a set of data that shows a strong linear correlation. What story do these data tell?

KEY CONCEPTS

- Linear regression is an analytic technique for determining the relationship between the independent and dependent variables. The line of best fit is the linear relationship that most closely matches the general trend shown by all points. For perfectly correlated data, the line of best fit can be determined exactly by calculating its slope and y-intercept using any two known points. When the correlation is not perfect, a more complex method must be used.

- The vertical distance between the line of best fit and any point is called that point's residual value. The line of best fit is sometimes called the least-squares line, because it minimizes the sum of the squares of the residual values of all points. What this means visually is that if a square is created between each point and the line, then the side length of each square represents that point's residual value. The sum of the areas of these squares is the least it can be.

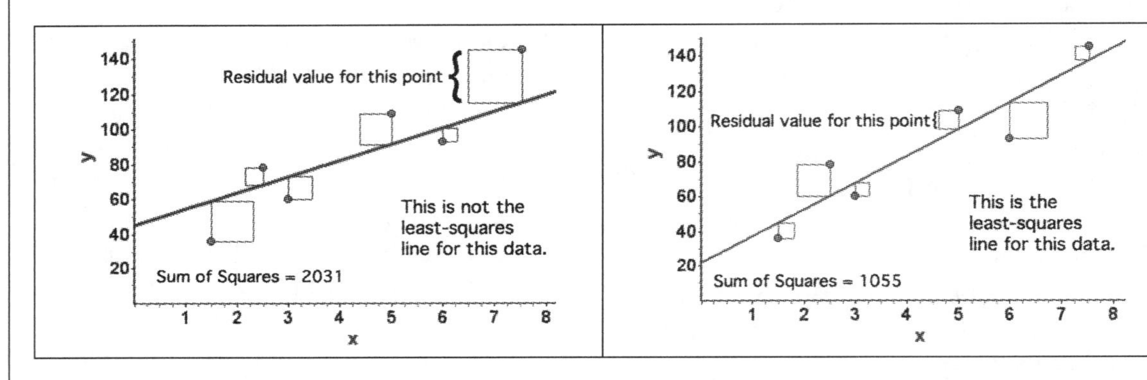

- The equation for the least-squares line (also known as the residual line, line of best fit, or trend line) describes the most appropriate average trend in the data. The equation can be used to predict values of one variable given values of the other.

- To calculate the equation of the line of best fit in the form $y = ax + b$, use

  $a = \dfrac{n(\Sigma xy) - (\Sigma x)(\Sigma y)}{n(\Sigma x^2) - (\Sigma x)^2}$ and $b = \overline{y} - a\overline{x}$

- The magnitude of the correlation coefficient measures how well a least-squares line fits a set of data.

- The linear-regression model for any set of multi-variable data is made up of the description of the correlation, the equation of the line of best fit, and the graph of the scatter plot.

- An outlier is a point that is far away from the line of best fit. When an outlier is removed from the data, the correlation should become noticeably stronger.

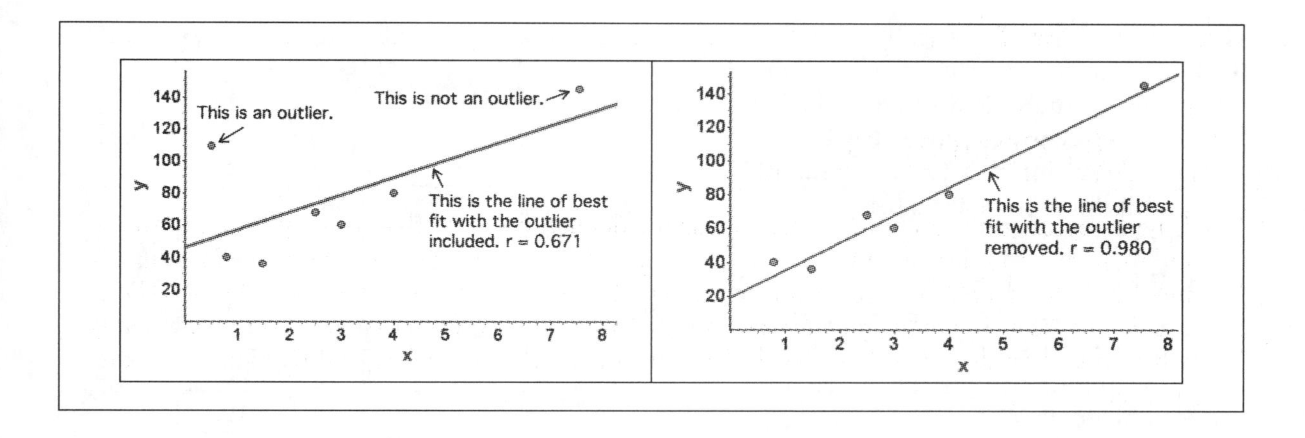

A

1. For each scatter plot, identify the outlier and draw the line of best fit
   (i) using all the data points
   (ii) using all the data points except the outlier

a)

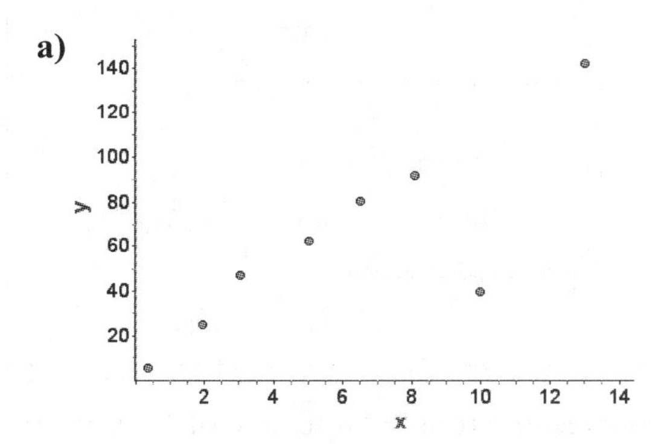

b)

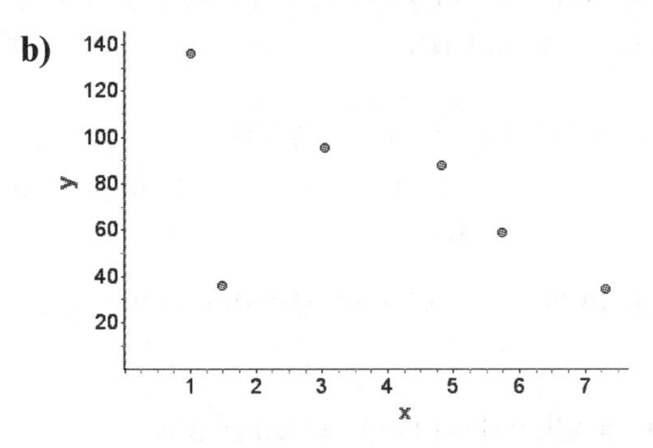

★2. For each set of data, determine the equation of the line of best fit, classify the relationship, and create a scatter plot using the indicated technology.

a) spreadsheet

x	y
156	63
119	57
100	34
177	71
160	42
183	58

b) graphing calculator

x	y
2.7	14
3.7	9.7
4.6	8.4
2.4	28
1.5	36
0.6	36

c) Fathom™

x	y
3.9	48
8.3	61
20	78
0.8	40
3.2	35
16.7	68

B

3. Go to *www.mcgrawhill.ca/links/MDM12* and select **Study Guide**. Follow the links to obtain a copy of the file **GasBill.csv**. This file has information about monthly gas bills for a home for a period of approximately two years.

   a) What relationship would you expect to see between the average temperature, in degrees Celsius, and the amount of gas consumed, in cubic metres?

   b) Determine the equation of the least-squares line and describe the relationship.

   c) What does the slope represent in the context of the question?

   d) Based on the data, how much gas would you expect to be consumed in a particularly cold month with an average temperature of −15°C?

   e) Based on the data, what would you expect gas consumption to be during a month when the average temperature is 25°C? Explain why the answer is not realistic.

4. In 1973, statistician F. J. Anscombe at Yale University created four sets of data with a very peculiar property:

Set I		Set II	
x	y	x	y
10	8.04	10	9.14
8	6.95	8	8.14
13	7.58	13	8.74
9	8.81	9	8.77
11	8.33	11	9.26
14	9.96	14	8.1
6	7.24	6	6.13
4	4.26	4	3.1
12	10.84	12	9.13
7	4.82	7	7.26
5	5.68	5	4.74

Set III		Set IV	
x	y	x	y
10	7.46	8	6.58
8	6.77	8	5.76
13	12.74	8	7.71
9	7.11	8	8.84
11	7.81	8	8.47
14	8.84	8	7.04
6	6.08	8	5.25
4	5.39	19	12.5
12	8.15	8	5.56
7	6.42	8	7.91
5	5.73	8	6.89

   a) Go to *www.mcgrawhill.ca/links/MDM12* and select **Study Guide**. Follow the links to obtain a copy of the file **Anscombe.csv**. Use a spreadsheet or graphing calculator to determine for each set

   i) the mean of x

   ii) the mean of y

   iii) the standard deviation of x

   iv) the standard deviation of y

   v) the correlation between x and y

   vi) the equation of the least-squares line

   b) Based on your answers in part a), what would you expect to be true about the relationships for each set?

   c) Create a scatter plot for each set. What do the scatter plots tell you about the importance of the visual representation of the data?

☆5. Your science teacher is demonstrating water displacement. She slowly adds beads to some water in a graduated cylinder. As beads are added, water is displaced and the overall level goes up.

   a) What type of relationship would you expect between the number of beads and the volume of the mixture? What about between the number of beads and the total mass?

b) You arrive at class late and manage to get only the following data:

Number of Beads	Volume of Mixture (mL)	Mass of Mixture (g)
10	38	172.0
15	41	182.1
25	49	202.5
35	58	222.7

Create linear-regression models for the relationship between the number of beads and the volume of the mixture, and between the number of beads and the mass of the mixture.

c) Use your models to determine the mass and volume of one bead.

d) What else could you determine from your regression models?

e) Is there a relationship between volume and mass? If so, which would be the independent variable?

6. An Internet video site had the following numbers of videos on the given dates.

Date in 2008	Day of Year	Number of Videos (millions)
Jan. 28	28	70.0
Mar. 12	72	77.4
Mar. 16	76	78.3

a) Create a linear-regression model for the day of the year and the number of videos.

b) What does the slope mean in the context of the question?

c) Based on this model, predict the number of videos the site will have by the end of 2009.

d) Based on this model, when will the site have a billion videos?

e) How accurate do you think these predictions are?

7. Go to *www.mcgrawhill.ca/links/MDM12* and select **Study Guide**. Follow the links to obtain a copy of the file **October2009Movies.csv**. This is the box office data for the top-grossing movies of the weekend of October 16, 2009.

a) Create a scatter plot using the **Gross** and **Theatres** attributes.

b) Describe the regression model of this data.

c) Are there any outliers? If so, remove them and describe how your regression model changes.

C

8. Go to *www.statcan.gc.ca/estat/licence-eng.htm* and click **Accept and enter** to access E-STAT. From the menu on the left, choose **Search CANSIM in E-STAT**. Enter **002-0011** in the **Search** field and click **Search**. In the **Food categories** list, select **Food available**. In the **Commodity** list, select **Soft drinks** and **Bottled water**. For the **Reference period**, choose 1970 to 2008. Click **Retrieve as individual Time Series**. Under **DOWNLOADABLE FILE**, select **CSV (comma-separated values) file: Time as rows**, and then click **Retrieve Now**. The file will be downloaded to a location you choose.

a) Create a linear-regression model for time and soft drinks. Describe any difficulties with this model.

b) Create a linear-regression model for time and bottled water. How would you compare the situations for soft drinks and bottled water?

c) Create a new linear-regression model for soft drinks. This time, use only data from the section of the graph where consumption is decreasing. Based on this model and the bottled water model, approximately when will bottled-water consumption surpass soft-drink consumption?

d) What story do these data seem to tell?

9. Go to *www.mcgrawhill.ca/links/MDM12* and select **Study Guide**. Follow the links to obtain a copy of the file **Library.ftm**. This represents the music files on an MP3 player.

   a) Create a linear-regression model for the **Size** and **Time** attributes. Describe any problems with the model.

   b) Drag the **BitRate** attribute to the centre of your graph. Describe how this affects your regression model. Research what the bit rate represents and describe how the model supports your research.

   c) What does the y-intercept mean in the context of the data? Compare the y-intercepts for both models. Why does this value make sense?

★ 10. Go to *www.mcgrawhill.ca/links/MDM12* and select **Study Guide**. Follow the links to obtain a copy of the file **Snowfall.ftm**. This file represents the total amount of snow that fell during a snowstorm.

   a) Produce a linear-regression model for the data.

   b) Sometimes even when a correlation is strong, the model used is not the correct one. One way to show that a set of data is linear is to look at its residual plot. For a set of linear data, this plot should look random. For non-linear data, a definite pattern should be seen. Right-click in your data and choose **Make Residual Plot** to show that this is non-linear data.

   c) To further confirm that this is a non-linear model, create a model and compare it using the sum of squares. To determine this, complete the following:

      i) Right-click in the graph. Deselect **Make Residual Plot** and choose **Show Squares**. What is the sum of the squares for the linear model?

      ii) Right-click in the graph again and choose **Plot Function**. In the function box, type $\boldsymbol{a} \times$ **time_h²** to add a quadratic model to the graph. In this case, a represents the coefficient of the general parabola $y = ax^2$.

      Expression for function
      Snowfall_cm = a time_h²

      iii) Move the slider for the a value until the parabola best fits the data. You can tell the best fit when the sum of the squares is as small as it can be. What is the curve of best fit?

   d) How do your sums of squares show that the non-linear model is better?

   e) Use both of your regression models to predict the amount of snow that will have fallen after 24 h. How do the two answers compare? Which would be the more reasonable?

11. Go to *www.mcgrawhill.ca/links/MDM12* and select **Study Guide**. Open the file **MeanFit.ftm**.

   a) Create a scatter plot with a least-squares line on it. Right-click in the plot. Choose **Plot Value** and type **mean(x)**. Right-click in the graph again, choose **Plot Function**, and type **mean(y)**.

   b) What does the intersection of the vertical line and horizontal line represent?

   c) Click and drag any point (as many as you wish) to change the original data and thus change the least-squares line. The point at the intersection found in part b) is called the mean fit point. What do you think the characteristics of this point are?

KEY CONCEPTS

- Correlation does not necessarily imply a cause-and-effect relationship. Correlations can also result from common-cause factors, reverse cause-and-effect relationships, accidental relationships, and presumed relationships.
- Extraneous variables can invalidate conclusions based on correlational evidence.
- Comparison with a control group can help remove the effect of extraneous variables in a study.

A

1. Identify the most likely type of causal relationship for each situation, assuming that the first variable is the independent variable.

a) Heart attack rates drop as fitness clubs bring in more revenue.

b) Toxic waste dumping increases in Lake Erie, and fish populations decline.

c) Two waffle-producing factories close due to fire, and the availability of waffles decreases.

d) You do well on tests when you wear a red shirt.

e) As ice cone sales increase, so do drowning incidents.

f) When you eat a lot of chocolate, you get a lot of acne.

g) As the number of police officers increases, so does the number of crimes.

h) The higher a ball is dropped from, the higher the ball bounces.

2. In each case, indicate what the common cause may be.

a) You notice that as a person's height increases, so does the person's arm span.

b) Children with more books in their houses do better in school.

c) The price of gas goes up as the prices of cars go up.

3. The graph shows some fictitious data about global warming and pirates.

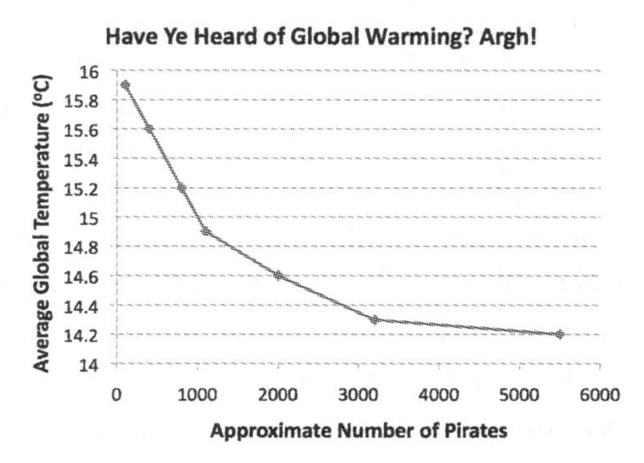

a) What is this graph trying to convey?

b) What type of causal relationship is it?

B

4. From the mid 1970s to the early 1990s, the number of crimes per year in Canada nearly doubled.

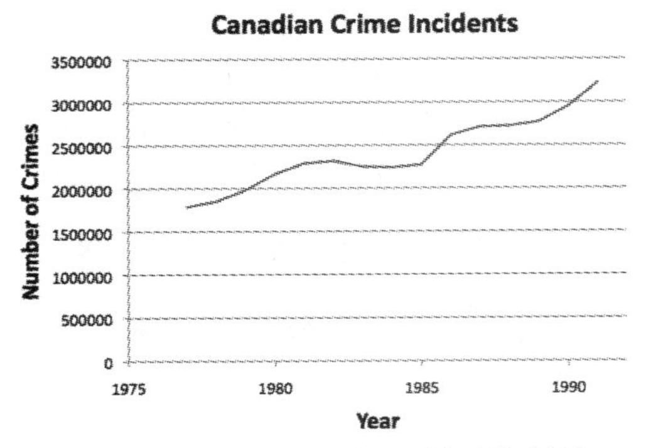

Source: Statistics Canada, Table 252-0013

A politician compared the number of police officers to the crime statistics for the years 1986–1991 to try to determine a cause. She noticed that as the number of police officers increased, so did the number of crimes.

Year	Number of Officers	Crime Incidents
1986	51 425	2 613 351
1987	52 510	2 712 389
1988	53 312	2 722 335
1989	54 211	2 769 367
1990	56 034	2 946 734
1991	56 768	3 218 778

The politician concluded that having more police officers was causing more crime.

a) What type of causal relationship was the politician using?

b) Go to *www.mcgrawhill.ca/links/MDM12* and select **Study Guide.** Follow the links to obtain a copy of the file **CrimePolice.csv**. Create a linear-regression model of the relationship between police and crime for the years up to 1991. Do you think this relationship accurately models this situation?

c) Create scatter plots that show the number of crimes occurring over time and the number of police officers over time. Compare the trends seen in these graphs and tell a data story.

5. The headline in a newspaper reads "Cable Profits Causing Death" and displays a graph.

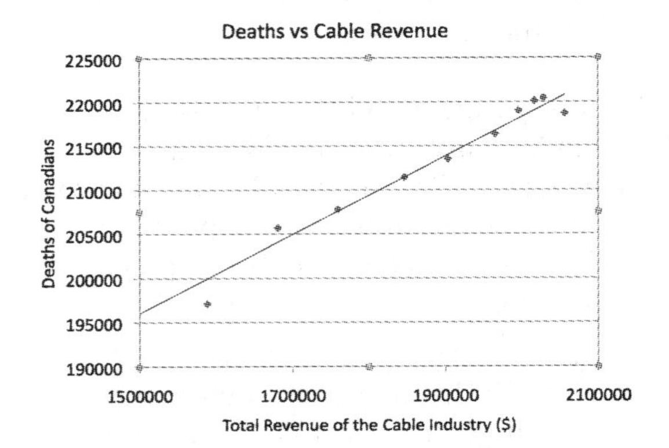

Source: Statistics Canada, Tables 101-0501 and 353-0001

a) What kind of causal relationship do you think this likely is?

b) Go to *www.mcgrawhill.ca/links/MDM12* and select **Study Guide.** Follow the links to obtain a copy of the file **Cabledeaths.csv**. Create scatter plots of deaths versus year, cable revenue versus year, and population versus year. What does this seem to say about the causal relationship between deaths and cable revenues?

6. Go to *www.mcgrawhill.ca/links/MDM12* and select **Study Guide**. Follow the links to obtain a copy of the file **InternetRollerCoasters.csv**. This file contains data about the number of broadband Internet connections per 10 people and the number of roller coasters per 10 million people in 30 countries.

   a) What type of causal relationship do you think these data have?

   b) Are there any outliers in this set of data? If so, identify them and indicate how they are outliers.

   c) How does Canada compare to the other countries?

7. A new cancer study has just come out linking cancer rates in females to pregnancy. The study claims that there is a strong correlation between the cancer rate for females and the number of pregnant women.

   a) Go to *www.mcgrawhill.ca/links/MDM12* and select **Study Guide**. Follow the links to obtain a copy of the file **PregnancyCancer.csv**. This contains data regarding the cancer rate in females (per 100 000) and the number of pregnancies. Create a linear-regression model that would agree with the claim of this study. What does the slope represent in the context of this question?

   b) What type of causal relationship is likely occurring here? Justify your answer.

★ 8. A study concludes that if car seats are used for children from two years to six years of age, there will be fewer fatalities from automobile accidents.

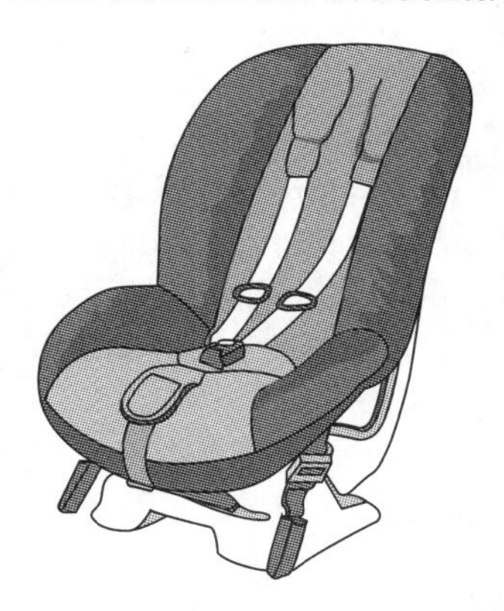

   a) What type of causal relationship do you think this represents?

   b) Go to *www.mcgrawhill.ca/links/MDM12* and select **Study Guide**. Follow the links to obtain a copy of the file **CrashData.csv**. This represents a summary of injuries sustained by children in the two-to-six-years age range in automobile accidents for the years 2005 to 2008. Determine the percent of fatal injuries for each protection system for the entire four years. Based on this, is the original claim correct?

   c) Who might not want this information to be known?

C

9. Go to *www.mcgrawhill.ca/links/MDM12* and select **Study Guide**. Follow the links to obtain a copy of the file **CauseAndEffect.csv**. This contains several data sets on various topics. Use these data sets to create some linear-regression models that are strongly correlated but are not cause-and-effect relationships. Describe what type of relationship each is.

KEY CONCEPTS

- Critical analysis of statistics can be challenging. Statistics that we see or hear are often presented only in summary form. This makes it difficult to know where the numbers originated, how the data were collected, whether any bias was present, or if other factors exist that could create a skewed perspective of the data. Often, very subtle wording can change the meaning of information.

Example

In February 2005, U.S. President George Bush announced a proposal to partially privatize a retirement-income program. The president stated that U.S. workers could "set aside four percentage points of their payroll taxes into private accounts." Many news stories reported that the proposal would redirect 4% of those taxes into private accounts, a figure that made the change appear to be small. However, if a worker paid a tax of 6.2%, redirecting four *percentage points* meant that $\frac{4}{6.2}$ or 65% of that payment could go to a private account!

- The use of large numbers can lead to misunderstandings about the significance of data and cause information to be taken out of context.

Example

Suppose that a news report states that annual health-care spending in Ontario will increase by $80 million. That appears to be a significant amount. However, because the population of Ontario is approximately 10 million, the increase works out to only about $8 per person.

- Be careful about comparing statistics about items that are not weighted equally.

Example

The bar chart shows unemployment statistics, by province, for Canada.

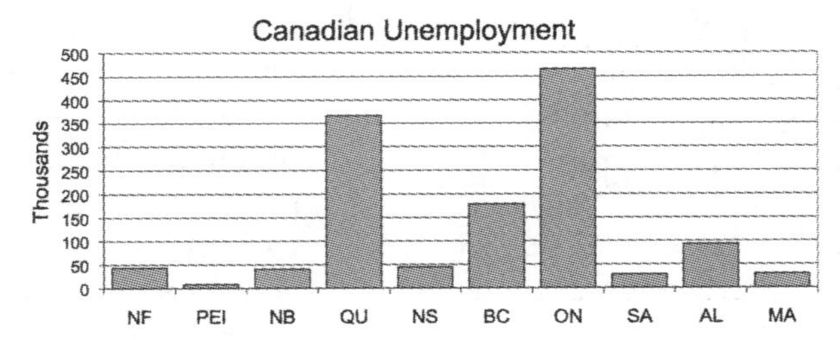

Based on this chart, a newspaper headline might be "Ontarians, Watch Out!" However, the fact that a large number of Ontario residents are unemployed should not be surprising given that Ontario has the greatest population. A more reasonable—and more accurate—depiction of unemployment would be a percent or per capita representation of the situation.

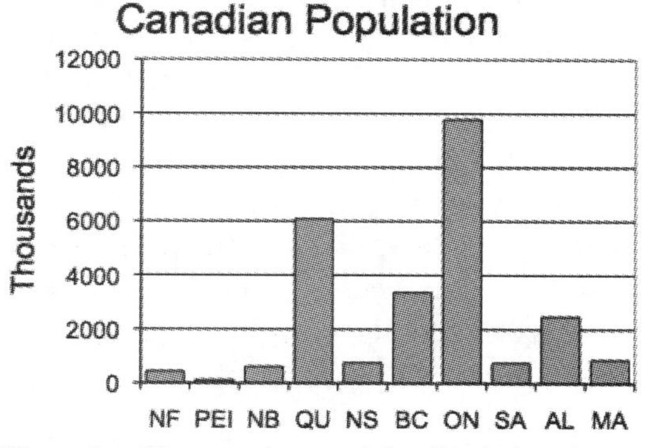

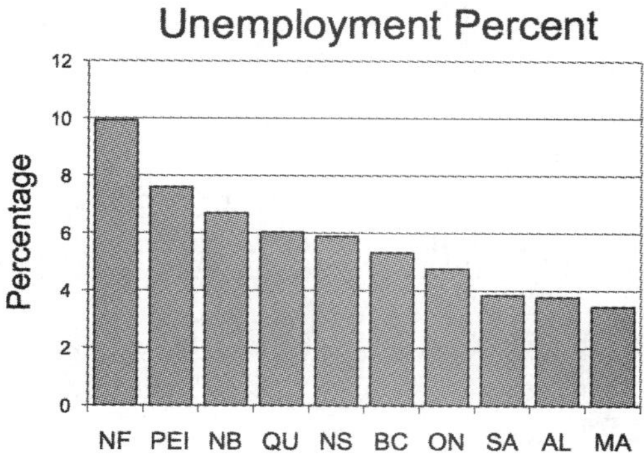

- When looking at data critically, it is often important to do more than a superficial examination. Looking much deeper into data may be required to find the correct meaning.
- Often, statistical mistakes (whether accidental or deliberate) can hide the real meaning of data. For this reason, when the media present information, you should be cautious about accepting any claim that does not include information about the sampling technique and analytical methods used.
- Sometimes small samples can be used to represent larger populations, and those small samples can distort the data.
- Extraneous variables must be eliminated or accounted for.
- A hidden variable can skew statistical results and yet be hard to detect.

A

1. In each case, determine the actual percent increase or decrease.

a) Last year, the unemployment rate was 8.5%. This year, unemployment has had a 1.5 percentage-point increase.

b) You are offered a bonus assignment that would increase your grade of 67% by 3 percentage points.

c) The percent of salmon returning to spawn in a particular stream was 45% last year. This year, there was a drop of 5 percentage points.

2. For each situation, manipulate the numbers to make better sense of the "big number".

a) A college diploma can help you to earn an additional $2 000 000 over your lifetime.

b) The estimated budget for the Canadian military in 2009 was $19.1 billion. The population of Canada is approximately 30 million.

c) The annual Super Bowl football championship has approximately 90 million television viewers and generates about $500 million in advertising revenue.

d) The 2009 annual revenue of the Canadian Cancer Society was about $200 million. The number of new cancer cases in Canada in 2009 was approximately 171 000.

e) Kentucky Derby–winning jockey Calvin Borel has competed in more than 25 000 races, winning more than 4300 of them.

f) Professional baseball player Alex Rodriguez has an annual salary of $25 million for playing 162 games.

B

3. Global temperature is often correlated with levels of carbon dioxide gas (CO_2) in the atmosphere. A large amount of data is being collected, and some information can be manipulated to show particular viewpoints.

   a) Dr. P. E. Semist collected data for almost a year and created the following graph:

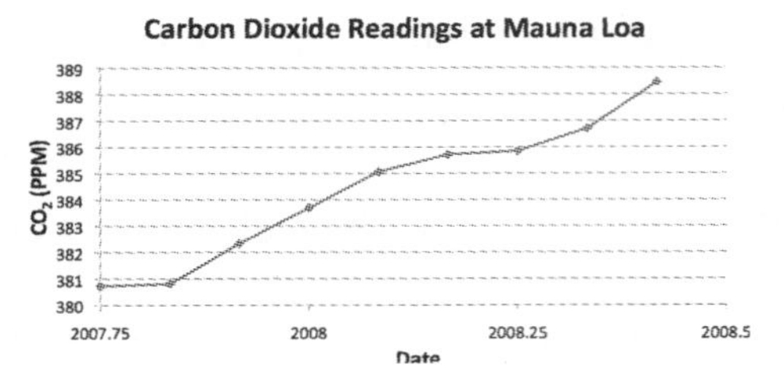

   What conclusion do you think Dr. Semist will make regarding CO_2 emissions?

   b) Dr. Izzy Nosobad decides to collect his own data. Results are displayed in the following graph:

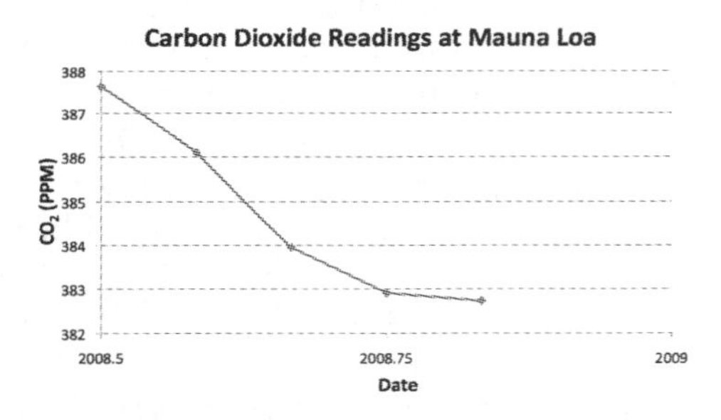

   What conclusion do you think Dr. Nosobad will make about CO_2 emissions?

   c) Dr. Beatrice Real decides to review the conclusions of the other scientists. Go to *www.mcgrawhill.ca/links/MDM12* and select **Study Guide**. Follow the links to obtain a copy of the file **BReal.csv** to see her work. What conclusions do you think Dr. Real will make?

   d) What does your answer in part c) tell you about sample size?

   e) Create a linear-regression model for this data. Based on your model, predict what the CO_2 level in the atmosphere will be at the beginning of 2015. Why might this not be a very accurate estimate?

4. The data below, from *www.quantcast.com*, show estimated traffic for some of the most popular web sites.

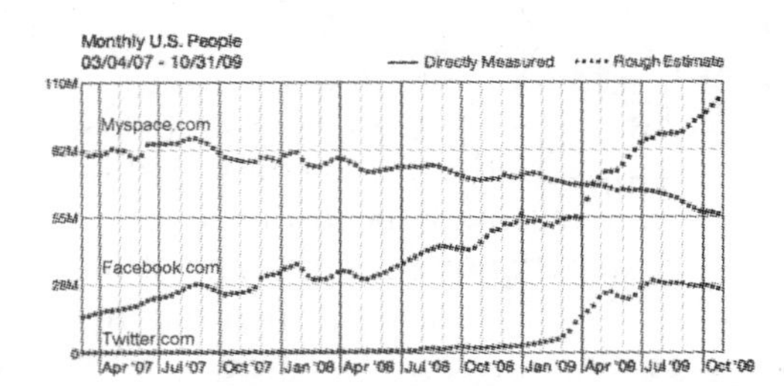

   a) Who might be interested in this data?

   b) What are some potential problems with the data?

   c) What kinds of stories could be written about this data?

5. Go to *www.mcgrawhill.ca/links/MDM12* and select **Study Guide**. Follow the links to obtain a copy of the file **NASCAR2009.csv**. This represents summary data for the 2009 NASCAR racing season.

   a) The driver with the most points at the end of the season wins the championship. Create linear-regression models of points versus each of pole positions, wins, top 5 finishes, and top 10 finishes to determine which has the strongest relationship to winning the championship.

   b) Even though your answer in part a) is a strong correlation, why would this linear model not be appropriate for this set of data? Research the NASCAR point structure to explain why the data are this way.

★ **6.** Go to *www.mcgrawhill.ca/links/MDM12*, select **Study Guide**, and follow the links to the article "Will Women Ever Outrun Men?" Read the article.

Then, open the data set **MarathonTimes.ftm**. This has the progression of world-record times for men and women in marathon running.

a) Create a linear-regression model for **Time** versus **Date**. Why is this model not appropriate for these data?

b) Drag the **Gender** attribute to the centre of the graph. Based on this information, will women ever be faster than men? Justify your answer.

7. Your local newspaper publishes the following graph under the headline "Ontarians Are the Laziest!"

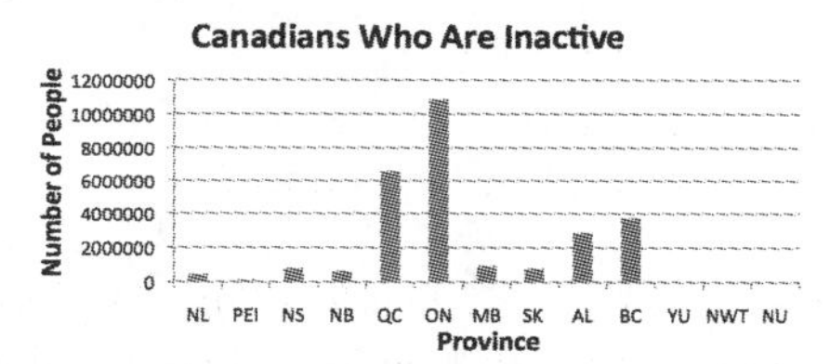

Go to *www.mcgrawhill.ca/links/MDM12* and select **Study Guide**. Use the data found in **Laziest.csv** to create a more realistic picture. What headline might be more appropriate?

C

8. Create an intentional bias about a statistical claim to try to convince your

a) principal

b) teacher

c) parent or guardian

Explain how you introduced the bias to manipulate the truth about the statistical claim.

9. a) Write down your best guess of the cost of each item in the list.

Item	Guessed Price ($)
Men's cologne	
12-pack AA batteries	
Newly released movie DVD	
10 Megapixel digital camera	
Pack of 8 razor blades	
Women's deodorant	
4GB SD memory card	
Box of premium chocolates	
8 GB MP3 player	
Newly released video game	
Leather gloves	
Leather handbag	
Flared jeans	
Leather boots	
D.J. style headphones	

b) Go to *www.mcgrawhill.ca/links/MDM12* and select **Study Guide**. Open the file **Prices.ftm**. Enter your data in the Guessed Price column. Create a linear-regression model of **Actual Price** versus **Guessed Price**. How good a guesser do you think you are?

c) What kind of correlation would you expect for a good guesser? What if you guessed exactly $10 more than the actual price for each item? What would your regression model be? What does this tell you about looking at the correlation alone to determine a good guesser?

d) What do you think the slope means in the context of the data? What if you guessed exactly double the actual price for each item? What would the characteristics of the regression model be in this case?

e) What do you think the characteristics would be of a regression model for a good guesser?

Chapter 4 Permutations and Organized Counting

4.1 Organized Counting

KEY CONCEPTS

- Combinatorics is the branch of mathematics dealing with the ideas and methods for counting, especially in complex situations.

- Tree diagrams are useful tools in organized counting.

Example

A golf club driver consists of the shaft, the grip, and the head. When ordering a custom club, you have your choice of steel, titanium, or carbon fibre for the shaft. For the head, you may select aluminum or titanium. For the grip, you have your choice of leather wrapped, rubberized, or silicone. How many different styles of clubs can be made?

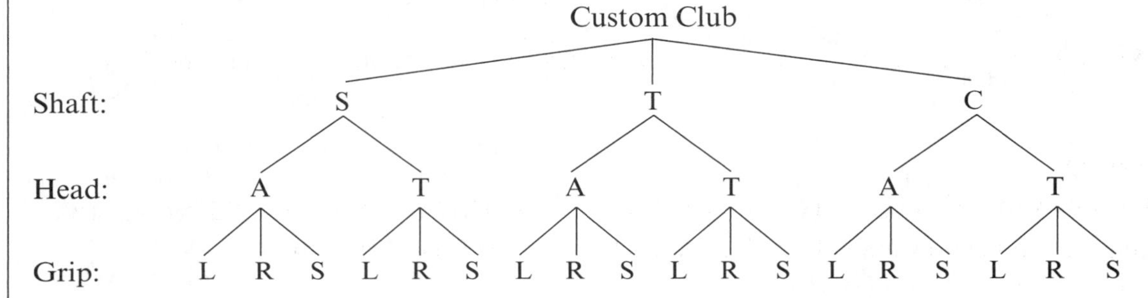

There are 18 different styles of club that can be made. For example, style TTL has a titanium shaft, a titanium head, and a leather grip.

- The fundamental counting principle (also called the multiplicative counting principle) states that if a task or process is made up of stages with separate choices, the total number of choices that can be made is the product of the number of possible choices at each stage.

Example

There are three choices in the previous example. Each choice has a number of possibilities:

Choice 1		Choice 2		Choice 3		
Shaft Type		Head Type		Grip Type		
3 Possibilities		2 Possibilities		3 Possibilities		
3	×	2	×	3	=	18 Clubs

- The additive counting principle (also called the rule of sum) must be used in a situation where some of the actions cannot occur at the same time. These actions are referred to as mutually exclusive events. Cases must be used to solve these types of problems. The results of each case are added together to get the total number of different results. Seeing the word "either" or "or" in a question is often a clue that this method should be used.

Example

A meal combo in the school cafeteria allows a student to choose one entree (sandwich, pasta, or chicken strips) and either a soup (tomato, bean, or vegetable) or a dessert (ice cream or fruit cup). How many different meal combinations are there?

Case 1: Soup Chosen			Case 2: Dessert Chosen		
Choice 1 Entree Type		Choice 2 Soup Type	Choice 1 Entree Type		Choice 2 Dessert Type
3	×	3	3	×	2

$$\begin{aligned}\text{Total} &= \text{Case 1} + \text{Case 2}\\ &= 9 + 6 \text{ or } 15 \text{ combos}\end{aligned}$$

- Sometimes it is easier to determine the complement of a situation to determine the answer.

Example

How many four-digit codes (which may begin with 0) do not have all four digits the same?

First, determine how many four-digit codes there are with no restrictions:

1st digit		2nd digit		3rd digit		4th digit		
10	×	10	×	10	×	10	=	10 000

Now determine what you do not want (the complement): there are exactly 10 codes with all digits the same (0000, 1111, … , 9999). Therefore, the total number of codes without four matching digits is 10 000 − 10, or 9990.

A

1. Create a tree diagram to determine all the ways you can arrange the letters of the word MATH.

2. Use the fundamental counting principle to answer question 1.

3. **a)** Create a tree diagram to represent how many outcomes there are when you flip a coin four times in a row.

   b) How many outcomes have exactly two heads?

4. Create a tree diagram to represent all the different outcomes of flipping a coin and rolling a standard six-sided die.

5. In the film *The Da Vinci Code*, the main character must determine the correct word in a five-letter cryptex. A cryptex is a kind of combination lock where five rings of 26 letters must be rotated to spell a word. How many different arrangements of letters are possible?

6. Jane, Hank, Aneesh, Grace, and Eve are sitting side-by-side in a row. How many different ways can they be positioned in the row?

B

7. How many different five-digit even numbers can be made from the digits 2, 3, 4, 5, 6, and 9?

8. In many driving-simulation programs, you can customize your vehicle. Suppose that a program has ten car models, and each model has the following options:

Option	Number of Choices
Colour	60
Wheels	64
Neon underglow	4
Spoilers	5
Bumpers	5
Decals	70
Window tint	2
Tail lights	4
Muffler tips	60
Headlights	4
Side mirrors	5
Doors	3

   a) How many possible car styles can be created?

   b) Why do you think game makers include this feature in their games?

   c) Automobile manufacturers often offer vehicle options, but usually make them available in packages rather than as individual options. Why do you think they do it this way?

9. The code on an electronic lock can be up to four digits long. How many different codes are there if a code must be at least two digits long?

10. Combination locks for lockers typically have three numbers in the combination and often have the numbers from 1 to 40 to choose from for each value. How many locks can be produced before a lock with a duplicate combination is made?

☆ 11. The diagram represents a simple dart board, with the points for each section shown. Use a tree diagram to determine all the possible point totals if you throw three darts and

   a) you hit the board each time

   b) you might not hit the board each time

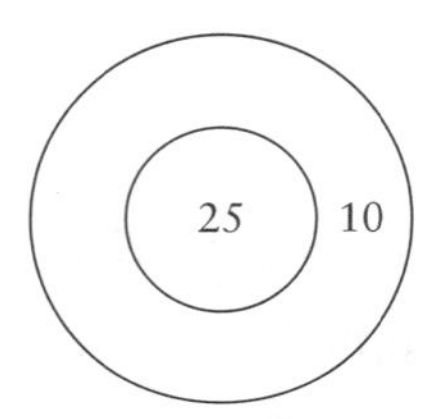

12. Carlita has the following uniform items in her closet:

Pants/skirts	grey pants, blue pants, blue shorts, plaid skirt
Shirts	white short-sleeve, white long-sleeve
Sweaters	blue cardigan, blue sweater, grey hoodie
Shoes	flats, sandals, running
Accessories	belt, scarf, headband

   a) How many ways can Carlita dress if she chooses one item from each category?

   b) How many days can Carlita go without wearing the same outfit if the only things she must wear are a shirt, pants or a skirt, and a pair of shoes?

13. Your sister has just received the latest toy craze, a stuffed animal called a Webpetz. The toy comes with an access code for the toy's online site where your sister can interact virtually with her pet.

a) The site currently has 1 000 000 users, and the access code is a four-character code that is not case sensitive but can have letters and numbers. How many more users can the site have based on the access code?

b) How many more users could the site have if the code was increased to six characters?

c) Your sister has forgotten her code after a couple of weeks. You help her find the original tag with the code, but the tag is partially ruined as shown:

You decide to help your sister by guessing the missing information. What is the total possible number of times you could try before getting the correct code?

14. Phone numbers in Canada and the United States consist of a three-digit area code, a three-digit exchange, and a four-digit number.

a) How many seven-digit phone numbers are possible if there are no restrictions?

b) How many three-digit exchanges are there if there are no restrictions?

c) If there are 210 exchanges in Ontario, what is the total possible number of seven-digit phone numbers?

d) If 11 area codes are in use in Ontario, what is the total possible number of ten-digit phone numbers?

15. Airlines use different rules and options to set fares. Because the choice of options affects the price of a ticket, it is possible for everyone on the same flight to have paid a different fare. Based on the following rules and options, how many different ticket prices could there be?

Rule	Options
Number of stopovers	0, 1, 2
Rewards member	yes, no
Number of seats	1, 2, 3, 4
Purchase location	airport, online, travel agent
Time of flight	morning, after-noon, evening, overnight
Type of seat	economy, busi-ness, first class
In-flight meal	yes, no
Number of extra bags	0, 1, 2, 3
Purchase date	more than two weeks in advance, within two weeks of flight, same day as flight
Time of day ticket purchased	regular business hours, evening, midnight to 8:00 a.m.
Discount code used	yes, no

☆ 16. Postal codes in Canada have six characters with alternating letters and digits in the form L#L#L#. How many postal codes do not have one letter repeated three times?

17. Many electronic adventure games require a player to go on a quest, moving from place to place. Suppose you begin a game in a room with four doors. Each door leads to new areas, each of which has three exits. As you move from area to area, you encounter either four exits or three exits, alternating from area to area. This continues for eight levels. Assuming that you can never enter the same area by more than one route, how many areas will there be at the eighth level?

C

☆ **18.** The song "A Hard Day's Night" by the Beatles begins with a unique and iconic guitar chord. Suppose that a guitar has 6 strings and 21 frets on the neck.

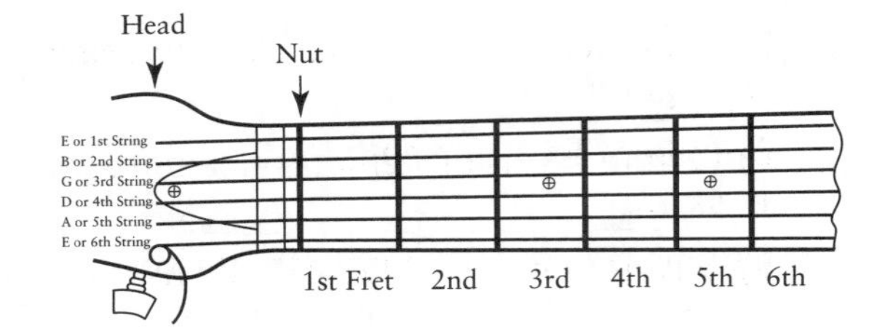

a) Determine the total number of possible chords that could be played assuming that

i) no fret may be used more than once in a chord (i.e., one finger per fret)

ii) frets must be used in blocks of four (one for each finger) and each finger must be used

b) Assuming that it takes 2 s to play a chord, how long would it take if every chord were played back to back?

c) Do you think the answer in part a) overestimates or underestimates the total number of chords? Explain.

19. A serial number is often needed to activate new computer software. For serial numbers to work, the software must store all of the possible correct serial numbers on the installation disk. Assume that a number has 24 characters (digits and letters) and that one byte of computer memory is needed to store each character.

a) How much memory would have to be used to store all possible serial numbers? Express your answer in terabytes, where 1 terabyte = 2^{40} bytes.

b) Samples of serial numbers for a particular software product are shown.

VMT4-57XK-9BBD-2B3Y-VQ34-BB46
VMG3-52XT-6BBD-2P2Y-YQ31-BB46
VMR1-58XQ-8BBD-2R6Y-RQ39-BB46

How much memory would you need to store all the possible serial numbers for this software? State any assumptions that you make.

20. The Build-Your-Own Ice Cream shop claims to offer more than 20 million different ice cream cone possibilities. Is this a correct claim based on the following information?

Build Your Own
1. Cone: plain, sugar, chocolate
2. Scoops: 1, 2, or 3
3. Flavours: 36
4. Toppings: 10 (choose up to 2)

KEY CONCEPTS

- A factorial indicates the multiplication of consecutive natural numbers.
  $n! = n(n-1)(n-2) \times \ldots \times 1$
- The number of permutations of n distinct items chosen n at a time in a definite order is $_nP_n = n!$
- The number of permutations of r items taken from n distinct items is. $_nP_r = \frac{n!}{(n-r)!}$.
- Another notation for $_nP_r$ is $P(n, r)$.
- When you see the word *arrangement*, the problem is dealing with a permutation. This means that order matters.

A

1. Evaluate.

a) $6!$ **b)** $12!$ **c)** $56!$

2. In each case, determine whether the expression is true or false.

a) $3! + 3! = 6!$ **b)** $(5!)^2 = 25!$

c) $(4 + 4)! = 8!$ **d)** $(6^2)! = 36!$

e) $5!6! = 30!$ **f)** $(4 \times 2)! = 8!$

g) $\left(\frac{15}{3}\right)! = 5!$ **h)** $\frac{15!}{3!} = 5!$

3. What is the greatest factorial that you can evaluate on your calculator?

4. Express each as a fraction involving only factorials.

a) $8 \times 7 \times 6 \times 5 \times 4$

b) $78 \times 77 \times 76$

c) $n(n-1)(n-2)(n-3)$

5. Simplify first and then, evaluate.

a) $\frac{32!}{25!}$ **b)** $\frac{89!}{84!}$

c) $\frac{120!}{118!}$ **d)** $10 \times 9!$

e) $16 \times 17 \times 15!$

6. Express each permutation statement using $_nP_r$ notation.

a) permutations of 12 items taken 3 at a time

b) permutations of 32 items taken 9 at a time

c) permutations of k items taken z at a time

d) $\frac{52!}{(52-8)!}$ **e)** $\frac{19!}{15!}$

f) $7 \times 6 \times 5 \times 4 \times 3 \times 2 \times 1$

g) $101 \times 100 \times 99$

B

7. a) Determine the number of possibilities for

i) the number of arrangements of four different coloured blocks

ii) the number of ways four people can be arranged in a line

iii) the number of anagrams (arrangements) of the word TIME

b) Many combinatorics problems are said to be the same problem in disguise. How does part a) support that statement?

8. Five people are lined up for a 100-m race.

a) Determine the number of different ways the first and second positions can be assigned by

i) writing out all the possibilities

ii) using a tree diagram

iii) using the fundamental theorem of counting

iv) using permutations

b) Which method requires the least amount of work?

c) Which method gives the most information?

9. A puzzle consists of 15 numbered tiles and 1 open space on a 16-square grid. A player uses the open space to slide tiles one at a time up, down, left, or right. The object of the game is to arrange the tiles so that the numbers are in order. How many scrambled versions of the puzzle are there?

2	7	3	10
5	8	1	11
9	14	6	4
13		15	12

1	2	3	4
5	6	7	8
9	10	11	12
13	14	15	

10. A baseball team is lining up for a team photo. There are ten players and two coaches. How many ways can they be arranged if

a) everyone is in one line?

b) everyone is in one line and the coaches are at each end?

c) the players line up in two rows of five with the coaches at each end of the back row?

☆**11.** You are seating people at a circular table.

a) How many different arrangements are possible if the number of people is

i) 3? **ii)** 4? **iii)** 5?

b) Determine a general method for solving this problem when n people are to be seated.

12. You disconnect the thermostat from your furnace but neglect to jot down how the wires connect to it. There are three wires coming from the wall (red, black, and white) but four terminals on the thermostat. How many different ways can the wires be connected?

13. For a holiday gift exchange, each of four friends buys and wraps one gift. The gifts are then distributed randomly, one to each person. How many different ways can the gifts be given out if

a) there are no restrictions?

b) no one is allowed to get his or her own gift?

14. How many ways can the vertices of the polygon be labelled using the letters shown?

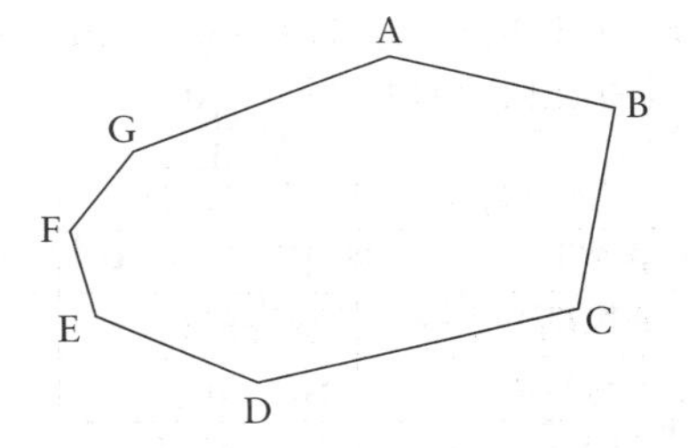

15. Your school is creating a new academic team, the Data Managers. The team will have four members to perform different tasks: collector, analyzer, grapher, and recorder. Ten people try out for the team. How many different teams can be made if

a) there are no restrictions?

b) the team is to have two alternate members?

c) Alonzo and Jen are the only two who can be collector or analyzer?

16. What does 0! equal? Explain why this is so.

17. You purchase 215 songs online and download them to listen to on your computer.

a) By using the shuffle feature of your music software, how many different playlists can your computer make if it uses all the songs?

b) Suppose the software will show only the next ten songs in the playlist. How many different arrangements of songs are possible to be shown?

c) Many software programs that play music learn your favourite music and will play those songs more often. How many different ten-song playlists can the software create if each one must start and end with one of ten songs by your favourite artist?

18. A student must choose one course for each of four periods. Based on the course selections below, how many different schedules can be made if

a) there are no restrictions?

b) English is during first period and a math course is in last period?

c) a student takes MCV4U and must also take MHF4U?

AMU4M	MDM4U
BAT4M	SBI4U
ENG4U	SCH4U
EBT4O	SPH4U
PSE4U	CGW4U
IDC4U	HZT4U
FSF4U	TGJ4M
MHF4U	ICS4U
MCV4U	

☆**19.** How many arrangements are there of the letters in FERMAT if

a) there are no restrictions and you use all the letters?

b) you use all the letters but each arrangement must end in a vowel?

c) you use only five of the letters?

d) you use only five of the letters but the first one must be T and the arrangement must end in a vowel?

20. How many four-digit numbers are there that

a) have no restrictions?

b) have four unique digits?

c) have unique digits and contain the number 21 somewhere in the arrangement?

C

21. For what values of n is $n!$ less than 3^n? Show your work.

22. How many zeros will there be at the end of 38!?

23. A double factorial, $n!!$, is defined as follows:

$$n!! = \begin{cases} n(n-2)\ldots 5\times 3\times 1 & n>0, \text{ odd} \\ n(n-2)\ldots 6\times 4\times 2 & n>0, \text{ even} \\ 1 & n=0 \end{cases}$$

Write out and evaluate the following based on the definition.

a) 10!! **b)** 17!! **c)** 9!!8!!

24. Determine 70!.

☆**25.** Simplify.

a) $\frac{(n+2)!}{n!}$ **b)** $\frac{(n+2)!}{n}$

c) $\frac{(2n)!}{(2n-2)!}$ **d)** $\frac{(n+2)!}{(n-2)!}$

26. A primorial, $n\#$, is like a factorial but uses only prime numbers. That is, $n\#$ is the product of all the prime numbers less than or equal to n. Evaluate the following primorials.

a) 5# **b)** 13# **c)** 17#

KEY CONCEPTS

- The number of permutations of n items that include a identical items of one type, b identical items of another type, c identical items of yet another type, and so on, is $\dfrac{n!}{a!b!c!\ldots}$

Example

How many anagrams are there for the word NOON?

Explictly:

NOON
NONO
NNOO
ONNO
ONON
OONN

Using Factorials:

$$\frac{n!}{a!b!c!\ldots} = \frac{4!}{2!2!} = \frac{4 \times 3 \times 2 \times 1}{2 \times 1 \times 2 \times 1} = 6$$

Using a Tree Diagram:

N	N	O – O		NNOO
	O	N – O		NONO
		O – N		NOON
O	N	N – O		ONNO
		O – N		ONON
	O	N – N		OONN

A

1. How many anagrams are there of each word?
   a) PASCAL
   b) TORONTO
   c) MATHEMATICS
   d) GOOGLE
   e) STATISTICAL

2. Consider the word *LOGO*.
   a) Determine the number of arrangements by
      i) writing them out explicitly
      ii) creating a tree diagram
      iii) using factorials
   b) Which method requires the least amount of work?
   c) Which method yields the most information?

B

3. For security, an e-mail program requires you to have a ten-character password and change it every three weeks. You like your current password, *XXQWERTYZZ*, because it is relatively easy to remember and type in. At the very least, you would like to keep the same letters when you change your password. How many different passwords can be created using your letters
   a) with no restrictions?
   b) keeping *QWERTY* together?
   c) keeping *QWERTY*, the *Z*s, and the *X*s together?

4. An interior designer is creating a pattern of tiles for a border. If the pattern is to contain an equal number of tiles of each of four colours, how many different patterns can be made if
   a) only four tiles are used in the pattern?
   b) eight tiles are used in the pattern?
   c) eight tiles are used and a green tile must begin and end the pattern?

☆**5.** The board game Mastermind® uses a four-colour code that can be made up of any of six colours: red, white, green, blue, black, and yellow. One player creates a code and the second player tries to solve it in as few attempts as possible.

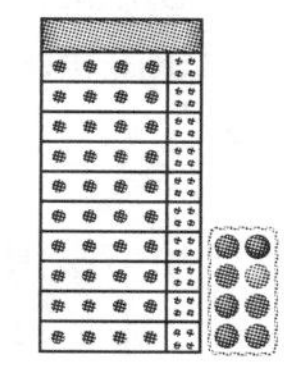

a) How many different codes are possible if

i) four different colours are used for the code?

ii) three different colours are used (with one colour repeated)?

iii) two different colours are used?

b) Using the fundamental counting principle, determine the total number of different codes possible. How is this number related to your answers in part a)?

c) On each turn, the code creator lets the code breaker know how close he or she is by showing up to four pegs. A white peg indicates a correct colour but not in the correct position. A black peg indicates a correct colour in the correct position. Assume the code maker shows three white pegs and one black peg. How many different possible next moves are there if

i) all the colours are different?

ii) there are three colours with one repeated?

6. In the movie *National Treasure*, the main character breaks into a room that has a password-protected lock. The password used the letters *AEFGLORVY*.

a) How many possible anagrams of the letters exist?

b) The actual password entered was *VALLEYFORGE*. How many more or fewer possible anagrams exist if two *E*s and two *L*s can be used?

c) How many more passwords exist if any one of the letters could have been used twice?

C

☆**7.** *Who Wants to Be a Mathematician?* is the latest reality show on the Math Network. On one segment, contestants are given five coloured blocks: two red, two blue, and one green. They are to create a code sequence to represent the letters of the alphabet and the ten digits.

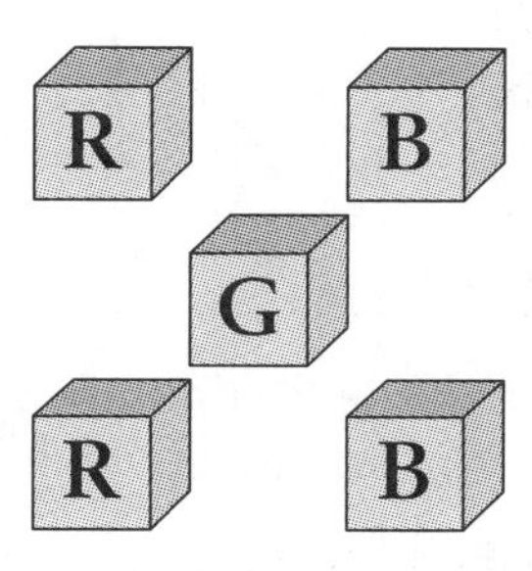

A sample of some of the possible codes is shown:

1 = R
2 = B
3 = G
4 = RB
5 = BR
6 = BG
7 = GB
8 = RG

Determine the greatest number of blocks needed in each code to be able to represent all the letters and digits.

KEY CONCEPTS

- Pascal's triangle is a pattern of numbers that, although named for Blaise Pascal, has existed for hundreds of years. It is a fairly simple number pattern that has many complex patterns embedded within it.

  The first seven rows of Pascal's triangle are shown. Note that although these are stated to be the first seven rows, there are actually eight rows. This is because the first row (which contains only a single 1) is considered the zeroth row. The second number in any row indicates the row number.

  1
  1 1
  1 2 1
  1 3 3 1
  1 4 6 4 1
  1 5 10 10 5 1
  1 6 15 20 15 6 1
  1 7 21 35 35 21 7 1

  Similarly, the first number in any row is considered to be the zeroth element. The second number in any row (which is the same as the row number) is considered the first element. For example,

  the third element in the sixth row is 20 and is denoted $t_{6,3} = 20$.

- Each row of Pascal's triangle begins and ends with a value of 1. To generate any other number in the table, Pascal's method is used. In Pascal's method, to determine any value, determine the sum of the two values above it.

  For example,

  the first value of 35 in the seventh row is the sum of the 15 and 20 above it.

- Many other patterns occur in Pascal's triangle. For example,

  the sum of the elements in any nth row will be 2^n

  the elements of the third diagonal are the triangular numbers

- Pascal's method can be used for counting paths in a number of arrays and grids, such as city streets and checkerboards.

A

1. Consider the part of Pascal's triangle shown.

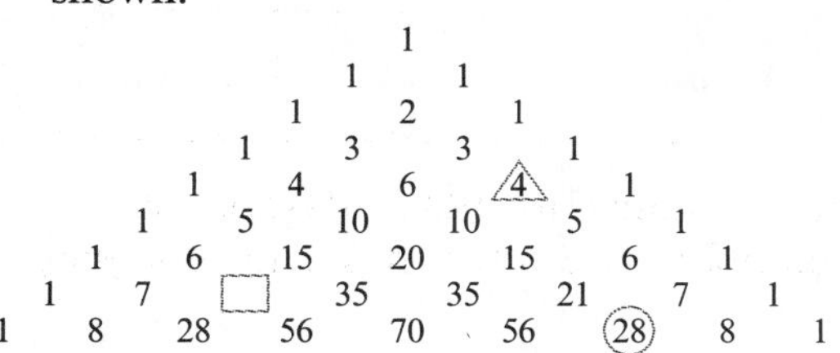

   a) Write out the next row. What row number is it?

   b) What row and element number is the value in the triangle?

   c) What is the value in the circle, in $t_{n,r}$ notation?

   d) What value belongs in the box? Give three reasons why you know.

   e) Describe the symmetry of the values in the triangle.

2. Consider the triangle below.

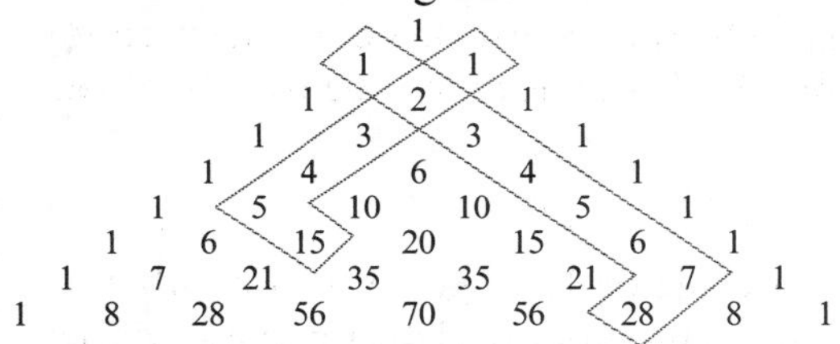

   a) The pattern is sometimes called the hockey stick pattern. Describe the pattern.

   b) Create your own hockey stick pattern.

3. For any row of Pascal's triangle where the first element is prime, what is true about all the other elements in that row (excluding the 1s)?

4. Determine the number of ways to spell each word.

   a) SAMPLE

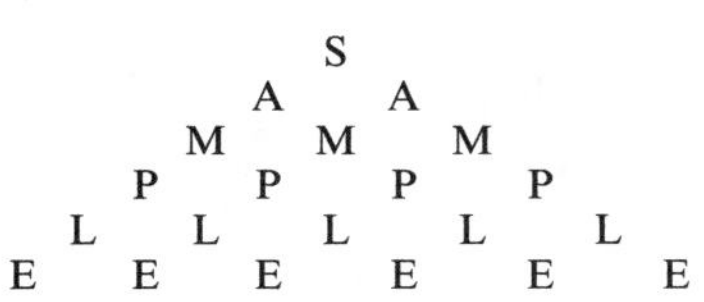

   b) CORRELATION

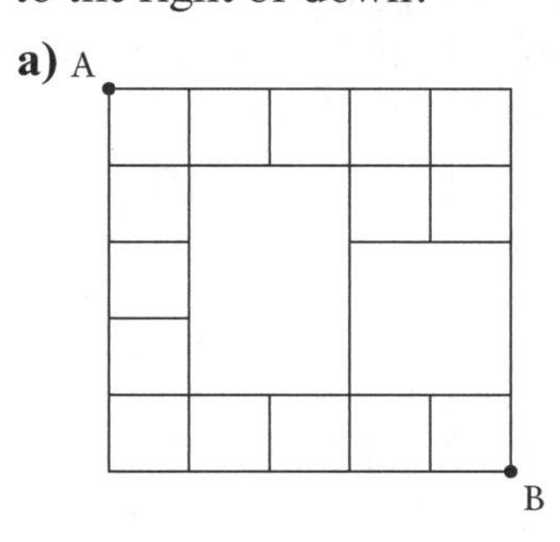

B

5. How many ways are there to go from A to B in each diagram if you can travel only to the right or down?

   a)

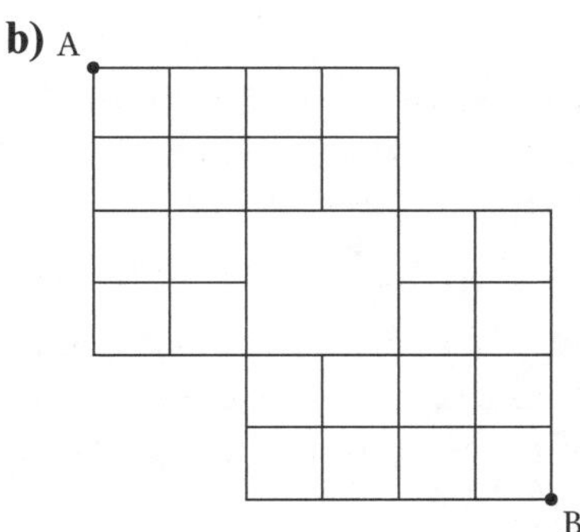

   b)

☆6. a) Create a tree diagram to represent the possible outcomes of flipping a coin five times. Complete the table based on your diagram.

Flip	Possible Outcomes	Number of Each Outcome
1	1H, 1T	1, 1
2	2H, 1H1T, 2T	1, 2, 1
3		
4		
5		

   b) How do these results relate to Pascal's triangle?

7. **a)** Complete the table indicating the number of each geometric feature formed by joining points on the circumference of the circle in each diagram.

Image	Circles	Points	Line Segments	Triangles	Quadri-laterals	Pentagons
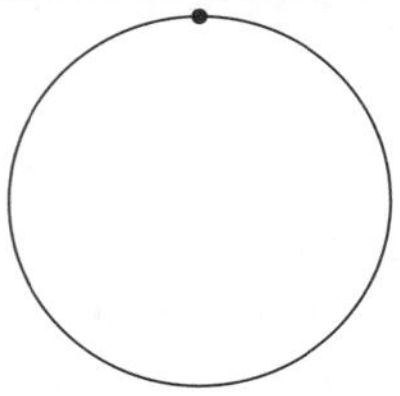	1	1				
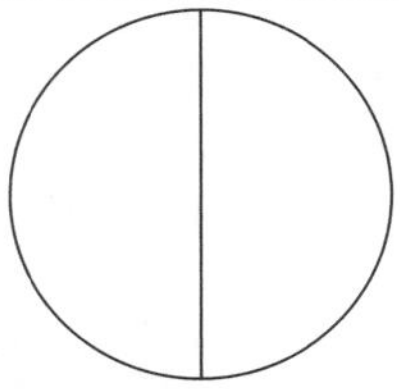						
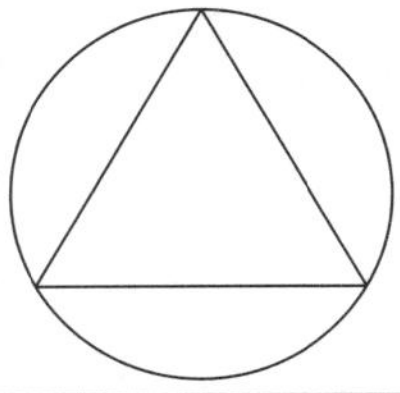						
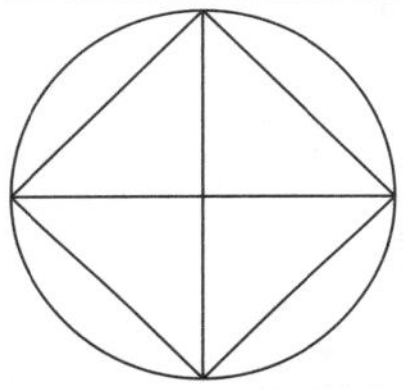						
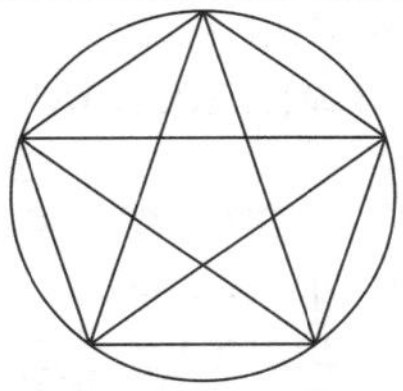						

b) How do your answers relate to Pascal's triangle?

c) How many hexagons would there be in a circle with eight points?

C

8. Create your own grid pattern and determine the number of ways to get from one corner to another.

9. Create your own word pattern and determine the number of ways to spell it.

Chapter 5 Combinations and the Binomial Theorem

5.1 Organized Counting With Venn Diagrams

KEY CONCEPTS

- Venn diagrams are a way to visualize sets of data where there may be overlapping classifications.

Example

Consider the following subsets of words:

Subset A: three-letter words {dog, hip, eye, bob, cat, why, zig, zag, gag}

Subset B: palindromic words {stats, eye, bob, gag, rotor, civic, radar}

Create a Venn diagram to represent these data.

Solution

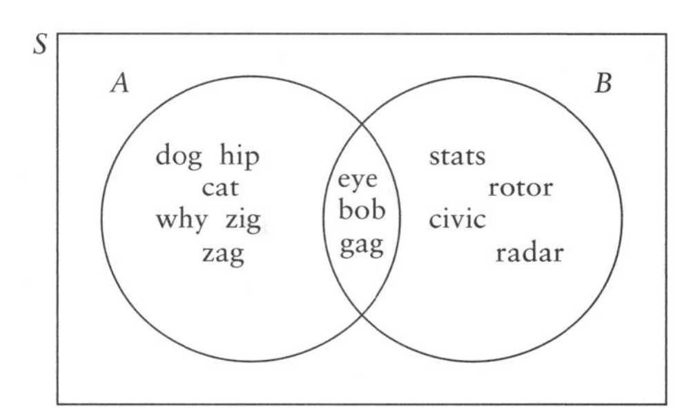

- In the previous example, the rectangle represents the universal set, S, of all words. Each circle represents a subset of the set S. The part where the two circles overlap is the section where elements of one subset are also elements of the other.
- When dealing with sets, certain notations are used.

Notation	Meaning
S	The universal set
A	The subset A
B'	The complement, or opposite, of subset B
$n(A)$	The number of elements in subset A
$A \cup B$	The union of sets A and B. This creates a set that contains all the elements that are in either set. This is sometimes called the "or" operation.
$A \cap B$	The intersection of sets A and B. This creates a set that contains only elements that are in both sets. This is sometimes called the "and" operation.
$\varnothing$	The null or empty set

Example

The following illustrate the use of notation rules using the sets in the first example.

i) $A \cup B$ = {dog, hip, eye, bob, cat, why, zig, zag, gag, stats, rotor, civic, radar}

ii) $n(A \cup B) = 13$

iii) $A \cap B$ = {eye, bob, gag}

iv) $n(A \cap B) = 3$

v) $(A \cap B)'$ = {dog, hip, cat, why, zig, zag, stats, rotor, civic, radar}

- The principle of inclusion and exclusion indicates how many elements are in the set that contains A or B:

  $n(A \cup B) = n(A) + n(B) - n(A \cap B)$

- For three subsets A, B, and C, the principle of inclusion and exclusion is

  $$n(A \cup B \cup C) = n(A) + n(B) + n(C) - n(A \cap B) - n(A \cap C) - n(B \cap C) + n(A \cap B \cap C)$$

- Sometimes Venn diagrams show only the number of elements in each section.

Example

Your school track team includes sprinters, jumpers, and hurdlers. Set A represents the sprinters, set B represents the jumpers, and set C represents the hurdlers.

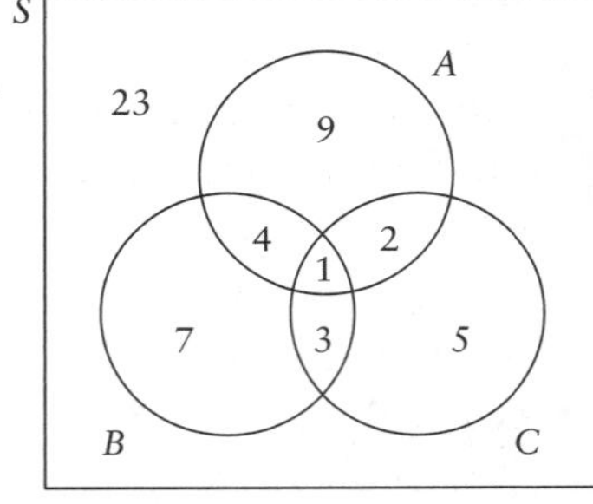

Determine the following and explain what each number represents.

a) $n(A)$

b) $n(S)$

c) $n(B) - n(A \cap B)$

Solution

a) $n(A) = 9 + 4 + 1 + 2$
$= 16$

This means there are 16 sprinters, some of whom are also doing other events.

b) $n(S) = 23 + 9 + 4 + 1 + 2 + 5 + 3 + 7$
$= 54$

This means there are 54 members of the entire track team. (Note that from the diagram, the value 23 represents the number of team members who are not sprinters, jumpers, or hurdlers.)

c) $n(B) - n(A \cap B) = (7 + 4 + 1 + 3) - (4 + 1)$
$= 10$

This means there are 10 members of the team who are jumpers but not sprinters.

A

1. Given a Venn diagram with two subsets A and B, shade the indicated areas.

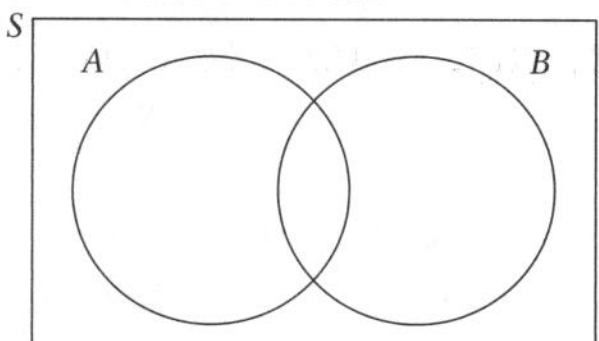

   a) A

   b) $A \cup B$

   c) B'

   d) $B \cap A'$

2. Consider the universal set of all the whole numbers from 1 to 20, inclusive. If set A consists of all the even numbers, set B is all the prime numbers, and set C is all the numbers divisible by 3, determine the following.

   a) $n(A)$
   b) $n(A \cap B)$
   c) $A \cup C$
   d) $n(B')$
   e) $A \cap C$
   f) $A \cap B \cap C$

B

☆3. A shoe manufacturer makes 60 models of shoes. Twelve models have the manufacturer's new JumpGoo soles. Fourteen models are basketball shoes, and four models are basketball shoes with JumpGoo soles.

   a) Create a Venn diagram based on what you know.

   b) How many models that are not basketball shoes do not have JumpGoo soles?

4. You are making up next week's work schedule. You ask each employee what shift they would like to work and compile the following information:
   - 3 people like to work midnights only
   - 5 people like to work afternoons only
   - 8 people like to work days only
   - 2 people like all shifts equally

   a) Create a Venn diagram to represent this situation.

   b) If one person works each shift each day, which people are more likely to get more hours next week?

5. Create a Venn diagram that shows how many cards in a deck are hearts or face cards.

6. You survey the students in your school about their primary means of electronic communication and find that
   - 38% use a cell phone
   - 53% use a social-networking site
   - 41% use e-mail
   - 23% use a cell phone and e-mail
   - 20% use a cell phone and a social-networking site
   - 22% use e-mail and a social-networking site
   - 15% use a cell phone, e-mail, and a social-networking site

   a) Create a Venn diagram to represent this information.

   b) What percent of students use only a cell phone?

   c) What percent of students use a cell phone or a social-networking site but do not use e-mail?

   d) What percent of students use a cell phone or e-mail?

C

☆7. Display in a Venn diagram the following information about student participation in a school's activity day.

	Males	Females
Participated	350	380
Did not participate	85	135

8. Go to *www.mcgrawhill.ca/links/MDM12* and select **Study Guide**. Follow the links to obtain a copy of the file **TechSurvey.csv** or **TechSurvey.ftm**. In this file are the results of a survey of 100 youths in an area. A local service provider wants to create an advertising campaign to attract people to its products. Use the data and a Venn diagram to determine who the campaign should be aimed at.

KEY CONCEPTS

- Chapter 4 examined arrangements (permutations) of items where the order mattered. For example, the number of ways to arrange three colours if there are four colours to choose from is ${}_4P_3$ or 24 ways. If the order does not matter, then the number of groups of items that can be made is different.

Example

Consider the colours blue, green, red, and yellow. Here are all of the arrangements of three of these colours.

BGR	BGY	BRY	GRY
BRG	BYG	BYR	GYR
GBR	GBY	RBY	RGY
GRB	GYB	RYB	RYG
RBG	YBG	YBR	YGR
RGB	YGB	YRB	YRG

If you look at the arrangements in each column, you will notice that they are made from the same group or set of colours. So, there are only four possible groups of three colours that can be made from four colours. In fact, the number of ways to group three of four elements is fewer than the number of ways to arrange three of four elements by a factor of 6, or 3!. Note that the number in the factorial comes from the number of colours being taken from the four colours.

- In general, the number of combinations of n elements taken r at a time is given by

$$\begin{aligned} {}_nC_r &= \frac{{}_nP_r}{r!} \\ &= \frac{n!}{(n-r)!r!} \end{aligned}$$

- The notations ${}_nC_r$, $C(n, r)$, and $\binom{n}{r}$ are different ways to express the same thing. Sometimes these are read "n choose r."
- The use of words such as *sets* or *groups* in a problem is usually a signal that the problem is dealing with combinations where order does not matter.
- The multiplicative and additive counting principles also apply to combinations.

Example

There are 15 males and 13 females in your class. How many ways can you make a group of 4 students if

a) there are no restrictions?

b) there must be two males and two females?

c) the group must be either all males or all females?

Solution

a) There are 28 students in total, so the number of groups, if there are no restrictions, is ${}_{28}C_4 = 20\,475$.

b) The number of ways to choose two males is ${}_{15}C_2$.

The number of ways to choose two females is ${}_{13}C_2$.

The total number of possibilities is ${}_{15}C_2 \times {}_{13}C_2$, or 8190.

c) Case 1: choose four boys, ${}_{15}C_4 = 1365$
Case 2: choose four girls, ${}_{13}C_4 = 715$

The two cases cannot happen at the same time, so the total number of possibilities is 1365 + 715, or 2080.

A

1. Evaluate.

a) ${}_{10}C_3$

b) ${}_{13}C_5$

c) ${}_{52}C_{13}$

2. Go to the Google™ Internet search engine. Repeat question 1 by typing each part in the search box in the form "*n* choose *r*" (e.g., "10 choose 3").

3. Represent each of the following using the notation $\binom{n}{r}$.

a) combinations of ten elements taken four at a time

b) groups of 8 made from 12 elements

c) ${}_{20}C_6$

d) ${}_{25}C_{18}$

e) $\frac{15!}{(15-6)!6!}$

f) $\frac{32!}{24!8!}$

4. Why might someone suggest that a combination lock is misnamed?

5. Determine a value for *r*, $r \neq 3$, that makes each statement true.

a) ${}_5C_3 = {}_5C_r$

b) ${}_8C_3 = {}_8C_r$

c) $\binom{12}{3} = \binom{12}{r}$

d) $\binom{20}{3} = \binom{20}{r}$

e) $C(35, 3) = C(35, r)$

f) $C(52, 3) = C(52, r)$

B

6. Explain why $n \geq r$ must be true for ${}_nC_r$.

7. You have lollypops of five different colours: red, green, blue, yellow, and orange. You want to make groups of three different-coloured lollypops to give out for Halloween.

   a) Determine the number of different groups you could make by

      i) writing them out explicitly
      ii) creating a tree diagram
      iii) using combinations

   b) Which method gives you the most information?

   c) Which method gives you the answer the quickest?

8. **a)** Consider the following questions.

      i) If there are ten people in a room, how many ways can a group of two people be made?
      ii) If there are ten people in a room, how many handshakes are made if each person shakes hands with each other person?
      iii) If ten points are arranged evenly around the circumference of a circle, how many line segments can be made joining any two points on the circle?

   b) How are the questions in part a) related to each other?

9. Use the definition of $C(n, r)$ to show that the following are true.

   a) $C(8, 5) = C(8, 3)$

   b) $C(13, 7) = C(13, 6)$

   c) $C(19, 2) = C(19, 17)$

   d) $C(n, r) = C(n, n - r)$

☆ 10. A standard deck of cards contains 52 cards.

   a) How many different 13-card hands are there?

   b) How many different 13-card hands can be dealt to 4 players?

   c) How are the questions in parts a) and b) different?

11. To make fruit juice, you buy apples, oranges, grapes, watermelon, lemons, pears, and bananas. How many different types of juice can you make if

   a) you choose any three types of fruit?

   b) you choose any three types and one of them is apple?

   c) you choose any number of different types?

12. A class has 7 males and 21 females. How many ways can your teacher select four students to go to the office to pick up boxes of textbooks if

   a) there are no restrictions?

   b) an equal number of males and females are selected?

   c) the sample is stratified (i.e., the sample has the same proportion of members from each group as the class has)?

   d) if there is at least one female?

13. There are 24 students in your class. Your teacher wants to put you in groups for your data-management project. How many ways can six groups of four be made if

   a) there are no restrictions?

   b) Paige, Jolene, Hans, Ahmed, Raul, and Paul have been chosen already to be the group leaders?

☆ 14. A researcher is studying the effects of exercise on concentration. She has 50 volunteers participating: 20 athletes and 30 non-athletes. How many groups of 10 volunteers can she make if

   a) there are no restrictions?

   b) half must come from each group?

   c) there must be at least three athletes?

15. How many ways can a company choose its 16 employees to occupy a row of five offices if

a) there are no restrictions?

b) one of the offices must be left empty for a photocopier?

c) the company president and two of the three vice-presidents each must have an office?

16. Comedian Jay Leno owns more than 80 cars. If he were to drive two cars each day, how long would it take him to drive every combination of pairs of his cars?

17. You and your friends are choosing teams for a road-hockey game. How many different ways can you create the teams if there are

a) ten people and no restrictions?

b) ten people and the only two goalies are Raj and Andre?

c) nine people and no restrictions?

☆ **18.** Your school's girls' basketball team is participating in a tournament. Use a spreadsheet to determine how many different sets of opening games are possible if

a) there are 16 teams

b) there are 15 teams, one of which has a bye (that is, does not play) in the first round

19. You are playing the online game SimHuman and have acquired enough credits to buy three items for your apartment. You may choose from

- three types of television
- four types of couch
- six types of bed
- two types of wall decoration

How many different ways can you choose your three items if

a) there are no restrictions?

b) you definitely want one television?

c) you do not want to purchase more than one type of any item?

20. When cribbage is played with two people, each player is dealt six cards, four of which the player keeps and two of which are put into the "crib."

a) How many ways can six cards be dealt?

b) How many ways can the two sets of six cards be dealt?

c) Once a player has six cards, the player must decide which four to keep and which two to put in the crib. How many ways can this occur?

d) One way to earn points is to get a pair. One pair is worth two points. Determine how many pairs can be made from having three jacks, and thus determine how many points you get for three of a kind.

e) Repeat part d) for four of a kind.

C

21. Consider the letters in the word *MISSISSAUGA*. How many different four-letter arrangements can be made if

a) each letter must be different?

b) there must be three *S*s (e.g., SSSM)?

c) there are two sets of double letters (e.g., SSII)?

d) there are no restrictions?

5.3 Problem Solving With Combinations

KEY CONCEPTS

- Use the formula $(p + 1)(q + 1)(r + 1) \ldots - 1$ to find the total number of selections of at least one item that can be made from p items of one kind, q items of a second kind, r items of a third kind, and so on.
- A set with n distinct elements has 2^n subsets including the null set.
- For combinations with some identical elements, you often have to consider all possible cases individually.
- In a situation where you must choose at least one particular item, either consider the total number of choices available minus the number without the desired item, or add all the cases in which it is possible to have the desired item.

A

1. How many sets can be made from six distinct items if there must be at least one item in each set?

2. How many sets can be made from three orange marbles, four green marbles, and two yellow marbles if there is at least one item in each set?

B

3. **a)** Determine all of the prime factors of 210.
   b) Determine all the ways the answers from part a) can be combined and multiplied to make factors of 210. Verify that each is a factor of 210.
   c) Use combinations to determine the total number of factors of 210.

4. Carlita has the following uniform items in her closet:

Pants/skirts	grey pants, blue pants, blue shorts, plaid skirt
Shirts	white short-sleeve, white long-sleeve
Sweaters	blue cardigan, blue pullover, grey hoodie
Shoes	flats, sandals, running
Accessories	belt, scarf, headband

   a) How many ways can Carlita dress if she chooses one item from each category?
   b) How many days can she go without wearing the same outfit more than once if the only items she must wear are a shirt, pants or a skirt, and a pair of shoes (assuming that if she chooses an accessory, she chooses only one)?

5. **a)** Determine all of the prime factors of 360.
   b) How many factors are there of 360?

☆6. There are seven males and five females on a cross-country team. Determine the number of four-member co-ed teams (at least one male and one female on the team) that can be made by
   a) using specific cases that are permissible co-ed teams
   b) subtracting the teams that are not permissible from the unrestricted case

7. At a party, each person shakes hands with every other person. If there are a total of 66 handshakes, how many people are at the party?

8. MegaBurger is a new restaurant. You can create your own custom hamburger using the guidelines below. How many different types of hamburger can you make at MegaBurger?

Required Elements	Optional Elements
Bun (7 types)	Condiments (10 types)
Patty (12 types)	Cheeses (8 types)
	Vegetables (6 types)
	Spicy bits (4 types)

9. The rules of a football league state that a team can score points in several ways.

Method	Points
Touchdown	6
Conversion	1
Field goal	3
Safety	2
Single	1

Assume that in a game a team gets at most seven touchdowns, seven conversions, six field goals, two safeties, and three singles.

a) How many total scores are possible for a single team in a game?

b) How many total scores are possible for two teams in a game?

10. A new pizza parlour offers five toppings and three pizza sizes to choose from.

a) How many different pizzas can be made from the menu?

b) The owner wants to name the restaurant Pizza 2000 to indicate all the different types of pizza that can be made. She increases the number of options to include three different cheeses, of which you can choose one, two, or three. If the owner wants to offer more than 2000 possible pizzas, how many topping choices must she make available?

11. The game of Go originated in China more than 2500 years ago. Players take turns placing black or white stones on the intersections of a 19-line × 19-line grid. The object of the game is to control more of the board than the other player does.

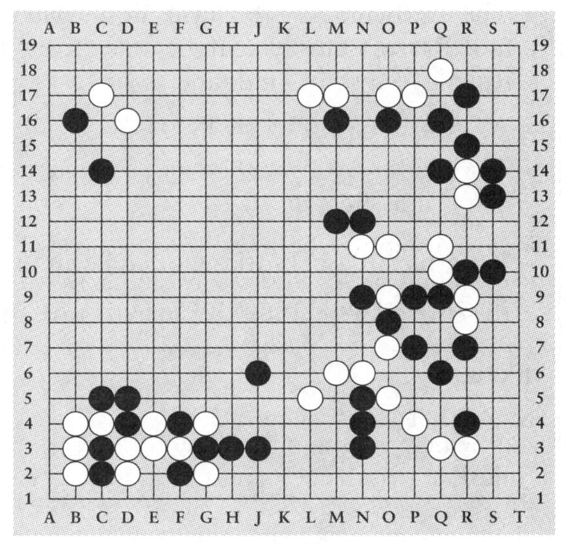

a) Use a spreadsheet to calculate the total number of ways in which stones can be placed on a Go board.

b) It is said that there are 4.63×10^{170} possible positions for a standard 19-by-19 Go board. Based on this, how many of the positions found in part a) are not possible?

12. You have three quarters, two loonies, and four toonies. How many different groups of coins can be made if

a) there are no restrictions?

b) there must be at least three coins in each group?

c) each group must have a different value?

C

13. How many perfect square factors does the number 5184 have?

☆ **14.** How many ways can you distribute ten apples among five people if each person gets at least one apple?

15. How many four-letter arrangements are there of the letters in the word *SUCCEED*?

5.4 The Binomial Theorem

KEY CONCEPTS

- The coefficients of the terms in the expansion of $(a + b)^n$ correspond to the terms in row n of Pascal's triangle.

$$
\begin{array}{ccccccccccc}
 & & & & & 1 & & & & & \\
 & & & & 1 & & 1 & & & & \\
 & & & 1 & & 2 & & 1 & & & \\
 & & 1 & & 3 & & 3 & & 1 & & \\
 & 1 & & 4 & & 6 & & 4 & & 1 & \\
1 & & 5 & & 10 & & 10 & & 5 & & 1
\end{array}
$$

$$
\begin{array}{ccccccccccc}
 & & & & & {}_0C_0 & & & & & \\
 & & & & {}_1C_0 & & {}_1C_1 & & & & \\
 & & & {}_2C_0 & & {}_2C_1 & & {}_2C_2 & & & \\
 & & {}_3C_0 & & {}_3C_1 & & {}_3C_2 & & {}_3C_3 & & \\
 & {}_4C_0 & & {}_4C_1 & & {}_4C_2 & & {}_4C_3 & & {}_4C_4 & \\
{}_5C_0 & & {}_5C_1 & & {}_5C_2 & & {}_5C_3 & & {}_5C_4 & & {}_5C_5
\end{array}
$$

- Patterns in Pascal's triangle can be summarized using combinatorial symbols.
- Pascal's formula is given by ${}_nC_r = {}_{n-1}C_{r-1} + {}_{n-1}C_r$.

Example

Show that ${}_9C_3 = {}_8C_2 + {}_8C_3$.

Solution

$$
\begin{aligned}
{}_8C_2 + {}_8C_3 &= \frac{8!}{(8-2)!2!} + \frac{8!}{(8-3)!3!} \\
&= \frac{8!}{6!2!} + \frac{8!}{5!3!} && \text{Simplify the brackets} \\
&= \frac{8!}{6!2!}\left(\frac{3}{3}\right) + \frac{8!}{5!3!}\left(\frac{6}{6}\right) && \text{Use a common denominator.} \\
&= \frac{3(8!)}{6!3!} + \frac{6(8!)}{6!3!} && \text{Simplify.} \\
&= \frac{9(8!)}{6!3!} && \text{Collect like terms and simplify the factorial.} \\
&= \frac{9!}{6!3!} \\
&= {}_9C_3
\end{aligned}
$$

- The binomial theorem can also be written using combinatorial symbols:

  $(a + b)^n = {}_nC_0a^n + {}_nC_1a^{n-1}b + {}_nC_2a^{n-2}b^2 + \ldots + {}_nC_nb^n$ or $(a + b)^n = \sum_{r=0}^{n} {}_nC_ra^{n-r}b^r$,

  where the rth term of the expansion is given by ${}_nC_ra^{n-r}b^r$.
- The degree of each term in the binomial expansion of $(a + b)^n$ is n.

A

1. Use Pascal's formula to rewrite each of the following.

a) ${}_{12}C_7$ **b)** ${}_{14}C_{11} + {}_{14}C_{12}$

c) $C(21, 8)$ **d)** $C(8, 3) + C(8, 4)$

e) $\binom{10}{7}$ **f)** $\binom{16}{5} + \binom{16}{6}$

2. Use ${}_nC_r = \frac{n!}{(n-r)!r!}$ to show that each equation is true.

a) ${}_{18}C_{15} = {}_{17}C_{14} + {}_{17}C_{15}$

b) $C(23, 10) = C(22, 9) + C(22, 10)$

c) $\binom{32}{25} - \binom{31}{24} = \binom{31}{25}$

3. Determine the number of terms in each expansion.

   a) $(x + y)^{11}$

   b) $(3a + b)^{16}$

   c) $(1 - 3x)^9$

4. Determine the value of k given the terms of the expansion of $(x + y)^{12}$.

   a) $C(12, k)x^7y^k$

   b) ${}_kC_9(x^3y^9)$

   c) $792(x^5y^k)$

   d) $66(x^{12-k}y^k)$

5. Consider $(x + y)^{15}$.

   a) Determine the expansion.

   b) What is true about the sum of the powers of x and y for each term?

   c) In general, what is true about the sum of the powers of x and y for each term in the expansion of $(x + y)^n$?

6. State the value of the coefficient of each term in the expansion of $(a + b)^n$.

   a) a^9b^4 **b)** $a^{12}b^2$ **c)** $a^{18}b^7$

B

7. Two people are sitting in a restaurant.

   a) How many different ways can a pair of these people be formed?

   b) If a third person joins, how many different pairs of people can be formed?

   c) Four more people come in, and after each one arrives, you calculate the number of different pairs of people that could be formed. What number did you get in each case?

   d) How do your answers from parts a) to c) relate to Pascal's triangle?

8. Expand and simplify.

   a) $(2x + y)^8$ **b)** $(x - y)^9$

   c) $(1 + 4x)^6$ **d)** $(3x - 5)^5$

   e) $(4x - 7y)^4$ **f)** $(3x^2 + 1)^9$

9. **a)** Determine how many ways there are to choose three of seven people in a room, using a direct method with combinations.

   b) Consider the following alternative method. Imagine that you are one of the seven people. When a group of three is made, you will either be in the group or not in the group.

   i) Case 1: You are not in the group. In this case, there are still three people to choose out of the six remaining. How many ways can this be done?

   ii) Case 2: You are in the group. In this case, there are only two people left to choose from the remaining six to make a total of three. How many ways can this be done?

   iii) Combine your cases to get the total number of ways to make a group of three from seven people.

   c) Parts a) and b) should yield the same answer. How do the methods relate to Pascal's method?

☆ 10. For each expansion, determine the sixth term.

   a) $(7x - 5y)^9$

   b) $(2x^3 + 1)^{13}$

11. Consider the diagram below.

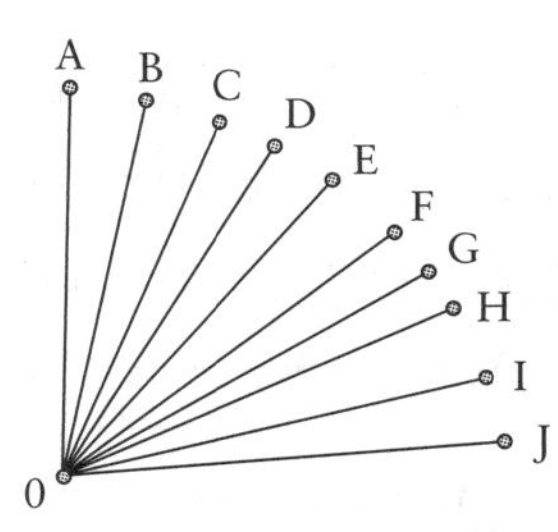

   a) How many acute angles are there?

   b) How does your answer relate to Pascal's triangle?

12. Use a graphing calculator to determine the eighth row of Pascal's triangle. Use the seq(Expression, Variable, Initial value, Final value, Increment) function.

- Press 2nd STAT for [List] , cursor over to OPS and select **5:seq(**
- Type **8**, then press MATH, cursor over to PRB and select **3:nCr**
- Press X,T,θ,*n* , X,T,θ,*n* , **0** , **8** , **1**)

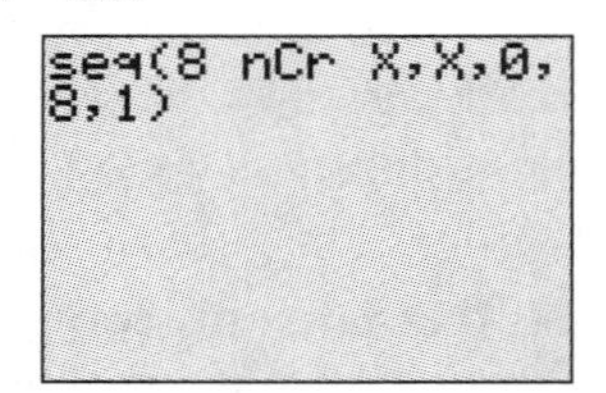

- Press ENTER.

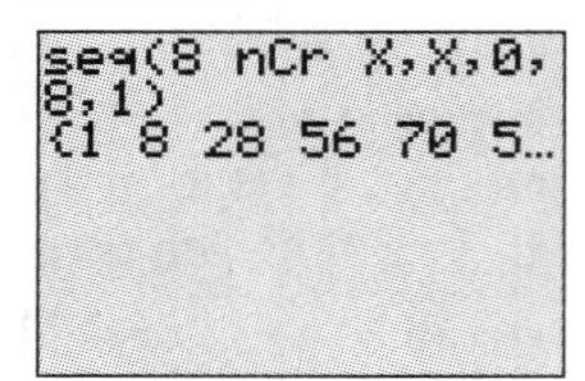

Use the arrow keys to see the entire sequence.

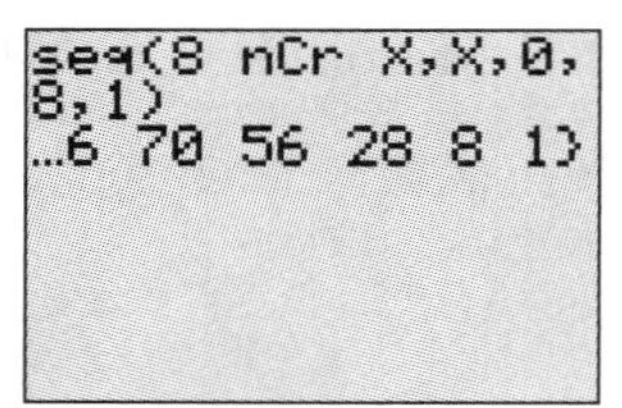

Repeat these steps to determine

a) the tenth row of Pascal's triangle

b) the first 5 terms of the 18th row of Pascal's triangle

c) the 15th through 18th terms of the 20th row of Pascal's triangle

13. Write each in the form $(1 \pm k)^n$. Then, determine the value of the first four terms. Round your answers to five decimal places.

a) $(1.02)^{10}$

b) $(0.999)^8$

14. In each case, write in the form $(a + b)^n$.

a) $x^5 + 5x^4y + 10x^3y^2 + 10x^2y^3 + 5xy^4 + y^5$

b) $1 + 7x + 21x^2 + 35x^3 + 35x^4 + 21x^5 + 7x^6 + x^7$

c) $16x^4 - 32x^3y + 24x^2y^2 - 8xy^3 + y^4$

d) $1 + 6(0.03) + 15(0.03)^2 + 20(0.03)^3 + 15(0.03)^4 + 6(0.03)^5 + (0.03)^6$

15. Manny has not studied for a test, so he is happy to see that there is a true-or-false section with five questions.

a) Use a tree diagram to determine all the possible outcomes if Manny guesses on each question. Complete the following table based on your tree diagram.

Outcome	Number of Ways
5 correct, 0 wrong	
4 correct, 1 wrong	
3 correct, 2 wrong	
2 correct, 3 wrong	
1 correct, 4 wrong	
0 correct, 5 wrong	

b) How do your answers in parts a) and b) relate to Pascal's triangle and the binomial theorem?

16. Use the binomial theorem to calculate the value of the indicated term(s), rounded to five decimal places.

a) $(0.8 + 0.2)^{10}$, term 3

b) $(0.25 + 0.75)^6$, term 4

c) $(0.9 + 0.1)^{50}$, terms 0 to 2

C

☆ 17. Consider the expansion of $(3x^2 - x^{-1})^8$.

a) Determine the term that is independent of x.

b) Which term has numerical coefficient −13 608?

18. Use the binomial theorem to determine the value of $(1.03)^9$, to four decimal places.

Chapter 6 Introduction to Probability

6.1 Basic Probability Concepts

KEY CONCEPTS

- Probability is one of the most important concepts in mathematics. Understanding the nature of probability can lead to a better understanding of the world and help promote more informed decision making.
- A probability event is a collection of outcomes satisfying a particular condition. The probability of an event can range from 0, or 0% (impossible), to 1, or 100% (certain).
- For any situation, the set of all possible outcomes is called the sample space, *S*. For example, for flipping a coin, the sample space is $S = \{\text{heads, tails}\}$. The sum of the probabilities of all possible outcomes in the sample space equals 1.
- A discrete sample space is one where the possible outcomes can be counted. For example, the sample space for rolling a standard die is discrete: $S = \{1, 2, 3, 4, 5, 6\}$.
- A continuous sample space is one where an outcome could be any value in a range. For example, the sample space for the time that guests could arrive at a party is $S = \{\text{any time between 6:00 P.M. and midnight}\}$.
- There are three main types of probability: subjective, experimental, and theoretical.
- Subjective probability is based on an educated guess and is the least reliable of probability types. It is usually the result of remembering experiences, though only informally. Since people tend to remember anomalies and forget the mundane, subjective probability is often skewed.
- Experimental probability (sometimes called empirical probability) is based on data recorded historically about a particular event. This is more reliable than subjective probability but only if a lot of data are collected in a large number of trials.

Example

A baseball player gets 15 hits in 50 times at bat. The probability of the batter getting a hit is

$$P(\text{hit}) = \frac{15}{50}$$
$$= 0.3 \text{ or } 30\%$$

In general, the experimental probability of any event is given by

$$P(\text{event}) = \frac{\text{number of times event occurs}}{\text{total number of trials}}$$

- Theoretical probability is based on the physical nature of the event and is the most reliable of all the probability calculations. In general, the theoretical probability of any event is given by

$$P(\text{event}) = \frac{n(\text{event})}{n(\text{sample})}$$
$$= \frac{\text{number of times event could occur}}{\text{total number of times any event could occur}}$$

Example

What is the probability of rolling 6 with a standard six-sided die?

$$P(\text{rolling 6}) = \frac{n(\text{rolling 6})}{n(\text{all possible rolls})}$$
$$= \frac{1}{6}$$
$$= 0.1\overline{6} \text{ or } 16.7\%$$

- Experimental probability is based on direct observation or experiment. For example, consider a coin flipped 12 times. Below are some possible outcomes. The theoretical probability of getting heads is $P(\text{heads}) = 0.5$. The experimental probability is the running probability as the coin is flipped.

Coin Flip	Number of Heads	*P*(Heads)	Coin Flip	Number of Heads	*P*(Heads)
T	0	$\frac{0}{1} = 0.00$	H	6	$\frac{6}{7} = 0.86$
H	1	$\frac{1}{2} = 0.50$	H	7	$\frac{7}{8} = 0.88$
H	2	$\frac{2}{3} = 0.67$	H	8	$\frac{8}{9} = 0.89$
H	3	$\frac{3}{4} = 0.75$	T	8	$\frac{8}{10} = 0.80$
H	4	$\frac{4}{5} = 0.80$	H	9	$\frac{9}{11} = 0.82$
H	5	$\frac{5}{6} = 0.83$	H	10	$\frac{10}{12} = 0.83$

Notice that the experimental probability rarely equals the theoretical probability. The experimental probability fluctuates quite a bit in the first 10 flips. However, as the number of flips increases, the fluctuation settles down around the theoretical probability (after about 60 flips). This illustrates that as the number of trials increases, the experimental probability approaches the theoretical probability.

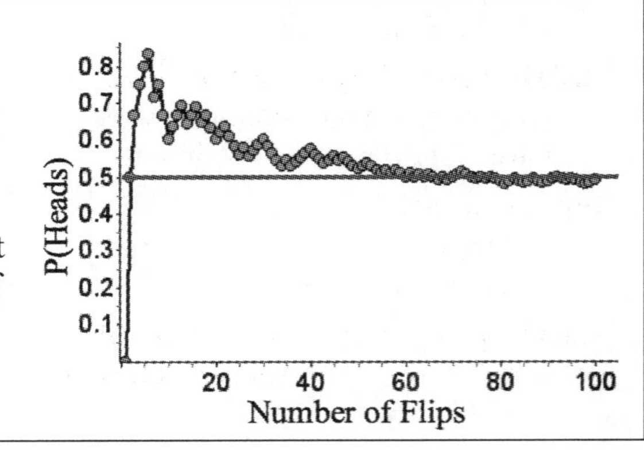

A

1. Identify the sample space for each event.
   a) drawing a card from a standard deck of 52 playing cards
   b) rolling a ten-sided die
   c) guessing a player's weight at a carnival midway game
   d) drawing a name from a hat
   e) the time when you leave home to catch the bus

2. For each event in question 1, indicate whether the sample space is discrete or continuous.

3. Determine whether each probability is subjective, experimental, or theoretical.
   a) The probability of rolling an even number with a standard die is 50%.
   b) The probability that your teacher will give a surprise quiz today is 75%.
   c) The probability that a baseball player will get a hit the next time at bat is 0.325.
   d) The probability of drawing a spade from a deck of playing cards is 0.25.
   e) A Toronto Maple Leafs fan says the probability of the Leafs winning the Stanley Cup this year is 60%.

4. Is each statement true or false? Explain your choice.
   a) The probability of rolling 7 with a standard die is 0.
   b) The probability of getting heads on your next flip of a coin is 1 when you have just flipped 15 tails in a row.
   c) The probability that your MP3 player will play the same song twice in a row when on shuffle is 0.
   d) The probability that you will roll a sum less than 12 with two standard dice is 1.

5. Determine the probability of each outcome. Indicate whether the probability is theoretical or experimental.
   a) *P*(red card) from a standard deck of 52 cards
   b) *P*(number divisible by 3) with a standard die
   c) *P*(green marble) from a bag of marbles consisting of four green, ten red, and six yellow
   d) *P*(goal in the next game) for a hockey player who has scored 55 goals in 80 games
   e) *P*(6) when you have rolled a die ten times and got {5, 1, 3, 6, 1, 3, 2, 4, 6, 3}

6. Suppose your school has the following numbers of students.

Grade	Male	Female	Totals
9	113	128	241
10	109	92	201
11	85	121	206
12	115	98	213
Totals	422	439	861

   Determine the probability of each of the following events.
   a) For a prize draw, the name of every student is placed in a hat. What is the probability that the name of a male student in grade 11 will be chosen?
   b) The grade 12 students are choosing a representative to sit on the Parent Council. What is the probability that the representative will be female?
   c) One male student and one female student will be chosen to ride on the school's float in the Thanksgiving Day parade. What is the probability that you will be selected?

B

7. Two standard dice are rolled and the sum of each roll is recorded.
   a) Create a tree diagram to represent all of the possible outcomes of rolling two dice.
   b) What is the sample space for these dice?
   c) What is $n(S)$?
   d) What is the probability of each sum occurring?
   e) Determine the sum of all of your answers in part d). Explain why this answer turns out as it does.
   f) What is the probability of rolling doubles with two dice?

8. A weather report states there is a 20% chance of rain tomorrow. Discuss what this value means and how it may have been determined.

9. A baseball player has had 14 hits in 60 times at bat in the last 20 games.
   a) What is the probability that the player will get a hit the next time at bat?
   b) What type of probability did you calculate in part a)?
   c) How many hits does the batter need in her next ten at bats to increase the probability of her getting a hit to 0.300?

☆ 10. A standard die is rolled 20 times with the following outcomes.
   1, 5, 2, 5, 3, 2, 4, 6, 4, 6,
   2, 3, 1, 6, 2, 6, 1, 5, 5, 1
   a) Create a table showing the running experimental probability of rolling 6.
   b) Graph your running probability from part a).
   c) Determine the theoretical probability of rolling 6. How does it compare to what you found in parts a) and b)?

11. In a dice game, you and a friend each roll one die and compare the outcomes. If the difference between them is 0, 1, or 2, then you win. If the difference is 3, 4, or 5, then your friend wins. Determine who has the advantage (if any).

12. A section of a test consists of four true-or-false questions.
   a) What is the sample space for all of the possible correct answers?
   b) Determine the probability of each of the following situations (where order does not matter). Show that the sum of each of the probabilities is 1. {4F0T, 3F1T, 2F2T, 1F3T, 0F4T}
   c) What is the probability that all the answers are the same?
   d) What is the probability that there are at least two answers that are true?

13. In the game Rock-Paper-Scissors, competitors simultaneously choose one of three hand positions. Rock (a closed fist) crushes scissors. Scissors (first two fingers in "V" formation) cut paper. Paper (hand held flat) covers rock.
   a) A round is best-of-three format. Assuming that each choice is equally likely, create a tree diagram to represent the sample space for one player's possible three throws.
   b) Statistically, in tournament play, scissors is thrown 29.6% of the time. How does this compare to the probability of scissors coming up in any particular game? at least once in three throws?
   c) Novice players tend to use rock for their first throw and avoid making the same choice twice in a row. What is the probability that if you are playing against a novice, you will be able to correctly guess what the player's three throws will be?

14. Jenna has 17 computer games. Of these games, 5 are music related, 7 are simulations, and 3 are both simulations and music related.

 a) Create a Venn diagram to represent this situation.

 b) Determine the probability that if Jenna randomly selects a game, the game is a simulation but not music related.

 c) Determine the probability that a game that Jenna selects will not be a simulation or music related.

15. Go to *www.mcgrawhill.ca/links/MDM12* and select **Study Guide**. Follow the links to obtain a copy of the following files to investigate the connection between experimental probability and theoretical probability.

	Situation	File
i)	Flipping a coin and getting heads	**Flip.ftm**
ii)	Rolling one die and getting 6	**Roll1.ftm**
iii)	Rolling two dice and getting a sum of 6	**Roll2.ftm**
iv)	Drawing a card from a standard deck and getting the 6 of hearts	**Draw.ftm**

 a) Determine the theoretical probability for each case.

 b) Open each indicated file and click the **Sample More Cases** button. Continue to click the button until the indicated variable's experimental probability settles around its theoretical probability. Repeat this five times in each case and get an average number of trials.

 c) What appears to be the connection between the size of the sample space and the number of trials needed for the experimental probability to settle around the theoretical probability?

16. Mack is attending the Harrow Fair, one of the longest-running country fairs in North America. He is playing a dart game where the small-prize area is a square with side dimensions of 10 cm. Within the square is the large-prize area, a circle having a diameter of 3 cm. If Mack hits the prize area, what is the probability that he will get a large prize?

C

17. What is the probability of rolling doubles or triples with three dice?

☆ 18. A game for two players has nine balls, each with a digit from 1 to 9. The object of the game is to be the first to choose three balls that have a sum of 15. Selection alternates between the players: after the first chooses a ball, the second makes a choice.

 a) What is the probability that if you choose any three balls, the sum will be 15?

 b) What number should you choose first in order to have the highest probability of winning?

 c) If you are the second player, how would it help to know all of the possible ways to get to a sum of 15?

KEY CONCEPTS

- The complement of any event is the opposite of that event. For example, the complement of rolling 1 with a standard die is rolling 2, 3, 4, 5, or 6. The complement of any event A is denoted by A'.
- If the probability of any event A is $P(A)$, then the probability of A' is $P(A') = 1 - P(A)$.
- Stating the odds of an event is another way to represent probability. There are two types of odds statements: the odds in favour of something and the odds against something.

The odds in favour of A occurring are given by the ratio $\frac{P(A)}{P(A')}$. Often the odds are stated as $P(A):P(A')$ or $P(A)$ to $P(A')$.	The odds against A occurring are given by the ratio $\frac{P(A')}{P(A)}$. Often the odds are stated as $P(A'):P(A)$ or $P(A')$ to $P(A)$.

Example

Consider rolling a standard die. The odds in favour of rolling 1 are given by

$$\frac{P(1)}{P(\text{not } 1)} = \frac{\frac{1}{6}}{\frac{5}{6}} \text{ or } \frac{1}{5}.$$

That is, the odds in favour of rolling 1 are 1 to 5.

- If the odds in favour of an event A are given by $\frac{h}{k}$, the probability can be obtained using the formula $P(A) = \frac{h}{h+k}$.

A

1. For each probability, determine the odds in favour.
   a) 0.25 **b)** $\frac{3}{5}$ **c)** 60%
   d) $\frac{1}{7}$ **e)** 35% **f)** 0.001

2. For each probability, determine the odds against.
   a) 0.98 **b)** 10% **c)** $\frac{10}{52}$
   d) 64% **e)** $\frac{6}{36}$ **f)** 0.05

3. Given the following odds in favour, determine the probability.
   a) 3 to 7 **b)** 5:35 **c)** $\frac{5}{2}$

B

4. Sometimes when odds are used for comparison, they are reduced so that the first number is 1 and the second number can be a decimal. Convert each of the odds in question 3 to be compared to 1.

5. Ali has 250 songs on her MP3 player. Of these, 150 are dance songs, 50 are R&B songs, and 30 songs are both dance and R&B.
   a) Create a Venn diagram to represent the songs on Ali's MP3 player.
   b) When Ali has her player on shuffle, what are the odds in favour of her hearing a dance song that is not an R&B song?
   c) What are the odds against her hearing a song that is both dance and R&B?

6. If you choose one card from a standard deck, what are the odds in favour of getting
   a) a 10?
   b) a spade?
   c) an ace or king?

7. In hockey, a team can win, lose, or tie any game. Assume that each outcome is equally likely. A team plays a three-game series.
   a) Create a tree diagram to represent the possible outcomes of the three games.
   b) What are the odds in favour of the team winning all three games?
   c) What are the odds in favour of the team winning the series by winning at least two games?
   d) What are the odds against the series ending in a draw (i.e., no distinct winner)?

8. A roulette wheel has 38 numbers (1 through 36, 0, and 00). As the wheel spins, a ball is rolled on the wheel and bets are placed as to where the ball will land. You can bet on a single number, as well as on 2, 3, 4, 5, 6, 12, or 18 numbers.

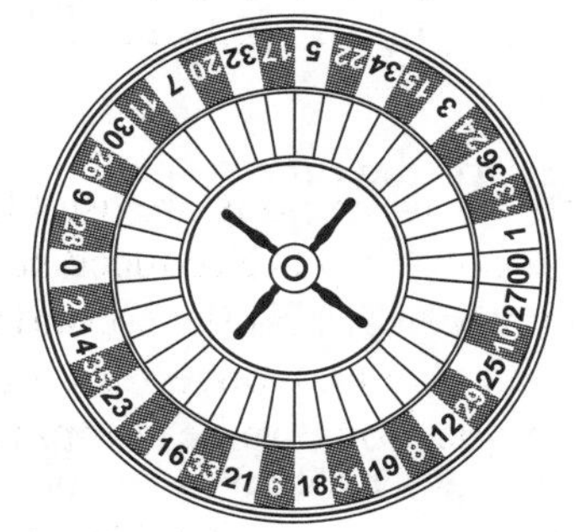

   a) What are the odds in favour of betting on an even number? Note: 0 and 00 are not considered even numbers.
   b) At a casino roulette table, a bet on an even number has 1 to 1, or even, odds of winning. That is, if you bet $5 on an even number and an even number comes up, you win $5 and get to keep your original $5. If you play 100 games and always bet on an even number, how much would you expect to win (or lose)?

9. Which would you rather play: a game with odds of 4 to 5 against winning or a game with odds of 3 to 4 in favour of winning? Explain.

C

10. A solid white cube measuring 5 cm on each side is painted so that each face is a different colour (red, green, blue, yellow, orange, and purple). The cube is then cut into 1-cm cubes that are put into a bag.

   a) What are the odds in favour of pulling out a cube with two sides painted?
   b) What are the odds against pulling out a cube with three sides painted?
   c) Which has the greater odds in favour: pulling out a cube with only one side painted or pulling out one that is entirely white?

11. You are playing darts on a board that is 30 cm in diameter. The inner ring is 6 cm in diameter and the middle ring is 18 cm in diameter.

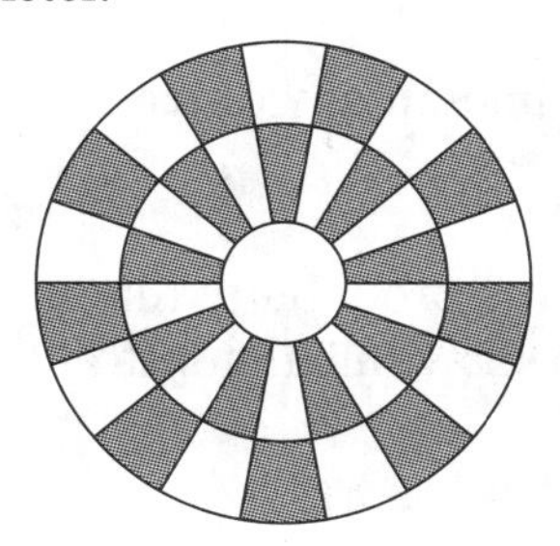

You throw a dart and hit the board.
   a) What are the odds against hitting the centre circle?
   b) What are the odds in favour of hitting a shaded section?

KEY CONCEPTS

- When using counting techniques with probability, the number of outcomes in the sample space, $n(S)$, often comes from the unrestricted case.

Example

What is the probability of getting three of a kind when dealt five cards from a well-shuffled regular deck?

Solution

In this case, the sample space is the unrestricted case of choosing any five cards from the deck: ${}_{52}C_5 = 2\ 598\ 960$. Now determine the number of ways to get three of a kind.

${}_{13}C_1$	$\times$	${}_{4}C_3$	$\times$	${}_{12}C_2$	$\times$	${}_{4}C_1$	$\times$	${}_{4}C_1$	$= 54\ 912$
First, choose the three-of-a-kind card. This card can be any one of the 13 number cards: A, 2–10, J, Q, or K.		Next, choose which three of four cards of the chosen number are in the hand.		There are two remaining cards in the hand. From the remaining cards, choose two. This guarantees they will be different.		Choose the suit of the first card.		Choose the suit of the second card.	

Note: A common mistake is to use ${}_{48}C_2$ for the two remaining cards, but these possibilities would include cases where two cards were paired and thus the hand would be a full house (three of a kind plus a pair) instead.

Thus, $P(\text{three of a kind}) = \dfrac{54\ 912}{2\ 598\ 960} = 0.0211$ or 2%. So, 2% of the time you will have three of a kind.

- Sometimes, many different techniques must be applied to solve a problem.

Example

What is the probability that if there are four people in a room, at least two of them will have the same birth month?

Solution

If there were 13 people in a room, it is guaranteed that at least 2 people in the room would have the same birth month. When the number of people is fewer than 13, the best way to approach the problem is to consider the complement. What is the probability that no 2 people have the same birth month?

First, consider the unrestricted case. There are 4 people, each of whom has 12 possibilities for his or her birth month. The total number of possibilities is $12^4 = 20\ 736$.

Next, determine the number of ways the four people would not have the same birth month:

12	×	11	×	10	×	9	= 11 880
The first person can have any birth month.		The next person can have any birth month except for the month of the first person.		The next person has a birth month different from the months of the first two people.		The last person must have a birth month different from those of the other three people.	

The probability of different birth months is $P(\text{different}) = \frac{11\ 880}{20\ 736} = 0.57$

Thus, the probability of having two birth months the same is $1 - 0.57 = 0.43$ or 43%.

A closer look at the answer yields $P(\text{different}) = \frac{11\ 880}{20\ 736} = \frac{12 \times 11 \times 10 \times 9}{12^4}$ or $\frac{_{12}P_4}{12^4}$

A

1. A bag contains four marbles of different colours, including red and blue.

a) How many ways can two marbles be pulled out at once?

b) What is the probability that the red and blue marbles are pulled out together?

2. Consider a standard deck of 52 playing cards.

a) How many ways are there to choose four cards?

b) What is the probability that all four cards you choose will be red?

c) What is the probability that all four cards you choose will be spades?

3. There are 20 people in your data-management class, each of whom has to make a presentation.

a) How many different orders are there for class members to make their presentations?

b) What is the probability that you will go first?

4. You have eight shirts, five pairs of pants, and ten pairs of socks.

a) How many different outfits can you create?

b) You buy clothes carefully so that everything matches, except that one shirt and one pair of pants do not go with each other. What is the probability that if you randomly select your clothes, your outfit will not match?

B

5. You are helping your sister to write addresses on the invitations to her wedding. You cannot make out the last two characters of one postal code. Knowing that postal codes have the format L#L#L#, what is the probability that if you guess the last two characters, you will be correct?

6. There are 25 candies in a bowl. Of these candies, 2 are orange, 5 are lemon, 8 are cherry, and the rest are grape.

a) If you choose one candy, what is the probability that it is grape?

b) Your friend chooses candy after you choose a grape candy. If your friend takes two candies, what is the probability that both will be cherry?

7. You have ten different pairs of socks mixed in a drawer.

   a) What is the probability that you will pull out two socks that match?

   b) What are the odds in favour of you not getting matching socks?

☆ 8. In Texas Hold'em, five cards are dealt face up and are considered community cards that any player can use. Each player also gets two cards face down that other players cannot see. With these seven cards, each player tries to make the best five-card poker hand.

   a) What is the probability of choosing five cards and getting a pair?

   b) Consider the following hand.

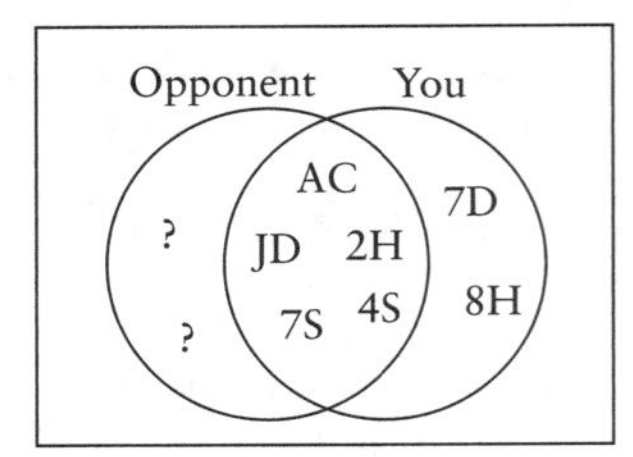

   What is the probability that your opponent has a higher pair than yours?

9. Your friend has a five-letter password for her e-mail account.

   a) What is the probability that if you guess her password, you will be correct?

   b) There are approximately 15 000 five-letter words in the English language. If your friend definitely used a word for her password, how much better is the probability of you guessing it correctly than if it were a random set of five letters?

10. Consider the second example in the Key Concepts on page 90. Determine a general formula for finding a pair of matching elements in a group of k elements when you have n possible different elements.

☆ 11. Your sister is celebrating her birthday by taking nine friends (three boys and six girls) to the movies. As a promotion, the theatre gives out three free movie passes randomly to people in the party.

   a) What is the probability that the three boys will get the passes?

   b) What is the probability that at least two girls get passes?

   c) If the theatre never gives a movie pass to the person celebrating the birthday, by how much would the probabilities in parts a) and b) change?

12. Sixty-four teams compete in a basketball tournament. Half the teams are eliminated in the first round of 32 games. Half of the remaining teams are eliminated in the 16-game second round, and so on until the final game determines the champion. Assuming that the teams are evenly matched so that each team is equally likely to win each game, determine

   a) the probability that any person will choose the winner correctly for every game

   b) the probability that the team you want to win the championship will do so

13. When your MP3 player is set to shuffle randomly, it plays songs by the same artist within a few songs of one another.

   a) Assuming that you have 10 songs by each of 20 artists, what is the probability that you will hear the same artist twice in any 8-song block?

   b) What is the least number of songs such that you still have greater than a 50% chance of hearing the same artist?

   c) Do the results from parts a) and b) help explain why you might hear the same artist more than once in a block of songs? Explain.

14. In a variation of Rock-Paper-Scissors, there are two added choices: fire and ice. Here are the outcomes for each possible game (excluding ties):

Rock smothers fire	Scissors cuts paper
Fire melts ice	Ice rusts scissors
Paper covers ice	Scissors divide fire
Paper covers rock	Fire burns paper
Ice freezes rock	Rock crushes scissors

a) Use combinations to verify the number of possible outcomes that are not ties.

b) If you throw ice, what is the probability that you will win?

c) Is there one choice that has a higher probability of winning than the others?

15. Theoretically, if there are 23 people in a room, there is a better than 50% chance that two of them will have the same birthday. Go to *www.mcgrawhill.ca/links/MDM12* and select **Study Guide**. Follow the links to obtain a copy of the file **Birthday.ftm**. In this file are birthdays of 800 students who have taken the Census at School survey.

a) From the **Birthdays** collection, create dot plots of the attributes **DayofMonth**, **Month**, and **DayofYear**. How do these plots suggest how the 800 birthdays are distributed throughout the year?

b) Follow the instructions in the file to repeatedly sample 23 of the students. Compare the number of trials where there is a pair of matching birthdays to the total number of trials to determine the experimental probability of two people in a group of 23 having the same birthday.

c) If the sample is 30 students instead of 23, how does the probability of getting a pair of matching birthdays change? To answer this, delete all of the data in the **Measures from Sample of Birthdays** collection. In the **Sample of Birthdays** collection, change the value 23 to 30 in the **Measures formula** as well as in the number of samples on the **Sample** tab.

d) How does your answer in part c) agree with its theoretical value?

C

16. Repeat question 15 to test the second example in the Key Concepts on page 90.

17. You have 12 balls in a bin. Each ball has one number on it as follows: {1, 2, 3, 1, 2, 3, 1, 2, 3, 1, 2, 3}. If you randomly select three balls, what is the probability that you will get balls whose numbers sum to 6?

☆ **18.** To play a lottery, you choose any six numbers from 1 to 49.

a) What is the probability that you will choose the winning set of six numbers?

b) You decide to buy 1000 tickets, each with different sets of six numbers. How much better is your probability of winning? Why does it not really matter?

c) You win another prize if you get five of the six numbers correct. What is the probability of winning this prize?

d) The easiest prize to win is for having three of the six numbers correct. What are the odds of winning this prize?

KEY CONCEPTS

- Compound events are composed of two or more simple events. The following are examples of compound events:
  a) flipping a coin and then rolling a die to see if you get heads and 6
  b) flipping a coin three times to see if you get three heads in a row
  c) drawing two cards from a deck one at a time to see if you get two hearts
- In general, when a compound event occurs, its probability is the product of the individual simple event probabilities.

Example

Consider the fictitious case of an army of 400 soldiers. When the soldiers reach a fork in the road, they always split up so that 25% go to the left and the rest go to the right. Thus, after two such forks, there will be four possible destinations for any soldier. Determine the probability that any soldier will be at each of the four locations.

Solution

You can determine the probability that any soldier will be at each of the four locations in two ways:

Method 1: Consider the Total Number of Soldiers at Each Fork	Method 2: Consider the Probability of Each Fork
400 → 100 → 25 at destination A $P(A) = \frac{25}{400} = \frac{1}{16}$	$\frac{1}{4}$ → $\frac{1}{4}$ destination A $P(A) = P(L) \times P(L) = \left(\frac{1}{4}\right)\left(\frac{1}{4}\right) = \frac{1}{16}$
400 → 100 → 75 at destination B $P(B) = \frac{75}{400} = \frac{3}{16}$	$\frac{1}{4}$ → $\frac{3}{4}$ destination B $P(B) = P(L) \times P(R) = \left(\frac{1}{4}\right)\left(\frac{3}{4}\right) = \frac{3}{16}$
400 → 300 → 75 at destination C $P(C) = \frac{75}{400} = \frac{3}{16}$	$\frac{3}{4}$ → $\frac{1}{4}$ destination C $P(C) = P(R) \times P(L) = \left(\frac{3}{4}\right)\left(\frac{1}{4}\right) = \frac{3}{16}$
400 → 300 → 225 at destination D $P(D) = \frac{225}{400} = \frac{9}{16}$	$\frac{3}{4}$ → $\frac{3}{4}$ destination D $P(D) = P(R) \times P(R) = \left(\frac{3}{4}\right)\left(\frac{3}{4}\right) = \frac{9}{16}$

Note that in both cases, the sum of the probabilities is 1.

- The events in examples a) and b) above, along with the army scenario, are pairs of independent events. They are independent because the outcome of one event does not affect the outcome of the other event.
- The events in example c) are dependent events. They are dependent because what happens in the first event affects what can happen in the second. Since one card is drawn out and kept out of the deck in the first event, the total number of possible outcomes changes from 52 for the first card to 51 for the second.
- In general, the following are true for independent and dependent events:

Independent Events	Dependent Events
$P(A \text{ and } B) = P(A) \times P(B)$	$P(A \text{ and } B) = P(A) \times P(B \mid A)$, where $P(B \mid A)$ is the conditional probability of B happening given that A has already occurred.

Example

Most tests to detect diseases are not 100% accurate. Sometimes a test can give a false positive (indicating that you have the disease or condition when you do not) or a false negative (indicating that you do not have the disease or condition when you do). Suppose a test to check donated blood for the virus that causes chickenpox is found to give a positive result 98% of the time when the blood contains the virus and to give a false positive result 7% of the time when the virus is not present. If one pint of blood in 1000 pints contains the virus, find the probability that

a) the blood has the virus and tests positive

b) the blood does not have the virus and tests positive

c) the blood does not have the virus and tests negative

d) the blood has the virus and tests negative

e) if the test is positive, it is correct

f) if the test is negative, it is correct

Solution

List the given facts:

$P(\text{positive result} \mid \text{virus is present}) = 0.98$

$P(\text{negative result} \mid \text{virus is present}) = 1 - 0.98$ or 0.02 (This is the false negative.)

$P(\text{positive result} \mid \text{no virus is present}) = 0.07$ (This is the false positive.)

$P(\text{negative result} \mid \text{no virus is present}) = 1 - 0.07$ or 0.93

a) Because there are two sets of results depending on whether the virus is present or not, these are dependent events. Therefore,

$$\begin{aligned} P(\text{virus and positive}) &= P(\text{virus}) \times P(\text{positive} \mid \text{virus}) \\ &= \left(\frac{1}{1000}\right)(0.98) \\ &= 0.000\,98 \end{aligned}$$

So, 0.098% of the time the test is positive and correct.

b)

$$\begin{aligned} P(\text{no virus and positive}) &= P(\text{no virus}) \times P(\text{positive} \mid \text{no virus}) \\ &= \left(\frac{999}{1000}\right)(0.07) \\ &= 0.069\,93 \end{aligned}$$

Then, 6.993% of the time the test is positive and incorrect.

c)

$$\begin{aligned} P(\text{no virus and negative}) &= P(\text{no virus}) \times P(\text{negative} \mid \text{no virus}) \\ &= \left(\frac{999}{1000}\right)(0.93) \\ &= 0.929\,07 \end{aligned}$$

92.907% of the time the test is negative and correct.

d)

$$\begin{aligned} P(\text{virus and negative}) &= P(\text{virus}) \times P(\text{negative} \mid \text{virus}) \\ &= \left(\frac{1}{1000}\right)(0.02) \\ &= 0.000\,02 \end{aligned}$$

0.002% of the time the test is negative and incorrect.

e) To answer this question, consider only the two ways the test could be positive: a true positive (0.098% of the tests) and a false positive (6.993% of the tests).

$$P(\text{positive and correct}) = \frac{\text{ways to be positive \& correct}}{\text{total number of ways to be positive}}$$
$$= \frac{0.098}{0.098 + 6.993}$$
$$= 0.014$$

This means if the test comes back positive, it is correct only 1.4% of the time.

f) $$P(\text{negative and correct}) = \frac{\text{ways to be negative \& correct}}{\text{total number of ways to be negative}}$$
$$= \frac{0.929\ 07}{0.929\ 07 + 0.000\ 02}$$
$$= 0.999\ 978$$

This means if the test comes back negative, it is correct 99.9% of the time.

A

1. Determine whether each event is compound or simple.
   - **a)** picking three cards from a deck
   - **b)** rolling a die
   - **c)** drawing two names from a hat
   - **d)** rolling a pair of dice
   - **e)** flipping a coin three times
   - **f)** flipping three coins

2. Determine whether the pairs of events are independent or dependent.
   - **a)** drawing one card from a deck, returning it, then drawing another
   - **b)** rolling two dice
   - **c)** flipping a coin twice
   - **d)** driving fast and getting a speeding ticket
   - **e)** placing first in a swimming competition and placing first in a running competition
   - **f)** staying up all night and failing a test next day

3. You have three toonies and two loonies in your pocket. You pull one coin out and then pull out another (without replacing the first).
   - **a)** Create a tree diagram to represent the probability of each outcome.
   - **b)** Show that the total of all the probabilities is 1.

B

4. Show using combinations that question 3 part b) is correct.

☆ 5. Determine the probability of drawing two cards from a deck (without replacement) and getting the same suit using
   - **a)** conditional probability
   - **b)** combinations

6. When Hannah wakes up on time, there is a 10% chance she will miss the bus for school. If she oversleeps, there is a 50% chance she will miss the bus. On average, Hannah oversleeps one morning a week.
   - **a)** What is the probability that Hannah will wake up late and miss the bus?
   - **b)** What is the probability that Hannah will miss the bus?

7. Several studies have been conducted on the effectiveness and safety of the H1N1 influenza vaccine. One study found that in a group that received the actual medication, the drug was 87% effective. In a group that received a placebo, there was a 10% effective rate. If the study used 400 people to receive the placebo and 600 people to get the actual drug, determine the probability that any person in the study will

   a) receive the drug and not get sick

   b) not get sick

8. On the television show *Numb3rs,* a mathematician helps to solve crimes. In one episode, the mathematician tested a person's claim to be a psychic by asking the person to guess the colour of 25 cards chosen from a standard deck.

   a) If a person were to guess the colour of each card, how well do you think the person would do? Explain.

   b) In the episode, the psychic guessed all of the 25 cards incorrectly. What is the probability of this happening?

   c) Would guessing all of them incorrectly strengthen or weaken the person's claim to be a psychic? Explain.

☆ 9. A basketball player has a basket-to-shot ratio of 40%, but after scoring two baskets in a row, her shooting percentage increases to 70%. This represents what many players perceive as a "hot hand" effect, a phenomenon such that when you have scored two or more shots in a row, you are more likely to continue to sink baskets.

   a) Create a tree diagram to represent the probability of all the possible outcomes of five shots.

   b) How many of the branches were affected by the hot hand?

   c) Look on the Internet to find out what researchers have determined to be true about the existence of the hot-hand effect?

10. In the game of Pig, players repeatedly roll a die to accumulate points. For every roll, a player adds that value to his or her point total. The catch is that if you roll 1, you lose all of your points. Players can roll as many times as they wish and quit whenever they want. The object is to be the person who quits with the highest point total.

   a) Determine the probability of not rolling 1 until

   i) the first roll

   ii) the second roll

   iii) the fourth roll

   iv) the sixth roll

   b) Conduct a physical simulation to develop a game strategy. Roll a die until you get 1. Once you get 1, repeat the experiment. Do this 20 times. Keep track of the number of rolls needed to get 1 in each trial. Use this information to help advise a player when the optimum time to quit would be.

   c) Go to *www.mcgrawhill.ca/links/MDM12* and select **Study Guide**. Follow the links to obtain a copy of the file **Pig.ftm**. With this file you will conduct a simulation in Fathom™ to conduct an experiment similar to part b) several times quite quickly. Follow the instructions in the file and compare your results to the simulation in part b). Would you change your advice?

★ **11.** A television game show allows contestants to choose one of three doors and keep whatever prize is behind that door. Usually, a valuable prize is behind only one of the doors, with token prizes behind the other two. When a contestant makes a selection and before that door is opened, the host shows that a token prize is behind one of the other two doors and offers the contestant the opportunity to switch to the remaining door.

Create a tree diagram to represent all of the possible different outcomes of the two choices (the original choice and the decision to switch or not) and determine whether switching or staying is the better strategy.

C

12. In a contest, five players qualify for a chance to win a car. To win, a contestant must choose from a group of five keys the only key that will start the car. Contestants are randomly selected to choose a key one at a time and try it. The game ends when someone starts the car. Determine which player, if any, has the highest chance of winning the car: the first to choose, the last to choose, or someone in the middle.

13. In your school, of all the students who take both Data Management and English, 90% pass Data Management, 85% pass English, and 82% pass both.

a) Determine the probability that if someone takes both Data Management and English, they will fail both.

b) Are passing English and passing Data Management independent events? Explain.

14. A manufacturer of mechanical pencils knows that 2% of its products are defective. You buy a box of pencils, pull out one pencil and try it, then try a second, then a third.

a) Create a tree diagram to represent all the different outcomes (good or defective).

b) What is the probability that one of the pencils you pulled out will be defective?

KEY CONCEPTS

- If two events are mutually exclusive, then the probability that either one will occur is given by $P(A \text{ or } B) = P(A) + P(B)$.

Example

A survey is conducted of 100 students about watching reality television shows.

	Males	Females	Totals
Watch Them	32	45	77
Do Not Watch Them	16	7	23
Totals	48	52	100

In two ways, determine the probability that if a person is selected at random from this sample, the person will be either a male or female who watches reality shows.

Solution

Let event A be male who watches and let event B be female who watches.

Method 1: Use Direct Determination

$$P(A \text{ or } B) = \frac{\text{number of ways a male or female who watches could be chosen}}{\text{number of ways anyone could be chosen}}$$
$$= \frac{77}{100} \text{ or } 0.77$$

Method 2: Add Simple Probabilities

Being a male and being a female are mutually exclusive events, so $P(A \text{ or } B) = P(A) + P(B)$.

$$P(A \text{ or } B) = P(\text{male who watches}) + P(\text{female who watches})$$
$$= \frac{\text{number of ways to choose a male who watches}}{\text{number of ways to choose anyone}} + \frac{\text{number of ways to choose a female who watches}}{\text{number of ways to choose anyone}}$$
$$= \frac{32}{100} + \frac{45}{100}$$
$$= \frac{77}{100} \text{ or } 0.77$$

- If two events are not mutually exclusive, then the probability that either one will occur is given by $P(A \text{ or } B) = P(A) + P(B) - P(A \cap B)$.

Example

A survey found that of 26 fast-food restaurants, 13 were pizza parlours, 9 did not sell alcohol, and 7 were pizza parlours that did not serve alcohol. Determine the probability that if one restaurant is chosen at random, it is a pizza parlour or does not serve alcohol.

Solution

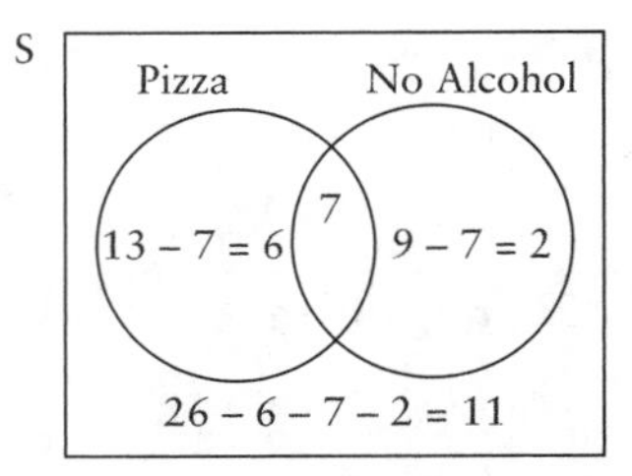

Method 1: Use a Venn Diagram

From the diagram, the number of ways to choose a restaurant that is either a pizza parlour or that does not serve alcohol is $6 + 7 + 2 = 15$. Therefore,

$P(\text{pizza or no alcohol}) = \frac{15}{26} = 0.58$

Method 2: Use Non–Mutually Exclusive Probabilities

The two events, being a pizza parlour or not serving alcohol, overlap (both can happen in the same restaurant). Thus, the probability of either happening is given by

$P(A \text{ or } B) = P(A) + P(B) - P(A \cap B)$. So

$$\begin{aligned} P(\text{pizza or no alcohol}) &= P(\text{pizza}) + P(\text{no alcohol}) - P(\text{pizza} \cap \text{no alcohol}) \\ &= \frac{13}{26} + \frac{9}{26} - \frac{7}{26} \\ &= \frac{15}{26} \text{ or } 0.58 \end{aligned}$$

There is a 58% chance of choosing a pizza parlour or a restaurant that does not serve alcohol.

- In general, if a question has the word *or* in it, you will have to do some sort of adding of probabilities.

A

1. Determine whether the events in each pair are mutually exclusive.

a) having red hair, having blue eyes

b) flipping a coin, rolling a die

c) drawing a spade from a deck of cards, drawing a jack from a deck of cards

d) celebrating a birthday, celebrating Thanksgiving

e) going on a roller coaster, going on a Ferris wheel

f) sleeping in, being late for school

2. You roll one die and flip one coin. Determine the probability of

a) rolling 6 or getting heads

b) rolling an even number or getting heads

c) rolling a number greater than 2 or getting tails

☆**3.** You roll two dice. What is the probability of getting

a) a pair or a sum of 8?

b) a 3 or a sum of 9?

c) a sum of 7 or a sum of 11?

4. You draw a card from a standard deck. What is the probability that the card is

   a) a jack or a heart?

   b) a face card or an ace?

   c) a spade or a heart?

5. Which events in questions 2, 3, and 4 are mutually exclusive events?

B

6. Create a scenario dealing with the rolling of two dice that will be

   a) mutually exclusive

   b) not mutually exclusive

   c) mutually exclusive and have a probability of $\frac{1}{4}$

   d) not mutually exclusive and have a probability of $\frac{11}{36}$

7. A newspaper surveyed 150 people about a change in its format. Of the people surveyed, 87 people like the change, 81 men participated in the survey, and 24 of those men like the change. One of the people who took the survey will win a year's subscription to the paper. What is the probability that the winner will be

   a) someone who likes the change in format?

   b) a man who does not like the change?

   c) a woman who likes the change?

8. A skateboard shop conducted a random survey of neighbourhood residents about a proposed skateboard park. Consider the following results:

	Males	Females
In Favour	142	45
Against	58	32
No Opinion	22	30

   What is the probability that a person randomly selected from this sample will

   a) be against the proposal or have no opinion?

   b) be a female or be in favour?

9. The Tri-Harder triathlon club has 98 members. Some of the members also compete in swimming, biking, and running competitions in addition to the regular triathlon races. Currently, no members compete in swim meets. There are 18 members who compete in races. Forty members compete in running, but 5 of them also compete in cycling. The club is asked to select a member to sit on a local roads and trails committee.

   a) What is the probability that the club will select someone who competes in races besides triathlons?

   b) Ten members who originally competed only in triathlons now decide to compete in swimming. Now what is the probability that the club will choose a person who competes in races besides triathlons?

10. The table has data about the fate of passengers in the sinking of the *Titanic* ocean liner in 1912.

	Men		Women		Children	
	Died	Lived	Died	Lived	Died	Lived
First Class	114	55	4	113	1	6
Second Class	139	13	13	78	0	25
Third Class	381	59	91	88	55	25

   a) What is the probability that if a person was from first class or second class, the person survived?

   b) What is the probability that if a person was from third class, the person survived?

   c) What is the probability a person survived if the passenger was

   i) in first class and not a man?

   ii) in second class and not a man?

   iii) in third class and not a man?

   d) Based on these data, how true do you think the practice of "women and children first" was adhered to when the *Titanic* was sinking?

11. The Parliament of Canada includes the House of Commons and the Senate. The table shows the political affiliations of the members of these chambers in 2009.

	Senate	House of Commons
Conservative	46	145
Liberal	49	77
Bloc Québécois	0	48
NDP	0	37
Progressive Conservative	2	0
Independent	3	1
Vacant	5	0

Determine the probability that a randomly selected public representative is

a) a Conservative member or a Liberal member

b) in the Senate or a member of the NDP

c) in the House of Commons or a Liberal member

d) not affiliated with any political party or in a party that is not in both chambers

C

12. Given three simple events A, B, and C, develop a formula to determine $P(A \text{ or } B \text{ or } C)$.

13. A data-management teacher has taken a survey of students and has created three categories that the students could be put into.

A) Think that the teacher is funny

B) Laugh at the teacher's jokes to be polite

C) Are excited about mathematics

The survey determines that

- 42% of the students are excited about math
- 41% of the students think the teacher is funny
- 55% of the students laugh to be polite
- 23% of the students are excited about math and laugh to be polite
- 18% laugh to be polite and think that the teacher is funny
- 10% think that the teacher is funny and are excited about math
- 3% think the teacher is funny, laugh to be polite, and are excited about math

a) What is the probability that if the teacher randomly selects a student, the student will be in only one of groups A, B, or C above?

b) What is the probability that if the teacher tells a clearly unfunny math joke, a student will laugh only if the student is excited about math?

Chapter 7 Probability Distributions

7.1 Probability Distributions

KEY CONCEPTS

- A random variable *X* has a single value for each outcome in an experiment. Discrete random variables have separated values, while continuous random variables have an infinite number of outcomes along a continuous interval.
- Randomness should not be confused with evenness. Of these two sets of dots, the scattering in Set 2 may appear more random than that in Set 1. In reality, when things are truly random, clustering appears. This means that Set 1 is closer to being random.

Set 1	Set 2
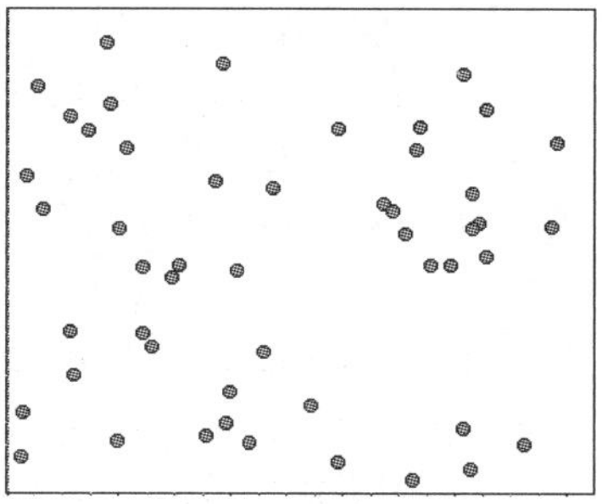	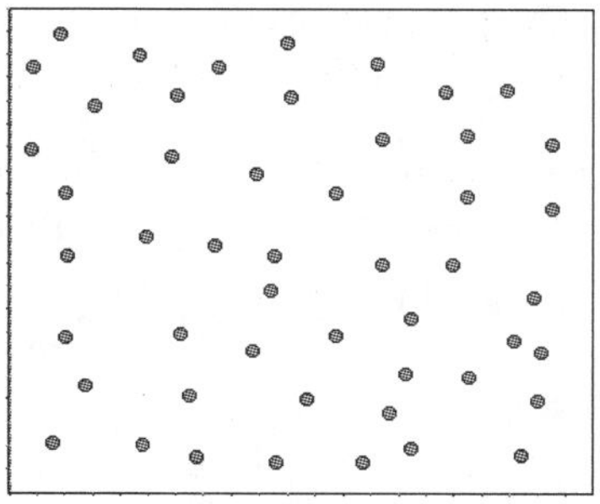

- A probability distribution shows the probabilities of all the possible outcomes of an experiment. The sum of the probabilities in any distribution is 1. Probability distributions can be displayed graphically in a probability histogram. A probability histogram has bars with heights equal to the probability of the outcome associated with each bar. Probability histograms can have many shapes.

Uniform	Structured	Non-Structured
Rolling a single die	Rolling two dice	Drawing different-coloured marbles from a bag
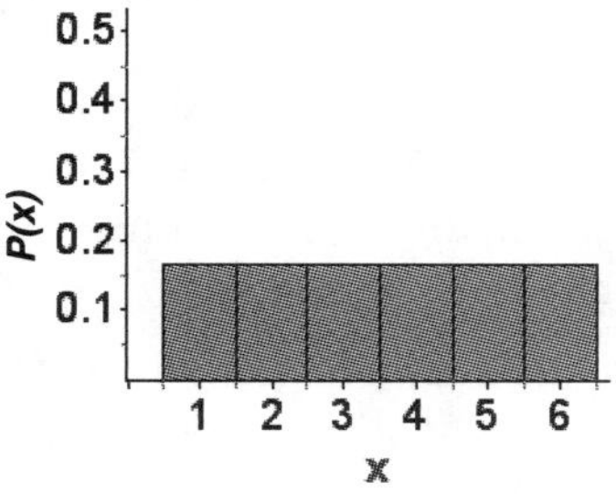	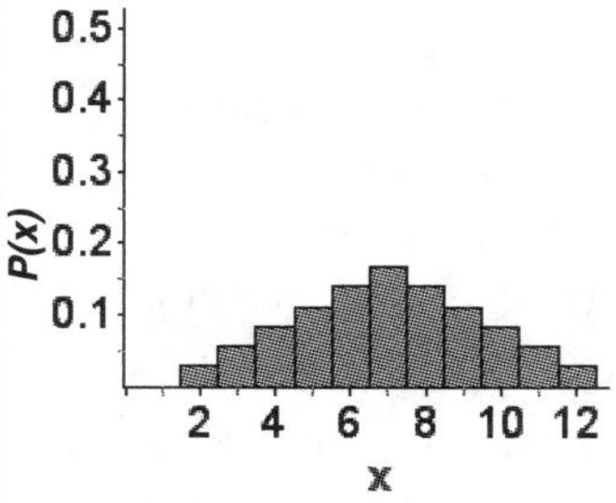	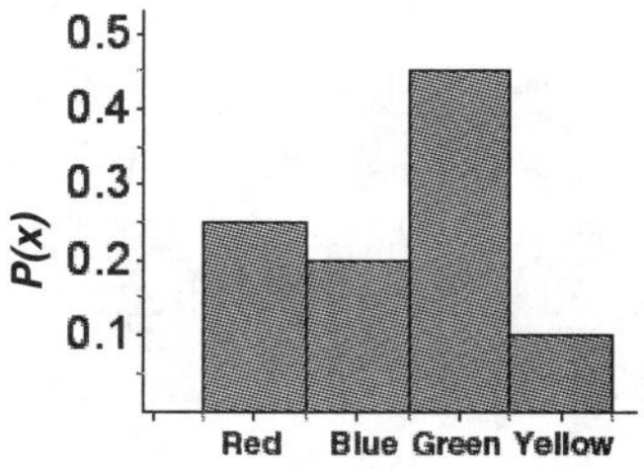

- Expectation, or expected value, is the predicted average of all possible outcomes of a probability experiment. It is calculated with the formula

  $$E(x) = x_1P(x_1) + x_2P(x_2) + ... + x_nP(x_n)$$
  $$= \sum_{i=1}^{n} x_iP(x_i)$$

  The expected value can be used to determine things like the cost of lottery tickets or the payout in games of chance.

Example

A school lottery has 5000 tickets for sale. There is one grand prize of $2000 and five second prizes of $100.

a) What is the random variable in this lottery?

b) Determine the expected value of the lottery and interpret what it represents.

c) If you buy a ticket for $1, what is your expected value?

d) How can the school use the expected value to help determine the ticket price?

Solution

a) The random variable, X, represents all the possible prize payouts. $X = \{\$2000, \$100, \$0\}$.

b) $E(X) = \$2000\left(\frac{1}{5000}\right) + \$100\left(\frac{5}{5000}\right) + \$0\left(\frac{4994}{5000}\right)$

$= \$0.50$

The expected value is the expected payout per ticket, assuming all tickets are sold.

c)

	Since the ticket costs $1, if you win the 1st prize, you actually win only $1999.		Similarly, if you win a 2nd prize, you win only $99.		If you do not win (which is the most likely case), then you actually lose $1.
$E(X)$	$= \$1999\left(\frac{1}{5000}\right)$	+	$\$99\left(\frac{5}{5000}\right)$	+	$(-\$1)\left(\frac{4994}{5000}\right)$

$= -\$0.50$

Thus, you would expect to lose $0.50 with each $1-ticket purchase. Stated another way, if you pay $1 to play the lottery, you can expect to win $0.50, which is a net loss of $0.50.

d) If the lottery sells all 5000 tickets, the payout per ticket is $0.50. In order to make money, the lottery needs to bring in more than that amount per ticket. An obvious choice would be to price the tickets at $1 each. However, at this price the lottery would need to sell at least half the tickets to break even, a better choice might be $2 per ticket.

- The expected value in a fair game is 0. The expected value can sometimes be misinterpreted. In the context of a game of chance, the expected value is the average payoff for a player if the player plays a large number of games. It has very little meaning in the context of the result of a single game. This is why casinos on average always win. For most of the games, the expected value is only slightly in their favour, but since they play substantially more games than any individual player, in the long run they will come out ahead.

- The outcomes of a uniform probability distribution all have the same probability, $P(x) = \frac{1}{n}$, where n is the number of possible outcomes in the experiment.
- You can simulate a probability distribution with manual methods, calculators, or computer software.

A

1. Classify each of the following random variables as continuous or discrete.
   a) whether it will rain tomorrow
   b) the spin of a roulette wheel
   c) the result of a lottery draw
   d) the length of time your computer will go without crashing
   e) the day of the week that you were born

2. Explain whether each experiment has a uniform probability distribution.
   a) selecting the month of the year
   b) selecting a weekday or a day on the weekend
   c) winning or losing in a lottery
   d) selecting a marble from a bag of six different-coloured marbles
   e) flipping a coin three times and counting the number of heads

3. In a restaurant promotion, customers are invited to spin a spinner after they order a meal. If the spinner lands on the month of the customer's birthday, the meal is free.
   a) Determine the probability distribution for this situation.
   b) Go to *www.mcgrawhill.ca/links/MDM12* and select **Study Guide**. Follow the links to obtain a copy of the file **Sample.ftm.** This file has the birthdates of 800 students. Create a frequency histogram and a relative frequency (probability) histogram of the birth month of these students. How does this agree or disagree with your answer in part a)?
   c) Describe how a frequency histogram and a probability histogram are related.

☆4. The table shows the theoretical frequency of each sum when two dice are rolled.

Sum	Frequency	Sum	Frequency
2	1	8	5
3	2	9	4
4	3	10	3
5	4	11	2
6	5	12	1
7	6		

   a) Create a frequency histogram for these data.
   b) Create a probability histogram for these data.
   c) How do the answers from parts a) and b) compare?

5. For each probability distribution, create a probability histogram and determine the expected value.

a)

x	$P(x)$
5000	$\frac{1}{1000}$
200	$\frac{10}{1000}$
50	$\frac{20}{1000}$
0	$\frac{969}{1000}$

b)

x	$P(x)$
1	0.7
10	0.25
50	0.05

B

6. A furniture shop is having a promotion.

LEO's Furniture
Roll the dice and save! With every purchase you get a chance to save.
• Roll 5, 6, 7, 8, or 9 and save $50
• Roll 3, 4, 10, or 11 and save $100
• Roll 2 or 12 and save $200

   a) Create a probability histogram with one bar for each prize.

   b) What is the expected prize?

   c) Do you think the person who set up the promotion knows anything about probability? Explain.

7. You are playing a dice game with a friend. A single die is rolled, and if 5 or 6 comes up, your friend pays you $6. Otherwise, you pay your friend $3.

   a) Who has the better chance of winning any particular game?

   b) What is your expected value?

   c) Is this a fair game? Explain.

8. A school is having a fundraising lottery. A total of 2000 tickets are available at a price of $25 each for these prizes.

Prize	Number
$10 000	1
$500	5
$25	494

   a) What is the probability of winning?

   b) What is the expected payout value for the school?

   c) What is your expected value if you buy one ticket?

   d) Determine the total of the prizes. How could this be used to answer parts b) or c)?

   e) How many tickets must be sold in order for the lottery to break even?

   f) Why do you think there are 494 prizes of $25 each?

9. A casino is developing a new game in which a coin is flipped three times and the number of tails is counted.

   a) Determine the probability histogram for this situation.

   b) A contestant will pay $2 to play this game and wins $10 if exactly two tails come up. Determine the expected value for the player.

   c) Explain why the casino may not want to use this game. Suggest how the game could be modified to benefit the casino.

☆ 10. In roulette, a player makes a bet on which of a wheel's 38 numbered spots a ball will land on. To increase the chance of winning, the player can bet on more than one number at once. If a person bets $1 and wins, the person is paid back the amount of the bet plus a multiple based on the following payout scale. If the player loses a bet, then the player loses his or her money.

Bet	Payout
1 number	35 to 1
2 numbers	17 to 1
3 numbers	11 to 1
4 numbers	8 to 1
5 numbers	6 to 1
6 numbers	5 to 1
12 numbers	2 to 1
18 numbers	1 to 1

   a) Determine the odds against winning for each bet. In each case, determine the odds as ▒ to 1.

   b) Compare the odds against winning to the payout. Based on this, is there any bet that is more advantageous to the player than any other?

   c) Determine the expected value for each bet. What does this information say about whether one bet is more advantageous than another?

11. In one version of the game show *Deal or No Deal*, a contestant chooses one of 26 briefcases. The case contains one of the amounts of money shown in the table below, but remains closed until the end of the game. The contestant owns the contents of the case unless he or she chooses to sell the case to a mysterious banker. As the game progresses, the amounts in the remaining cases are revealed and the banker makes cash offers to the contestant, which the contestant can accept and make a "deal" or refuse and make a "no deal". The game continues until the contestant either makes a deal or all the cases are opened.

$0.01	$1 000
$1	$5 000
$5	$10 000
$10	$25 000
$25	$50 000
$50	$75 000
$75	$100 000
$100	$200 000
$200	$300 000
$300	$400 000
$400	$500 000
$500	$750 000
$750	$1 000 000

a) What type of probability distribution will this game have?

b) What is the expected value at the beginning of the game?

12. A salesperson selling cars for $20 000 each has found that if a customer is offered a $1000 option package for free, the chance of making a sale is 70%. If no incentive is offered, then the chance of making a sale is only 60%. Use the expectation of each situation to determine which is more advantageous.

13. A football team keeps statistics about the probability of getting a touchdown or a field goal from various field positions.

Field Position	Probability of Event	
	Touchdown (7 points)	Field Goal (3 points)
5-yd line	60%	99%
10-yd line	50%	98%
20-yd line	40%	97%
30-yd line	25%	90%
50-yd line	10%	60%

Based on the points available, determine the expected value of each event and determine when it is more advantageous to choose one event over the other.

C

14. Design a fair game for two players who each roll one die where

a) the players have an equal chance of winning each game

b) the players do not have an equal chance of winning each game

c) the players have an equal chance of winning each game, and there is a way to tie

☆ **15.** Economists have studied the game *Deal or No Deal*. Among their findings is that one strategy contestants can use to maximize their winnings is to hold out on making a deal until the banker's offer gets close to or is better than the current expected value. A contestant has cases with $75, $500, $750, $5000, $10 000, and $400 000 in them, and the banker offers $35 000.

a) Explain why the contestant refuses to sell and opens one case worth $5000.

b) Suppose the banker's new offer is $75 000. Use mathematical reasoning to determine whether the player should make a deal.

KEY CONCEPTS

- A binomial distribution has a specified number of independent trials in which the outcome is either success or failure. The probability of success is the same in each trial.
- The probability of x successes in n independent trials is $P(x) = {}_nC_x p^x q^{n-x}$, where p is the probability of success on an individual trial and q is the probability of failure on that same individual trial ($p + q = 1$).
- The expectation for a binomial distribution is $E(X) = np$.

Example

A basketball player has five shots in a free-throw competition. Historically, the player has a probability of 35% of scoring on a free throw.

a) Assuming this probability stays constant, create a table showing the probability distribution and expectation for each number of baskets the player might make. Graph the probability distribution.

b) What is the probability of the player scoring more than two baskets?

c) What is the expected number of baskets the player will make?

Solution

a) The probability of success is $p = 0.35$ and the probability of failure is $q = 0.65$.

Number of Baskets, x	$P(x)$		$xP(x)$
0	${}_5C_0(0.35)^0(0.65)^5$	$= 0.1160$	$= 0$
1	${}_5C_1(0.35)^1(0.65)^4$	$= 0.3124$	$= 0.3124$
2	${}_5C_2(0.35)^2(0.65)^3$	$= 0.3364$	$= 0.6728$
3	${}_5C_3(0.35)^3(0.65)^2$	$= 0.1811$	$= 0.5434$
4	${}_5C_4(0.35)^4(0.65)^1$	$= 0.0488$	$= 0.1951$
5	${}_5C_5(0.35)^5(0.65)^0$	$= 0.0053$	$= 0.0263$
	Totals	$= 1.0000$	$= 1.75$

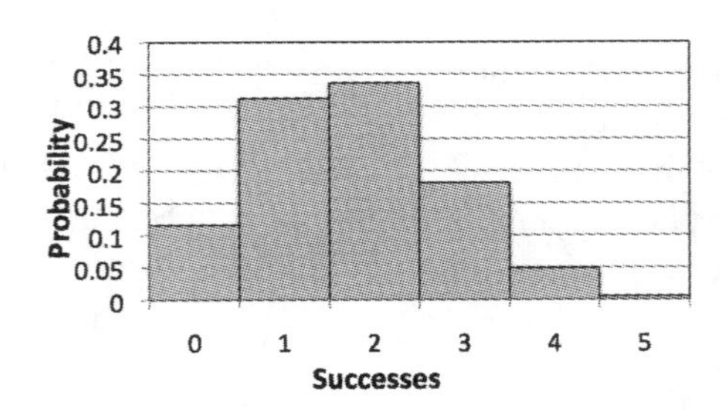

b) $P(\text{more than 2 baskets}) = P(3 \text{ baskets}) + P(4 \text{ baskets}) + P(5 \text{ baskets})$
$= 0.1811 + 0.0488 + 0.0053$
$= 0.2344$
$= 23.4\%$

c) $E(X) = np$
$= 5(0.35)$
$= 1.75$

The expected number of baskets is 1.75.

- To simulate a binomial experiment
  - choose a simulation method that accurately reflects the probabilities in each trial
  - set up the simulation tool to ensure that each trial is independent
  - record the number of successes and failures in each experiment
  - summarize the results by calculating the probabilities for r successes in n trials (the sum of individual probabilities must equal 1)

A

1. Determine which of the following situations could be represented by a binomial distribution. Justify your answers.

a) A coin is flipped seven times and the number of heads is recorded.

b) Five hundred people are playing bingo and the number of wins is counted.

c) The failure rate of a computer hard drive is 2.3%. The number of customer complaints is tracked by the manufacturer.

d) Health Canada tracks the infection rate for people who received the H1N1 vaccine.

e) Five cards are pulled one at a time without replacement from a standard deck to see if a spade is chosen.

☆ **2.** For each situation, determine the probability distribution, histogram, and expected value.

a) $p = 0.02, n = 6$

b) $p = 0.7, n = 10$

3. For $p = 0.25$ and $n = 9$, determine the probability distribution using

a) a graphing calculator

b) a spreadsheet

c) Fathom™

4. Describe the binomial situation in each of the following.

a) ${}_{10}C_3(0.91)^3(0.09)^7$

b) ${}_{6}C_5(0.15)^5(0.85)^1$

c) ${}_{25}C_0(0.1)^0(0.9)^{25} + {}_{25}C_1(0.1)^1(0.9)^{24} + {}_{25}C_2(0.1)^2(0.9)^{23}$

B

5. Evelyn and Aislin play basketball. Evelyn has a 45% success rate for free throws. Aislin's success rate is 60%.

a) If both players get five free throws, what is each player's probability of sinking exactly three?

b) In a particular game, Evelyn gets ten free throws while Aislin is awarded seven. Who has the greater probability of sinking at least four free throws?

6. A coin is flipped ten times. Determine the probability of getting
   a) at least four heads
   b) either three heads or three tails

7. You roll a die 20 times.
   a) What is the probability that you will get your favourite number, 3, as often as all the other numbers combined?
   b) Why can a binomial distribution be used in this question?

8. A person chosen at random is asked the day of the week on which he or she was born. Determine the probability that
   a) any person chosen at random was born on a Monday
   b) exactly one person out of seven chosen was born on a Monday

9. Traffic lights in your town operate on a short cycle. Red lights stay on for 30 s, amber lights for 5 s, and green lights for 25 s. There are eight sets of lights on your route to school. You want to assess the probability of being able to drive through an intersection without having to stop because the light is amber or red.
   a) How can this situation relate to a binomial distribution?
   b) Determine the probability histogram for being able to drive through any intersection without stopping.
   c) How many green lights should you expect to get on any particular day?

10. You and your friends are evaluating new gaming consoles to decide what to buy. You find the following data about power-supply failures.

Console	Failure Rate (%)
Wee	2.7
XStation	10.0
PlayBox	23.7

If you each decide to buy the same model of console, determine whether there is a system that has more than a 50% chance of at least one of you having a failure.

11. Consider that $p = \{0.2, 0.4, 0.6, 0.8\}$.
   a) For each probability of success for ten trials
      i) determine the probability distribution
      ii) create a probability histogram
      iii) determine the expected value
   b) Compare your answers from part a) for each of the given probabilities. How do you think the distributions would change if there were 100 trials instead of 10?

12. Use technology to compare the probability histograms of rolling a die and getting 6 in
   a) 10 rolls
   b) 20 rolls
   c) 50 rolls

13. Suppose that the probability that a child will have blue eyes is 70% if both mother and father have blue eyes. How many children with blue eyes can be expected if a mother and father both with blue eyes have
   a) 3 children
   b) 5 children
   c) 19 children

14. The electronics store where you work sells MP3 players with scroll wheels that fail in 0.5% of the units. You just received a shipment of 50 players.
   a) What is the probability that one of the units will be defective?
   b) What is the most likely number of returns you should expect?
   c) Should you be surprised if one unit is found to be defective?

★ **15.** The table has data about Canadian students from a Census at School survey.

	Responses by Boys (%)	Responses by Girls (%)
Which is your dominant hand?		
Right	82.1	87.5
Left	10.2	8.1
Ambidextrous	7.7	4.4
How do you get to school?		
Bus	35.7	39.7
Walk	22.4	20.4
Car	33.6	35.1
Cycle	3.3	0.6
Other	5.0	4.2
How many cigarettes do you smoke a day?		
0	85.4	91.5
1–25	7.2	5.4
26–50	2.6	1.7
51 or more	4.8	1.4

Source: Census at School 2008/2009

You survey 15 boys and 12 girls at your school.

a) What is the probability that two or more boys cycle to school?

b) What is the probability that exactly three boys smoke cigarettes?

c) Of the girls you interview, four are not right handed. Should this be surprising? Explain your answer.

16. How many cards would you have to draw (with replacement) from a standard deck so that there is a 50% chance of getting at least one ace?

C

17. Each box of your favourite cereal has one of five secret decoder rings inside. You like only one of the rings and would like to have one for each finger on one hand.

a) What is the probability that if you buy six boxes, you will get at least four of your favourite ring?

b) How many boxes do you need to buy to have a probability of more than 80% of getting at least four of your favourite ring?

c) How many boxes would you have to buy to have an expected value of four of your favourite ring?

18. Go to *www.mcgrawhill.ca/links/MDM12* and select **Study Guide**. Follow the links to obtain a copy of the file **CardBinomial.ftm** and use it to verify experimentally your answer to question 16.

★ **19.** A company is accused of discrimination in its hiring practices. Of the company's last 15 new hires, only 3 are women.

a) The people who claim that the company discriminates state that approximately half of the new hires should be women. What assumptions are they making and how can a binomial distribution confirm this?

b) The company claims that only 30% of female applicants were actually qualified. Is the company's hiring record in line with this claim?

KEY CONCEPTS

- A geometric distribution has an infinite number of independent trials, each with two possible outcomes, success or failure. The random variable is the number of unsuccessful outcomes before a success occurs.
- The probability of success after a waiting time of x failures is $P(x) = q^x p$, where p is the probability of success in each single trial and q is the probability of failure.
- The expectation of a geometric distribution is $E(X) = \frac{q}{p}$.

Example

An assembly-line robot installs a CD player in each car produced on the line. For quality control, the units are tested to make sure that they are installed properly. The robot has a probability of malfunctioning of 0.2.

a) Determine the probability distribution of the number of tests performed before a malfunction is found.

b) What is the expected number of tests before a malfunction is found?

Solution

a) In this case, a "success" is a malfunction, so $p = 0.2$ and $q = 1 - 0.2$ or 0.8.

Number of Tests Before a Malfunction (Waiting Time), x	$P(x) = q^x p$
0	$(0.8)^0(0.2) = 0.2$
1	$(0.8)^1(0.2) = 0.16$
2	$(0.8)^2(0.2) = 0.128$
3	$(0.8)^3(0.2) = 0.1024$
4	$(0.8)^4(0.2) = 0.08192$
.....	.....

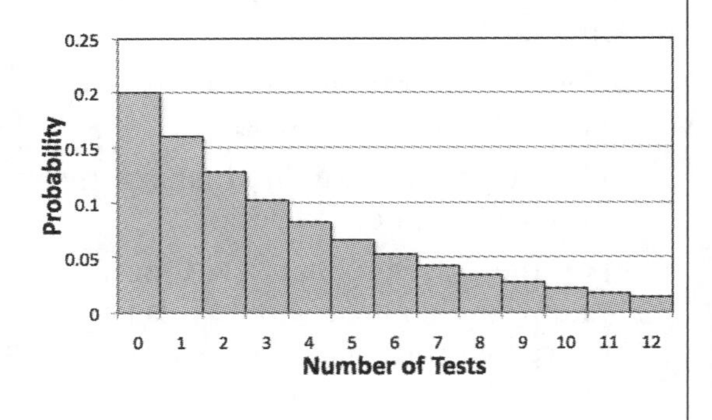

b) The expected value is given by $E(X) = \frac{q}{p} = \frac{0.8}{0.2} = 4$. Thus, you would expect to perform four tests before a malfunction was found.

- To simulate a geometric experiment, you must ensure that the probability on a single trial is accurate for the situation and that each trial is independent. Summarize the results by calculating probabilities and the expected waiting time.

A

1. Indicate whether a geometric distribution model can be used. If not, state why.
   a) A coin is flipped until a head appears.
   b) An NHL goalie is having shots taken on net in a shootout.
   c) Bart Simpson is opening cereal boxes in a grocery store, marking each one that does not contain the prize he is looking for as "Damaged—sell at reduced price!"
   d) You draw a card from a standard deck (without replacement) to see how soon you get an ace.
   e) Your little brother sees how many times he can jump on a frozen puddle before the ice breaks.

☆ 2. Given each probability of success, determine the probability distribution and histogram for up to six trials and determine the expected wait time.
   a) $p = 0.5$
   b) $p = 0.25$
   c) $p = 0.9$

3. For the probabilities in question 2, recalculate the probability distributions using
   a) a graphing calculator
   b) a spreadsheet
   c) Fathom™

4. Compare the geometric probability histograms for six trials using $p = \{0.2, 0.4, 0.6, 0.8\}$.

B

5. A card is drawn from a well-shuffled deck to see if it is an ace, and then it is replaced.
   a) Determine a probability distribution for drawing up to six cards.
   b) How long should it be before an ace is drawn from the deck?

6. An experienced darts player has about a 15% chance of hitting triple-20 on any individual throw.
   a) Determine a probability distribution of missing the triple-20 for throwing up to six darts.
   b) What is the expected number of throws before there would be a miss?

7. A manufacturing company had a failure rate of 30% when their compact discs were used in tests.
   a) Determine the expected wait time until you pull out a bad disc from a new box.
   b) The company improved product quality and this year achieved a failure rate of 13.5%. Compared to your answer in part a), how much longer will it be now before you expect to pull out a bad disc?

8. You work for a telemarketing company making calls to promote a new gadget. Based on company data, the following are the likely outcomes of each call.

Outcome	Percent
People hang up right away	45
No answer	38
People listen to you but do not buy	16
People purchase	1

   a) How many calls would you have to make before you get to speak to someone?
   b) Assuming that you can make 25 calls per hour and work an 8-hour shift, how many sales could you expect to make on any particular shift?

9. Sometimes statistics can be biased. A particular statistician knows that approximately 35% of people do not tell the truth when polled. If 400 people were polled in a survey, how many would be expected to answer before someone answers untruthfully?

10. Your aunt plays the 6/49 lottery, each time picking the same set of six different numbers out of 49 in the hope that those numbers will come up. You have told her the probability of her winning is quite low, but she plays twice a week, claiming that by the end of the year, her numbers "are sure to come up." How much better is the probability of her winning at the end of the year compared to the probability at the beginning of the year?

11. Your local hockey team has a contest in which spectators are chosen at random to take a single shot on net to try to score. The catch is that most of the net is blocked by a board. Organizers know that an average contestant has about a 1 in 100 chance of succeeding.

a) What is the probability that the tenth contestant gets the puck in?

b) What is the probability of a winner among the first ten people to play?

12. The failure rate for a professionally packed parachute is approximately 0.001.

a) How many jumps would a person have to make before his or her probability of having a parachute failure exceeded 50%?

b) If the failure rate increases to 1%, how many fewer jumps would a parachutist have to make before the probability of having a failure is over 50%?

13. In the game of Pig, players repeatedly roll a die to accumulate points. Players may roll as many times as they wish and quit whenever they want. The object is to be the person who quits with the highest point total. The value of each roll is added to a player's points. The catch is that if you roll a 1, you lose all your points. On average, what is the most strategic way to play this game? Justify your answer.

★ 14. A cereal company has a promotion offering one of five prizes in each box. Assuming that the probability of each prize is the same, determine the expected number of boxes you would have to buy to get

a) any prize

b) two different prizes

c) three different prizes

d) four different prizes

e) five different prizes

C

15. In the Prisoners game, six prisoners are locked in separate cells. The cells are numbered 0, 1, 2, 3, 4, and 5. To get any prisoner out of jail, a player rolls two dice and calculates the difference in values. The prisoner in the cell with the number matching the difference is set free. What is the expected number of rolls needed to release all of the prisoners?

16. Go to *www.mcgrawhill.ca/links/MDM12* and select **Study Guide**. Follow the links to obtain a copy of the file **Prizes.ftm** to experimentally simulate the situation in question 14. How does the simulation support your theoretical answer?

17. A manufacturer of precision parts for the automotive industry has a prototype of a new parts machine. The machine is tested for several days and the operator keeps track of the number of parts it makes before it makes a defective one. The results are as follows: 73, 60, 70, 60, 42, 64, 61, 59, 65, 68, 105, 71, 76, 59, 69, 65, 70, 120, 74, 62, 79, 79, 57, 76, 78. How could a geometric probability distribution be used to help estimate the probability of a defect for this machine?

7.4 Hypergeometric Distributions

KEY CONCEPTS

- A hypergeometric distribution has a specified number of dependent trials having two possible outcomes: success or failure. The random variable is the number of successful outcomes in the specified number of trials. The individual outcomes cannot be repeated within these trials.
- The probability of x successes in r dependent trials is $P(X) = \frac{({}_aC_x)({}_{N-a}C_{r-x})}{({}_NC_r)}$, where N is the population size, and a is the number of successes in the population. (Note that the Student Edition uses the variable n to represent the population size. However, here N is used to represent population size; otherwise the formulas for $P(x)$ are identical. This change is made because n is sometimes used to represent a sample size; here r is used for the sample size.)
- The expectation for a hypergeometric distribution is $E(X) = \frac{ra}{N}$.
- To simulate a hypergeometric experiment, ensure that the number of trials is representative of the situation and that each trial is dependent (no replacement or resetting between trials). Record the number of successes and summarize the results by calculating probabilities and expectation.

Example

A bag contains six red marbles, five yellow marbles, and four black marbles. Marbles are selected from the bag one at a time without replacement until four marbles have been taken.

a) Determine the probability distribution for choosing red marbles using

i) a tree diagram

ii) combinations

b) Create a probability histogram.

c) Determine the expected number of red marbles.

Solution

Since the marbles are taken without replacement, the draws are not independent events. Since we are concerned only with red marbles, there are only two possible outcomes for each draw: getting a red marble or not getting a red marble. Thus, this will be a hypergeometric distribution.

a) The probability of getting a red marble (success) is $p = \frac{6}{15}$, and the probability of anything else (failure) is $q = \frac{9}{15}$.

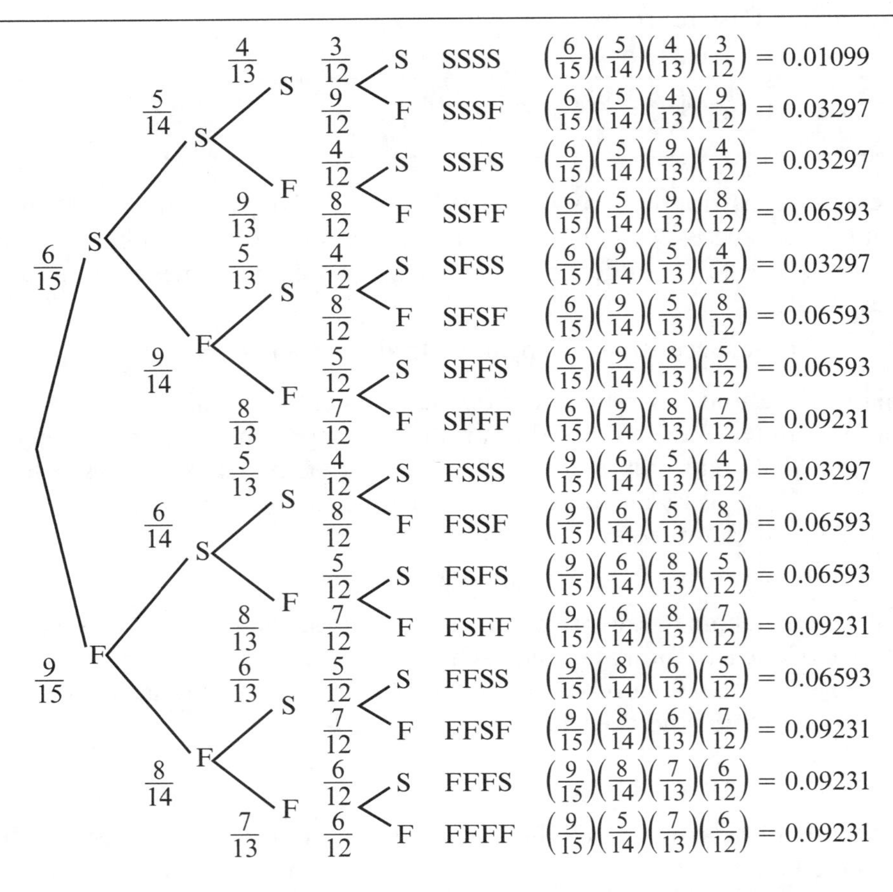

Number of Red, x	i)	ii)
	$P(x)$	$P(x)$
0	$1(0.09231) = 0.09231$	$\frac{({}_6C_0)({}_9C_4)}{({}_{15}C_4)} = 0.09231$
1	$4(0.09231) = 0.36923$	$\frac{({}_6C_1)({}_9C_3)}{({}_{15}C_4)} = 0.36923$
2	$6(0.06593) = 0.39560$	$\frac{({}_6C_2)({}_9C_2)}{({}_{15}C_4)} = 0.39560$
3	$4(0.03297) = 0.13187$	$\frac{({}_6C_3)({}_9C_1)}{({}_{15}C_4)} = 0.13187$
4	$1(0.01099) = 0.01099$	$\frac{({}_6C_4)({}_9C_0)}{({}_{15}C_4)} = 0.01099$

b) 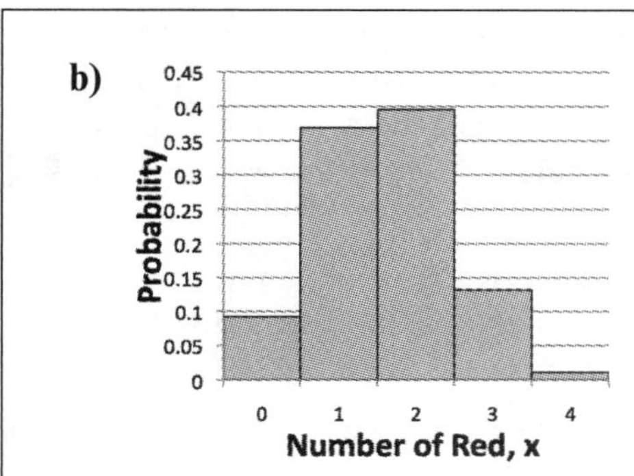

c) Total number of outcomes: $N = 15$;
number of successful outcomes: $a = 6$;
number of trials: $r = 4$

$$E(X) = \frac{ra}{N} = \frac{(4)(6)}{(15)} = 1.6$$

The expected number of red marbles to be chosen is 1.6.

A

1. Can each situation be modelled by a hypergeometric probability distribution? Justify your answer.

a) flipping a coin ten times and counting the heads

b) testing a part by putting it through 100 fatigue tests in a row

c) drawing a card with replacement from a deck and looking for a 10

d) choosing a co-ed team of 5 students from a class of 20 and counting the number of girls on the team

e) drawing five names at once from a hat to see who wins a prize

f) removing seven chips from a bag that contains 30% red chips and 70% blue chips and noting the number of red chips

g) selecting 6 marbles without replacement from a bag that contains 14 red marbles, 35 green marbles, and 26 blue marbles, and counting the number of blue marbles

☆**2.** Consider a hypergeometric situation with $N = 30$, $r = 6$, and $a = 10$.

a) Determine the probability distribution.

b) Show that $E(X) = \sum_{i=0}^{r} x_i P(x_i) = \frac{ra}{N}$.

3. For each situation, determine the probability histogram and expected value.

a) $N = 12$, $r = 4$, $a = 5$

b) $N = 52$, $r = 5$, $a = 13$

c) $N = 150$, $r = 13$, $a = 20$

4. Repeat question 3 using

a) a graphing calculator

b) a spreadsheet

c) Fathom™

B

☆**5.** A bag of 80 candies contains 25% each of orange and blue candies and 12.5% each of brown, red, yellow, and green. Eight candies are taken from the bag.

a) Explain why the number of green candies selected can be modelled with a hypergeometric distribution.

b) Determine the distribution and create a probability histogram.

c) What is the probability that exactly half of the candies taken from the bag will be green?

d) What is the probability that at least half of them will be green?

e) What is the expected number of green candies?

6. Explain what each combination represents in $P(X) = \frac{({}_aC_x)({}_{N-a}C_{r-x})}{({}_NC_r)}$.

7. You are in a small class consisting of 4 girls and 11 boys. Your teacher is making up a group of 4 students to conduct a survey at school.
   a) Use a tree diagram to determine the probability distribution for the number of girls in the group.
   b) Recalculate your answers in part a) using combinations.
   c) Discuss the methods used in parts a) and b). Include advantages and disadvantages of each.

8. A bag of marbles has 40% blue and 60% red. Five marbles are chosen at random, and the number of red marbles is noted.
   a) Explain why there is not enough information to determine a probability histogram for a hypergeometric distribution but there is enough information for a binomial distribution.
   b) Use technology to generate a table and probability histogram for both a binomial and a hypergeometric distribution given the following population sizes:
   i) 15 **ii)** 50 **iii)** 100
   c) Use your results to discuss when a binomial situation could be used to approximate a hypergeometric situation.

9. You have a bag with three red marbles and seven black marbles. You choose four marbles without replacement and count the number of red marbles.
   a) Discuss any problem(s) with the hypergeometric distribution in this situation.
   b) Do any of the problems from part a) disappear if you count the black marbles instead?

10. At an automotive test facility, a researcher prepares a set of 100 air bags to test. She makes sure that there are 6 defective bags in the batch. A second researcher tests 4 of the air bags.
   a) What is the probability that at least one of the air bags tested will not deploy (i.e., is defective)?
   b) What is the expected number of defects?

11. A company produces equal numbers of white, yellow, and green bars of soap. One day, the labelling machine malfunctions and mislabels the packages of 3000 bars. Rather than discard the bars, the company offers cases of 24 to its employees.
   a) Determine the expected number of green bars of soap in each case.
   b) Determine the probability histogram of getting green bars of soap.
   c) You really like green soap. What is the probability that at least half of the mislabelled packages contain green bars?

12. A company produces engine piston rings. A box of 50 is known to contain 3 defective rings. Ten rings are sampled.
   a) What is the probability of getting exactly one defective ring?
   b) What is the probability of getting any defective rings?
   c) The actual sample taken has two defective rings. Why should this not be surprising?

13. A convenience store is having a contest. Gold tickets worth $100 are hidden in 5 of 500 coffee cups. If you buy two coffees per day over the course of two weeks, what is the probability that you will win at least one ticket?

★**14.** A biologist is preparing bacteria cultures. Over the time of their preparation, it is possible that some may become tainted. A batch containing 64 cultures will be disposed of if more than two become tainted. It is necessary to destroy a culture in order to test it, so the biologist decides to sample the batch instead.

a) Suppose a batch contains three tainted cultures (and thus would be disposed of). Determine the expected value for the following sample sizes:

i) 2 **ii)** 5 **iii)** 10

b) Based on your answers from part a), should a batch be disposed of if one of the samples comes back tainted? Explain.

c) For each of the cases in part a), what is the probability that none will come back tainted? The biologist knows that it is possible that no sample will come back tainted even though there are three tainted cultures in the batch. Why is this a problem?

d) How many need to be sampled to have a 50% chance of finding at least one tainted culture if there are three bad ones in the batch?

15. To promote tourism, your town stocks a pond with 500 bass to add to the existing population of other species. Later, a researcher samples 50 of the fish to make an estimate of the total population.

a) What is the estimate of the fish population if the researcher catches ten bass?

b) Using the researcher's data, the town council decides to promote fishing by claiming that "one in five fish caught is a bass" and, if this is not your experience, offering to return your fishing licence fee. What is the expected value of the fees collected for this promotion if a fishing licence costs $50?

C

16. Consider question 13.

a) How many more coffees would you have to buy in order to increase your probability of winning to 50%?

b) How many more gold tickets would there have to be if you bought only two coffees a day for two weeks but your probability of winning was 50%?

17. You are trying to impress your friends by guessing the colour of a card before they draw it from a deck.

a) If you guess and the result is not told to you, what is the expected number of correct guesses if 20 cards are drawn one at a time?

b) If you are told the result of each guess, how could this affect the expected number of correct guesses?

c) If you are told the result of each card draw, develop a strategy to maximize the probability of guessing correctly.

18. Go to *www.mcgrawhill.ca/links/MDM12* and select **Study Guide**. Follow the links to obtain a copy of the file **ESP.ftm.** This is a simulation to test the strategy you developed in question 17c). Explain how the simulation works and determine whether it verifies that your strategy will work.

Chapter 8 The Normal Distribution

8.1 Continuous Probability Distributions

KEY CONCEPTS

- Chapter 7 dealt primarily with discrete probability distributions, where the possible values of the random variable are natural numbers. Chapter 8 deals with continuous probability distributions. Any random variable that can have fractional or decimal values can be a continuous variable. Height of a person, duration of a phone call, mass of an object, and diameter of a piston are examples of such random variables.

- Since any value is possible, some ways of representing a distribution are to use a frequency table, histogram, or frequency polygon. For example, here are representations of race times for males aged 40–44 in the New York City Marathon.

New York Marathon Times (male, age 40−44)	
Interval (h)	**Frequency**
2.0−2.5	4
2.5−3.0	194
3.0−3.5	820
3.5−4.0	1590
4.0−4.5	1424
4.5−5.0	869
5.0−5.5	349
5.5−6.0	180
6.0−6.5	69
6.5−7.0	31

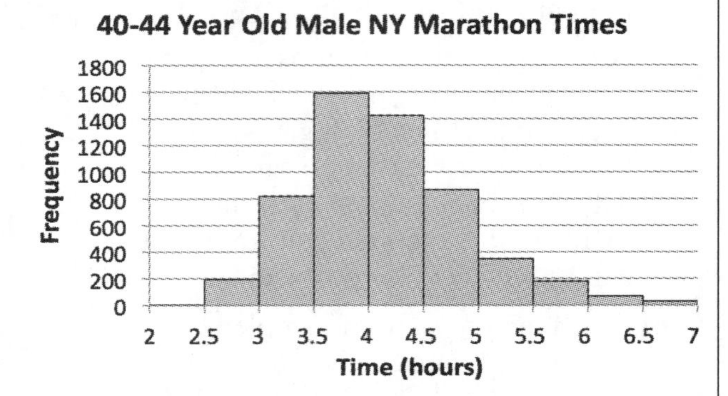

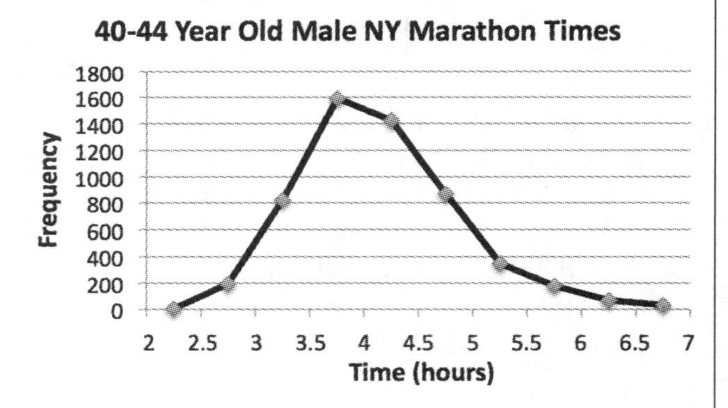

- The simplest continuous distribution is a uniform distribution. For uniform probability distributions, a constant function is used.

• Here are three different views of the same set of data.

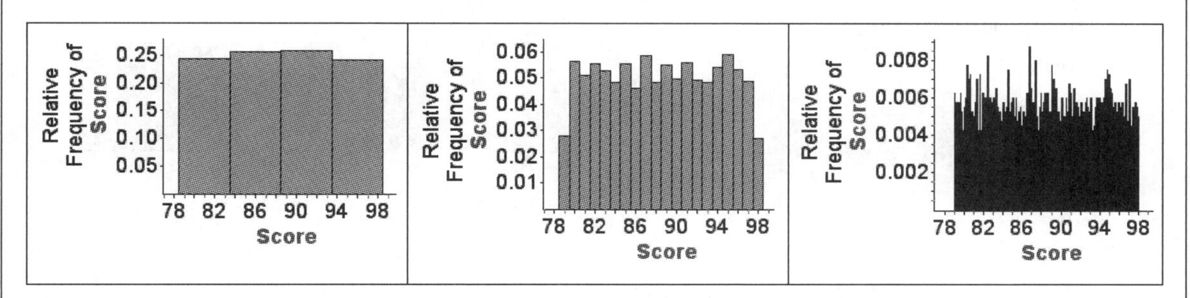

Notice that the shape of the histogram becomes erratic and non-uniform as the bin width becomes smaller. Since any possible value can occur, not all possible values will occur at the same frequency until you have an infinitely large sample size. Because of this, continuous data are often modelled by a function instead. The function represents the probability density, which is the probability per unit area. This is called a probability distribution function, sometimes referred to as a PDF.

• For any probability distribution, the total area under the curve equals 1, since it represents all possible outcomes. The probability of any single event is 0, because there are infinitely many possible events.

Example

For the previous set of sample data, the scores are known to have a uniform distribution over the range 78.5–98.5.

a) Determine the probability density function.

b) What is the probability that a score will be between 90 and 94?

Solution

a) To graph the probability density for a uniform function, the height of the graph must be known. Since the total area under the curve must be 1, the height is given by solving the equation $h(b - a) = 1$, where h is the height and b and a are the end points of the range of the distribution.

$$\begin{aligned} h &= \frac{1}{b-a} \\ &= \frac{1}{98.5-78.5} \\ &= 0.05 \end{aligned}$$

The probability density function is 0.05 and is represented in the graph.

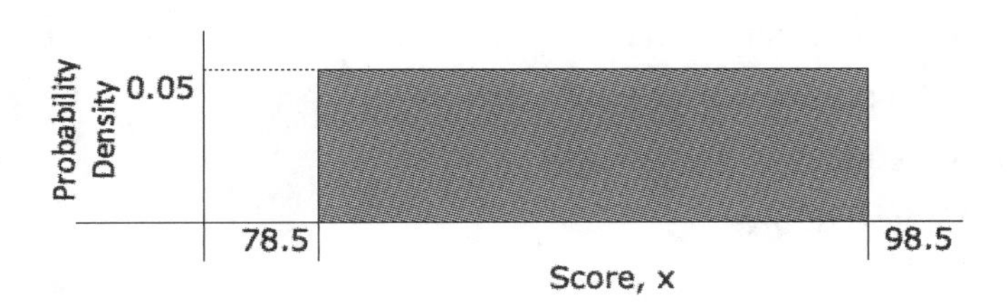

b) The probability of a score between 90 and 94, $P(90 < x < 94)$, is the area of the density between 90 and 94.

$P(90 < x < 94) = 0.05(94 - 90)$

$= 0.2$

The probability is 0.2, or 20%.

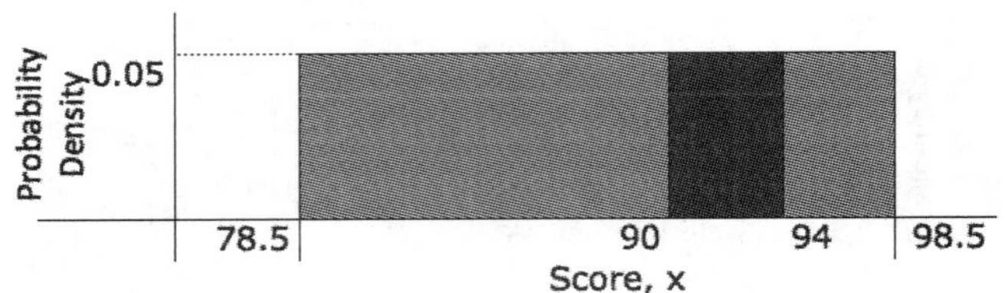

- Another common density function is the normal density function. This has a symmetrical shape with the mean centred within it. You will study its characteristics in this chapter.

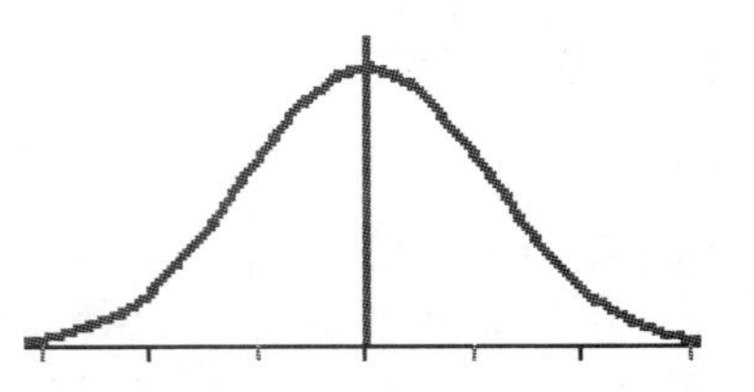

- Other distributions include bimodal, multimodal, and asymmetrical. The exponential density function $y = ke^{-kx}$ can be used to predict waiting times, where $k = \frac{1}{\mu}$ is the number of events per unit time, and $e \doteq 2.718\ 28$.

Bimodal	Negatively Skewed	Exponential Positively Skewed

A

1. Which will have a continuous probability distribution?

a) the length of a cut piece of wood

b) the number of marbles in a bag

c) the number of children in a family

d) the mass of fish caught

e) a golf score

f) a score on a BMX vertical competition

2. For each uniformly continuous distribution, determine the height of the graph.

a) a range of data of 155–210

b) a range of data of 0.52–0.57

c)

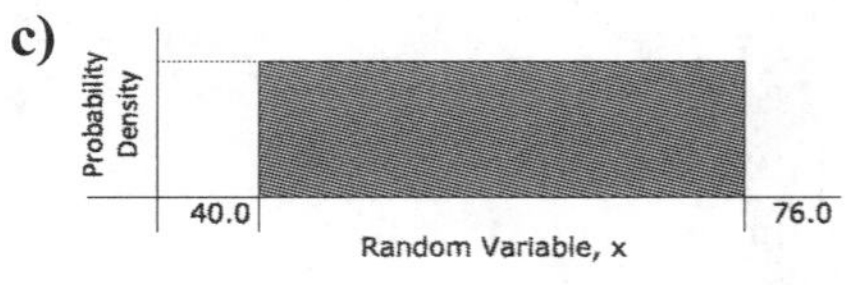

d)

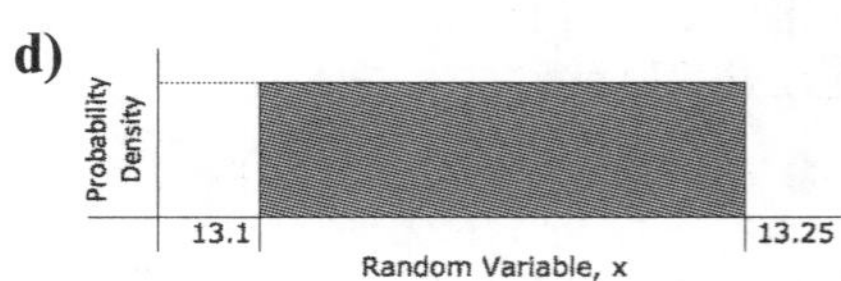

3. Consider the uniform distribution function shown in the graph.

Determine

a) $P(X < 130)$

b) $P(X \geq 135)$

c) $P(122 < X < 128)$

d) $P(129.5 < X < 130.3)$

4. Describe how to answer probability questions for uniform distributions without having to make a graph.

B

5. Consider the probability density function represented in the graph.

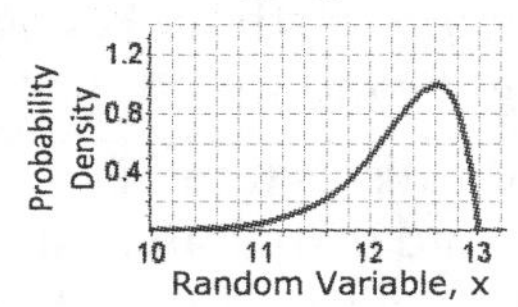

a) What should be true about the area under the curve?

b) Determine approximately the following probabilities.

i) $P(X < 12)$

ii) $P(X > 11.4)$

iii) $P(12 < X < 12.4)$

☆6. Consider a uniform probability distribution for $55 < x < 63$.

a) Determine the following probabilities.

i) $P(57 < X < 59)$

ii) $P(57.5 < X < 58.5)$

iii) $P(57.9 < X < 58.1)$

iv) $P(57.99 < X < 58.01)$

v) $P(57.999 < X < 58.001)$

b) How do your answers in part a) help show that $P(X = x) = 0$ (i.e., the probability of any individual value is zero)?

☆7. The table is a summary of data from an analysis of 250 magazines.

Price Interval	Frequency
\$1–\$2	5
\$2–\$3	8
\$3–\$4	69
\$4–\$5	88
\$5–\$6	51
\$6–\$7	16
\$7–\$8	8
\$8–\$9	5

a) Create a frequency table (including relative frequency), histogram, and frequency polygon of these data.

b) For an approximate density plot, recall that the total area of all of the bars of the histogram must equal 1. What would be the height of each bar for this plot?

c) What is the area of each bar of the density plot? How do these compare to the relative frequency? Do you think this will always be true?

d) Determine the probability that if any magazine is chosen at random, its price will be between \$3 and \$5.

8. A web server receives thousands of requests per hour. Typically, the arrival times of these requests follow a uniform distribution.

a) Create a graph of this distribution for any particular hour.

b) What is the probability that a request will come in the first 5 min?

c) What is the probability that a request will come between the 15th and 25th minutes?

d) Go to *www.mcgrawhill.ca/links/MDM12* and select **Study Guide**. Follow the links to obtain a copy of the file **WebServer.ftm**. Use the file to discover how to solve these problems using Fathom™.

9. Sailboat racing uses a handicap system to adjust for differences in types of boats. The Offshore Racing Rule system gives each boat a rating based on its measurements and past performance. Typical ratings follow a uniform distribution between 0.5 and 2.0.

a) Create a graph of this distribution.

b) What is the probability that any particular boat will have a rating greater than 1.4?

c) To make things fair, sub-classes are created containing boats within a range of ratings. What is the probability that a boat will be in the class with ratings between 1.0 and 1.1?

d) In any particular race, the distribution of ratings may not be uniform. With a course covering more than 500 km, the race from Chicago to Mackinac, Michigan (called "The Mac"), is the longest freshwater yachting race in the world. Go to *www.mcgrawhill.ca/links/MDM12* and select **Study Guide**. Follow the links to obtain a copy of the file **ChicagoMac.ftm**. Determine the type of distribution the boat ratings follow. Between which two ratings is the middle 50% of the data?

10. Go to *www.mcgrawhill.ca/links/MDM12* and select **Study Guide**. Follow the links to obtain a copy of the file **Heights.csv**. The data are the heights of 500 students compiled by Census at School.

a) Use a spreadsheet to create a frequency table (including relative frequency), histogram, and frequency polygon of these data.

b) Create an approximate probability density graph for these data. Hint: see question 7 for instructions.

c) Determine the probability that the height of any student chosen at random will be less than 170 cm.

d) A clothing company is using this data to help decide what sizes of clothes to produce. The company wants to produce sizes that will fit the greatest number of people but does not want to make every size. It decides to produce sizes only for heights within a 30-cm range. Where should this range be in order for the company to maximize its chances of making clothes for the greatest number of people? What is the probability that any person will fit into the company's clothes?

11. Go to *www.mcgrawhill.ca/links/MDM12* and select **Study Guide**. Follow the links to obtain a copy of the file **Earthquakes .ftm**. This file has data on all earthquakes that occurred in January 2010.

a) How many earthquakes occurred during the month?

b) Create a histogram of the **Mag** (magnitude) attribute using interval widths of 0.5. What type of distribution does this represent? Justify your answer.

c) Create four more histograms with the following interval widths: 0.25, 0.1, 0.05, 0.01. If these were frequency polygons instead, comment on how good they would be as an approximation of the probability distribution of these data.

12. The Canadian Community Health Survey was first conducted in 2000 and given to a sample of 130 000 Canadians. The survey consisted of questions related to each person's health obtained in a 45-min interview. The table has the results of two similar questions: *If you are a smoker, at what age did you start?* and *If you have quit smoking, at what age did you start?*

If you wish to use a spreadsheet, go to *www.mcgrawhill.ca/links/MDM12* and select **Study Guide**. Follow the links to obtain a copy of the file **Smokers.csv**.

Age Interval	Canada		Ontario	
	Smokers	Quitters	Smokers	Quitters
5–12	145 738	105 612	44 258	29 866
12–15	1 026 210	899 168	315 047	279 720
15–20	3 141 313	3 611 030	1 157 858	1 298 666
20–25	766 994	1 152 493	300 816	441 849
25–30	208 166	277 474	75 318	108 349
30–35	90 048	116 200	37 900	38 157
35–40	39 656	47 413	12 885	18 471
40–45	26 270	29 579	11 671	9 019
45–50	13 565	14 507	5 783	5 523
Over 50	9 094	11 334	3 231	3 030

a) What does the number 145 738 represent?

b) For each region, create a third column that represents the total of the smokers and the quitters. Explain what this column represents.

c) Create four frequency polygons, two for the smokers and two for the quitters.

d) Create the relative frequency for each set of data. For each of the following probabilities, compare your answers for all of Canada to those for Ontario. Determine the probability that someone who

i) is still smoking started before they were 20

ii) has quit started before they were 20

iii) is still smoking started in their 20s

iv) has quit started in their 20s

e) Create a new set of data that represents the percent of people in each age group who have quit. At what age could you start smoking and be the least likely to quit? At what age could you start smoking and be the most likely to quit?

13. An article critical of Canada's health-care system reports that the average wait time to see a specialist in Canada is 18.3 weeks. You have a pain in your knee and your doctor is sending you to a specialist.

a) Create a probability distribution graph using a spreadsheet.

b) Determine the probability that you will have to wait less than 10 weeks to see the specialist.

c) Go to *www.mcgrawhill.ca/links/MDM12* and select **Study Guide**. Follow the links to obtain a copy of the file **KneeWait.ftm** to learn how to repeat parts a) and b) using Fathom™.

14. At a busy radio station, the average time between request calls is 125 s. What is the probability that the next call will occur after a wait of 4 min?

C

15. Go to *www.mcgrawhill.ca/links/MDM12* and select **Study Guide**. Follow the links to obtain a copy of the file **SmokersProvinces.csv**. This file has each province's smoking statistics from the survey described in question 12. Complete an analysis similar to that in question 12d) and e) to determine if there are any anomalous provinces.

16. For the Chicago to Mackinac race data from question 9, determine as accurately as you can the approximate probability that any boat chosen at random has a rating between 0.75 and 1.00. Explain how you ensured the accuracy of your answer.

8.2 Properties of the Normal Distribution

KEY CONCEPTS

- The most commonly used distribution is the normal distribution. This distribution is one that models data that vary symmetrically about the mean. This is often referred to as the bell curve.

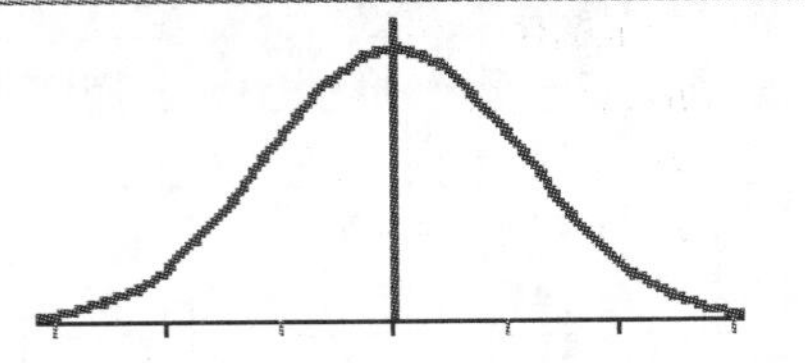

- The characteristics of a normal distribution rely directly on the mean, μ, and standard deviation, σ, of any particular data set. However, for all sets of normally distributed data, 68% of the data lie within one standard deviation of the mean, 95% within two standard deviations, and 99.7% within three standard deviations. This holds regardless of the mean and standard deviation of the data.

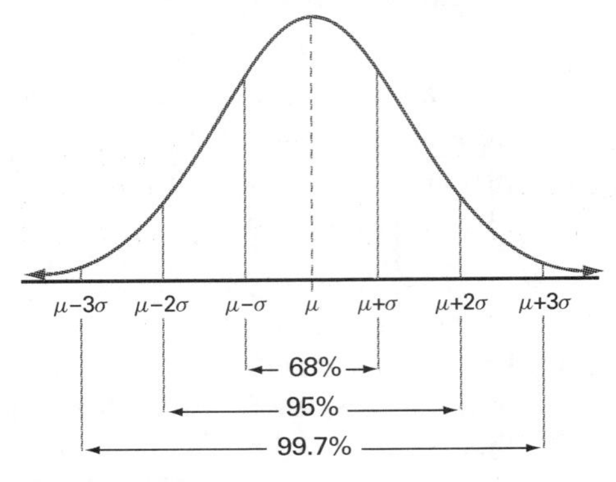

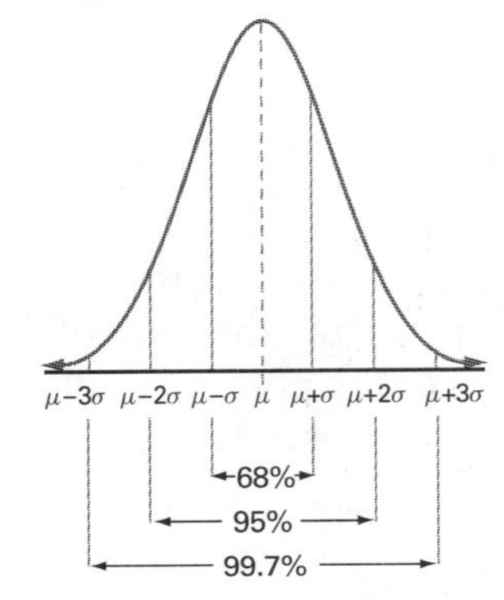

- The standard normal curve will have $\mu = 0$ and $\sigma = 1$. As with other probability distributions, the area under the curve equals 1. There are several methods for determining the probability from a normal distribution.

- Another useful measure for describing normal distributions is the z-score, $z = \frac{x - \mu}{\sigma}$. The z-score measures the number of standard deviations any point is away from the mean. By using the z-score, any normally distributed set of data can be converted to the standard normal distribution (with $\mu = 0$ and $\sigma = 1$). When the z-scores are used, another way to determine probabilities is by using tables of areas under a normal distribution curve. Refer to pages 606 and 607 of *Mathematics of Data Management* student edition, or go to *www.mcgrawhill.ca/links/MDM12*, select **Study Guide**, and follow the links to obtain a copy of the table.

Example

An assembly-line worker installs seats in cars. An ergonomics expert is analyzing the amount of time that it takes to do each job. She finds that the data are normally distributed with a mean of 43.5 s and a standard deviation of 4.5 s.

a) Generate a graph of the distribution.

b) Determine the probability that any particular job will take less than 49 s.

c) Determine the probability that any particular job will take between 40 s and 50 s.

Solution

a) Method 1: Use a Graphing Calculator

To graph the distribution, press [Y=], then [2nd], [VARS]. From the **DISTR** menu, select **1:normalpdf(**. To graph it, use the following parameters: **X,43.5,4.5)**. You will need to adjust the window so that the mean is roughly centred and the width is at least three standard deviations on each side. For the vertical scale, start at 0 and go to 0.5 (and adjust if necessary).

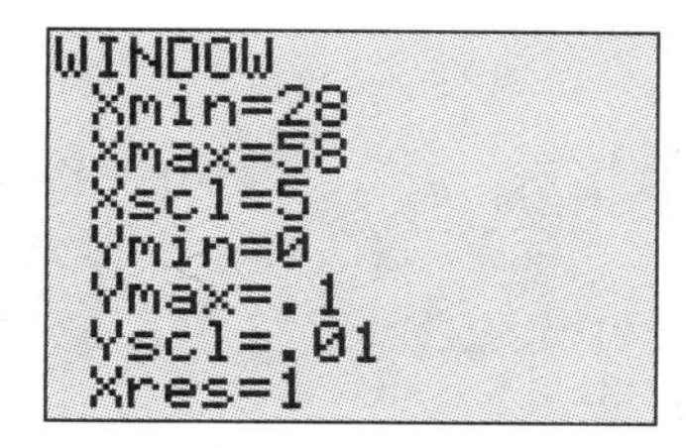

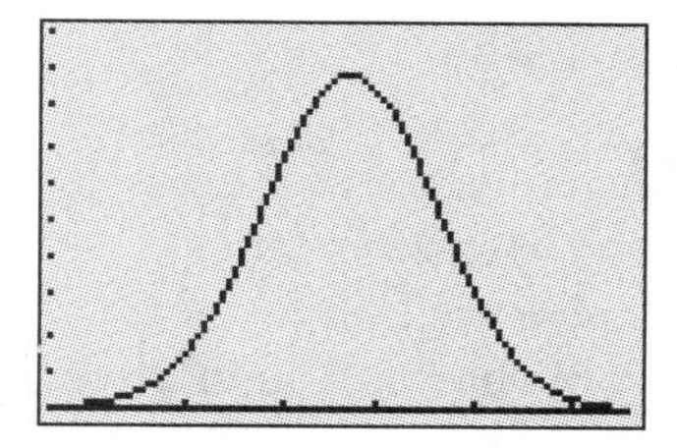

Method 2: Use Fathom™

Open a new file. Drag an empty graph to the workspace. At the top right of the graph, click the drop-down menu and choose **Function Plot**. Right-click the graph and choose **Plot Function**. In the function window, type **normalDensity(x,43.5,4.5)**. Adjust the scale so that the horizontal axis is centred at approximately 43.5 and there are at least three standard deviations on each side. Similarly, the vertical scale should start at 0 and end at about 0.5. Adjust the scales after the graph is displayed.

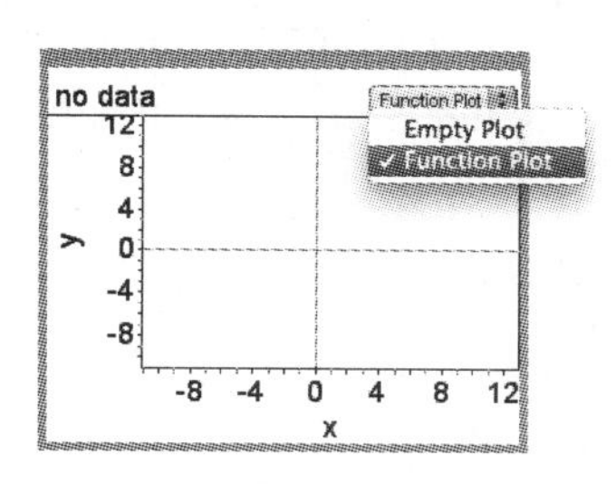

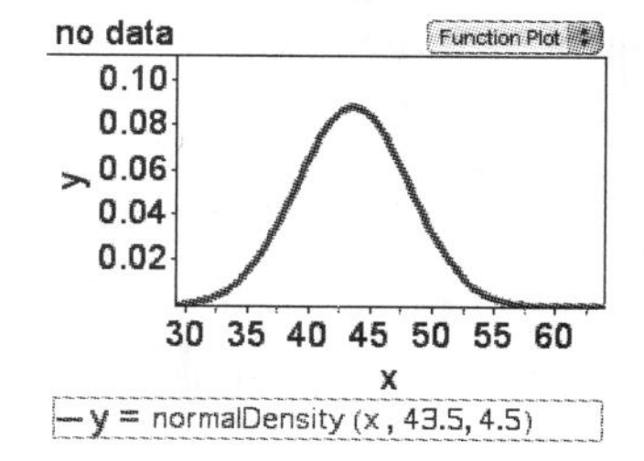

Method 3: Use a Spreadsheet

Create a column of numbers to represent your X-values for the range desired. The smaller the increments, the smoother the curve will be. In this example, the range starts at 30 and goes to 60 in intervals of 2. Next, calculate the distribution values by using the function **NormDist(A2,43.5,4.5,0)**. Finally, create a scatter plot with a smooth curve through the points.

	A	B	C
1	Time	Density	
2	32	=NORMDIST(A2,43.5,4.5,0)	
3	34	NORMDIST(x, mean, standard_dev, cumulative)	
4	36	0.02210603	

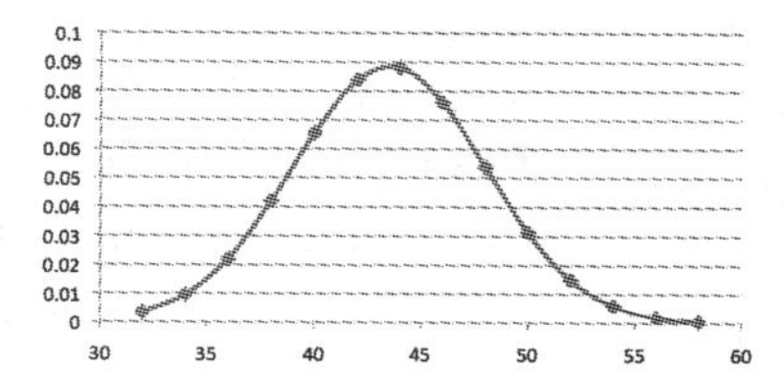

b) To determine the probability, use the cumulative normal density.

Method 1: Use a Graphing Calculator

From the **DISTR** menu, choose **2:normalcdf(**.
Use the following parameters: **−1e99,49,43.5,4.5)**.
Notice that −1e99 represents the left bound of this probability (the least value that can be entered).

Method 2: Use Fathom™

Open up a new collection, double-click it, and go to the **Measures** tab.
Create a new measure and in the formula box, type **normalCumulative (49,43.5,4.5)**.

Inspect Collection 1

Cases | Measures | Comments | Display | Categories

Measure	Value	Formula
prob_b	0.889188	normalCumulative (49, 43.5, 4.5)

Method 3: Use a Spreadsheet
In any cell, use the **NormDist** function with the **cumulative** parameter set to **1**.

0.8891882 =NORMDIST(49,43.5,4.5,1)
NORMDIST(x, mean, standard_dev, cumulative)

Method 4: Use Tables
Determine the z-score for a time of 49 s: $z = \frac{49 - 43.5}{4.5} \doteq 1.22$. Use probability tables to determine the cumulative probability. Note that in this case, the value is 0.8888, which is slightly different from the technology-based methods. This is because the table method is not as accurate, since it uses a z-score that is accurate to only two decimal places.

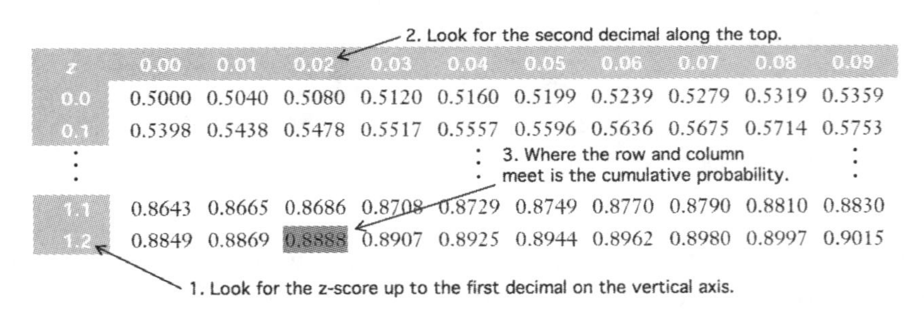

z	0.00	0.01	0.02	0.03	0.04	0.05	0.06	0.07	0.08	0.09
0.0	0.5000	0.5040	0.5080	0.5120	0.5160	0.5199	0.5239	0.5279	0.5319	0.5359
0.1	0.5398	0.5438	0.5478	0.5517	0.5557	0.5596	0.5636	0.5675	0.5714	0.5753
⋮					⋮					⋮
1.1	0.8643	0.8665	0.8686	0.8708	0.8729	0.8749	0.8770	0.8790	0.8810	0.8830
1.2	0.8849	0.8869	0.8888	0.8907	0.8925	0.8944	0.8962	0.8980	0.8997	0.9015

c) Method 1: Use a Graphing Calculator

Use the **normalcdf(** function, now with a lower limit of 40 and an upper limit of 50.

Method 2: Use Fathom™
Because the cumulative density function only calculates the probability below a given parameter, a slightly different method must be used. By calculating the probabilities below each boundary and then subtracting them, you get the probability between the boundaries.

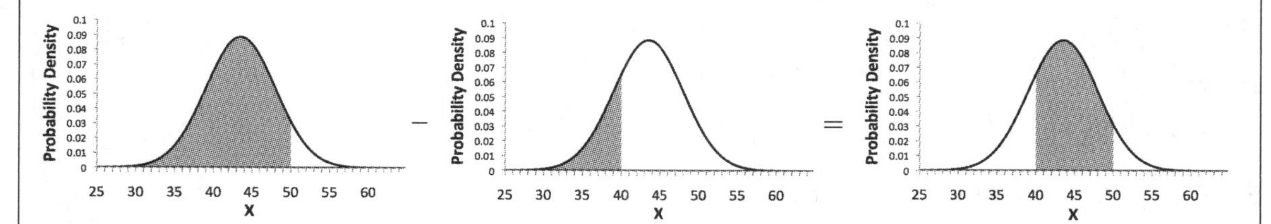

Again, use the **Measures** tab, but this time, subtract the two values.

Inspect Collection 1

Cases **Measures** Comments Display Categories

Measure	Value	Formula
prob_b	0.889188	normalCumulative (49, 43.5, 4.5)
prob_c	0.707343	normalCumulative (50, 43.5, 4.5) – normalCumulative (40, 43.5, 4.5)

Method 3: Use a Spreadsheet
Similarly, subtract the densities at each of the boundaries.

0.70734299	=NORMDIST(50,43.5,4.5,1)-NORMDIST(40,43.5,4.5,1)
	NORMDIST(x, mean, standard_dev, cumulative)

Method 4: Use Tables
First, determine the z-scores: $z_{40} = \dfrac{40 - 43.5}{4.5}$, $z_{50} = \dfrac{50 - 43.5}{4.5}$

$\doteq -0.78 \qquad \doteq 1.44$

Then, determine the cumulative probabilities from the table and subtract them.
$P(Z < 50) - P(Z < 40) = 0.9251 - 0.2148$
$= 0.7103$

A

1. Determine the z-score, to three decimal places.
   a) $\mu = 15.2, \sigma = 3.1, x = 11.3$
   b) $\mu = 129.1, \sigma = 15.7, x = 142.3$
   c) $\mu = 0.586, \sigma = 0.013, x = 0.580$

2. Determine $P(X < x)$ using the tables of areas under a normal distribution curve. Refer to pages 606 and 607 of *Mathematics of Data Management* student edition or go to *www.mcgrawhill.ca/links/MDM12*, select **Study Guide**, and follow the links to obtain a copy.
   a) $\mu = 88.7, \sigma = 5.6, x = 84.3$
   b) $\mu = 1952.7, \sigma = 15.6, x = 1981.4$
   c) $\mu = 13.716, \sigma = 0.015, x = 13.742$

3. Repeat question 2 using the technology of your choice.

☆4. For each interval, determine $P(a < X < b)$ using any method.

	μ	σ	a	b
a)	19.2	3.5	17	18
b)	245.8	22.9	210	230
c)	3.56	0.31	3.8	4.0

5. a) Use technology to create the graph of the normal probability distribution and cumulative probability distribution for data with $\mu = 73$ and $\sigma = 9$.

   b) What does the cumulative probability distribution graph represent?

B

6. Explore the connection between the normal distribution, mean, and standard deviation.

   a) Create the graph of the normal probability distribution for each of the following using technology. Use the same axis scales, for comparison purposes.

      i) $\mu = 50, \sigma = 10$

      ii) $\mu = 50, \sigma = 5$

      iii) $\mu = 50, \sigma = 2$

      iv) $\mu = 70, \sigma = 10$

      v) $\mu = 70, \sigma = 5$

      vi) $\mu = 70, \sigma = 2$

   b) Go to *www.mcgrawhill.ca/links/MDM12* and select **Study Guide.** Follow the links to obtain a copy of the file **NormalPlay.ftm**. Use this file to investigate the connection between the normal density curve, the mean, and the standard deviation by using the dynamic sliders in Fathom™.

   c) Describe how the graph of the normal distribution, the mean, and the standard deviation are connected.

7. Approximate the mean and the standard deviation for each distribution.

   a)

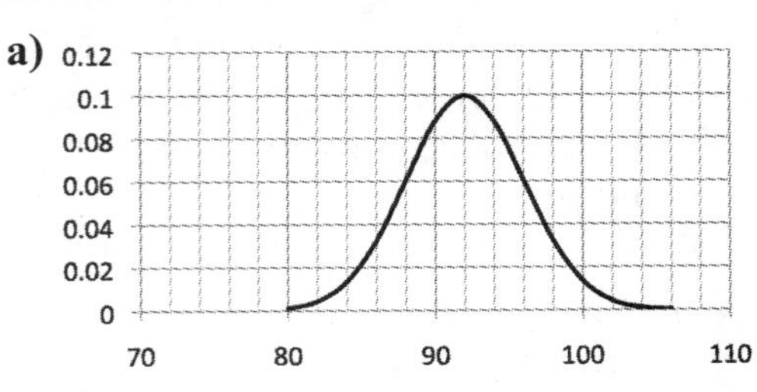

   b)

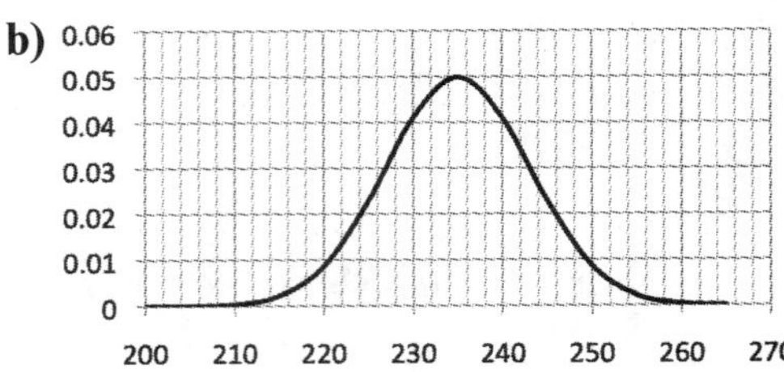

8. The mean of a set of data is 83 and the standard deviation is 3.5.

   a) Determine the probabilities between

      i) 79.5 and 86.5

      ii) 76 and 90

      iii) 72.5 and 93.5

   b) How do the answers from part a) relate to the characteristics of a normal distribution?

9. Historically, the results of a final exam given by a particular university professor have a mean of 65.3% with a standard deviation of 8.5.

   a) Calculate the probability of a student getting a mark between

      i) 75 and 85

      ii) 79 and 81

      iii) 79.9 and 80.1

   b) Based on these results, a student says, "It's almost impossible to get an 80 on the exam." Comment on whether this is true.

10. An agency is conducting performance tests of different brands of skateboards. Tests show that the pressure, in pounds per square inch, required to break the boards is normally distributed with a mean of 54.7 and a standard deviation of 6.2.

a) What percent of the boards have breaking pressures between 50 and 60?

b) What is the probability that any board will require at least a pressure of 70 to break?

c) In what range, centred on the mean, will 80% of the data lie?

11. A doctor is researching the effects of weight on people who have heart conditions. She has recorded the data, but some of the printed results are damaged and cannot be read. Below is the table of what the doctor has been able to recover.

	Percent of Data	Range (kg)	μ (kg)	σ
Males	95	62.2-107.0	84.6	[illegible]
Females	90	102.7-142.9	122.7	[illegible]

Assuming the data are normally distributed and the ranges are centred on the mean, determine the missing standard deviation values.

12. On an assembly line, each vehicle goes by in 54 s. Managers are analyzing a task that requires an operator to install a driver-side door panel. Since the plant produces cars around the clock, there are three shifts and thus three different people to perform the task. Results of the analysis are shown.

	μ (s)	σ (s)
Shift 1 Worker	41	9.4
Shift 2 Worker	45	2.1
Shift 3 Worker	43	4.2

How can these data be used to decide which of the three workers need retraining?

13. An outlier is defined as a point that is more than 1.5 times the interquartile range before the 1st quartile or after the 3rd quartile. A set of data has a mean of 115.8 and standard deviation of 9.6. For what values would any particular piece of data be considered an outlier?

☆ **14.** German engineer Friedhelm Hillebrand, known as the father of text messaging, proposed in 1985 that a total of 160 characters is all that is needed for most communications. Today, most SMS and Twitter messages are limited to 140 characters (with 20 reserved for the sender's address information).

a) Assuming that the number of characters in a text message fits the characteristics of a normal distribution, create three scenarios of mean and standard deviation that would have 95% of all messages less than 140 characters.

b) Explain why someone might think an answer of $\mu = 70$ and $\sigma = 42.4$ for part a) could be correct but really could not be.

c) List reasons why you think a normal distribution model might not fit this situation.

C

15. Two teachers have classes of similar sizes. After final exams, the mean of the grades in each class is 73%. However, one class has a standard deviation of 4% while the other's is 8%. In which class would a mark of 90% be more meaningful?

16. A medical journal reports that for a given surgery, the mean recovery time is 250 days with a standard deviation of 150 days. Describe why this is likely not a normally distributed set of data and what might be true about the distribution instead.

8.3 Normal Sampling and Modelling

KEY CONCEPTS

- If a population is known to be normal, then predictions can be made from a sample taken from that population.

Example

Consider a set of data from an experiment to measure the speed of light. Due to the magnitude of its speed, it is possible to conduct several measurements per second. Note how the histogram and values of the mean, $\overline{x}$, and standard deviation, s, for the sample change as the number of samples decreases.

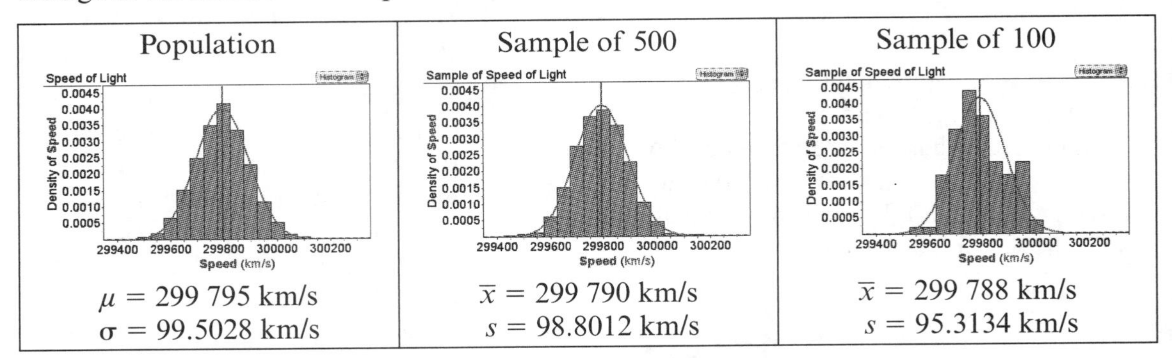

Although the values change, they are all close to the population mean, μ, and standard deviation, σ. It is possible to see the visual nature of the data start to deteriorate as the number of samples decreases, but the general shape remains.

- There are many samples of data that appear to be normal in nature. Often, around the mean, the data will fit the normal distribution well. However, it is often the tails (low and high extremes) that do not follow the normal distribution. As the sample becomes smaller, this difference is amplified.

- One way to help identify whether a set of data is normally distributed is to see if 68% of the data are within one standard deviation of the mean, 95% of the data are within two standard deviations of the mean, and 99.7% of the data are within three standard deviations of the mean.

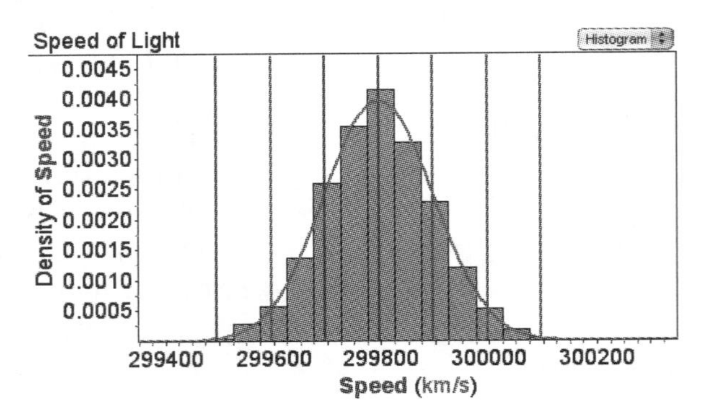

A

1. For each set of data, assume the data are normally distributed. Determine the range of data within each number of standard deviations of the mean

   i) one **ii)** two **iii)** three

	μ	σ
a)	45	8
b)	87.6	15.2
c)	13.785	0.023

2. Determine the standard deviation, given the mean and the range of data within two standard deviations of the mean, for each normally distributed set of data.

	μ	Range
a)	1250	1084–1416
b)	35.78	25.18–46.38
c)	1.235	0.433–2.037

B

☆3. The table shows data collected regarding the speed of cars on a local highway.

Interval (km/h)	Frequency
67.5–72.5	1
72.5–77.5	1
77.5–82.5	4
82.5–87.5	14
87.5–92.5	19
92.5–97.5	13
97.5–102.5	11
102.5–107.5	5
107.5–112.5	1
112.5–117.5	1

a) Assume the data are normally distributed. Determine the approximate mean and standard deviation.

b) Go to *www.mcgrawhill.ca/links/MDM12* and select **Study Guide.** Follow the links to obtain a copy of the file **SpeedData.ftm**. Use the relative frequencies, along with the midrange value of each interval, to verify your answers to part a) using a normal density function graph.

4. The application store for the latest 3G phone is reporting that it has average daily sales of close to $65 million with a standard deviation of $12.5 million. The collection of almost three years of data is shown.

Interval ($ million)	Frequency
22.5–27.5	2
27.5–32.5	7
32.5–37.5	11
37.5–42.5	28
42.5–47.5	46
47.5–52.5	76
52.5–57.5	115
57.5–62.5	148
62.5–67.5	164
67.5–72.5	128
72.5–77.5	102
77.5–82.5	85
82.5–87.5	45
87.5–92.5	24
92.5–97.5	12
97.5–102.5	4
102.5–107.5	2
107.5–112.5	1

a) Determine whether these data are normally distributed or not.

b) Determine the probability that they will have sales of less than $70 million in any day.

☆**5.** A sample of 50 students were measured and the following heights, in centimetres, were recorded. Go to *www.mcgrawhill.ca/links/MDM12* and select **Study Guide**. Follow the links to obtain a copy of this data in the file **50Heights.csv**.

160, 150, 139, 165, 160, 183, 165, 160, 147, 141, 143, 135, 150, 164, 158, 174, 157, 130, 180, 165, 144, 136, 140, 139, 158, 144, 170, 146, 150, 120, 139, 155, 157, 171, 149, 173, 158, 186, 158, 174, 148, 199, 161, 158, 155, 150, 135, 141, 142, 128

a) Create a frequency table and histogram for this data.

b) Do these data have a normal distribution? Justify your answer.

c) Go to *www.mcgrawhill.ca/links/MDM12* and select **Study Guide**. Follow the links to obtain a copy of the file **500Heights.ftm**. This file has the population from which the sample was taken. Create histograms of both the population and the sample. Change the interval sizes (bin widths) and positions (bin alignment) for each a few times. How does changing these values affect the look of the histograms?

d) Determine the mean and standard deviation of the population and sample, and graph the normal density curve for each. How well do you think the sample represents the population?

6. A sample of 40 students kept track of the number of hours per week that they watch television. Go to *www.mcgrawhill.ca/links/MDM12* and select **Study Guide**. Follow the links to obtain a copy of this data in the file **TelevisionUse.csv**.

16.3, 11.7, 14.3, 13.3, 9.1, 9.3, 11.7, 13.2, 13.5, 11.6, 6.7, 12.0, 24.2, 11.3, 13.6, 13.8, 14.7, 5.8, 16.6, 10.0, 11.0, 8.4, 13.0, 17.2, 10.4, 19.4, 15.2, 13.4, 12.9, 11.1, 7.2, 9.4, 10.6, 13.1, 13.1, 14.2, 9.3, 17.2, 10.5, 6.9

Assume that the data are normally distributed.

a) Determine the mean and standard deviation.

b) Determine the probability that the number of hours of television watched will be less than 8.

c) Determine the probability that the number of hours of television watched will be between 10 and 15.

7. Hamilton's Around the Bay 30-km Road Race is the oldest long-distance running race in North America. Go to *www.mcgrawhill.ca/links/MDM12* and select **Study Guide**. Follow the links to obtain a copy of the file **AroundtheBay .ftm** to get the winning times for each of the races since 1894. The data includes both male and female finishers as well as times from before and after 1982 (when the course was shortened slightly to be certified to be 30 km).

a) Create a histogram of the winning times for this race. Give reasons why the distribution may not have a normal shape.

b) Create histograms and graph the normal distribution curve for the male finishers before and after 1982. Discuss how well these fit a normal distribution.

c) Assuming both could be modelled with a normal distribution, what is the probability that any particular male would win the race with a time less than 100 min? Why do the values from before and after 1982 make sense when compared to each other?

C

8. Go to *www.mcgrawhill.ca/links/MDM12* and select **Study Guide**. Follow the links to obtain a copy of the file **CanadianPennies .ftm** to get the mass of each penny in a sample of 1750 pennies. Over the years, many different metals have been used to make the coins. The majority of the older pennies are said to be heavier than the newer ones.

a) Explain why a simple normal distribution model cannot be used in this case.

b) Use the sliders in Fathom™ to approximate the mean and standard deviation for the older and newer pennies by matching the histogram to the normal distribution curves.

c) Determine the probability of finding a penny that has a mass of less than 2.3 g. Hint: recall independent and dependent events.

d) Determine the probability of finding a penny that has a mass of more than 2.5 g.

9. In the Ironman triathlon, competitors must first swim 3.8 km, then cycle 180 km, and finally run a marathon of 42.2 km. In order for any particular athlete to qualify for the world championships in Kona, Hawaii, they must first be one of the top finishers in their division in another Ironman competition earlier in the year. Go to *www.mcgrawhill.ca/links/MDM12* and select **Study Guide**. Follow the links to obtain a copy of the file **IronmanQualifiers.ftm**, which has a sample of some of the qualifiers for the 2008 world championship.

a) Each person who qualifies does so from within his or her own category. The most elite of the categories is the Male Pro category. These are the fastest of all the competitors. Determine the probability that a qualifier chosen at random will have a qualifying time faster than the slowest pro male.

b) One of the most competitive categories is the male 40−44 age category. Determine the probability that a 42-year-old athlete who finishes his qualifying race in a time longer than 10 h will qualify for the world championships. Assume that in general, the population of 40- to 44-year-old males have times that are normally distributed.

8.4 Normal Approximation to the Binomial Distribution

KEY CONCEPTS

Discrete Data and the Normal Distribution

- Although the normal distribution is usually discussed in terms of continuous data, large data sets of discrete data can be modelled by a normal distribution.

Example

Consider a situation where a person wins a particular game at an amusement park. The winner's prize consists of a handful of pennies that the winner can grab from a jar. Over the course of a day, the game is played several times and the number of pennies grabbed per win is recorded.

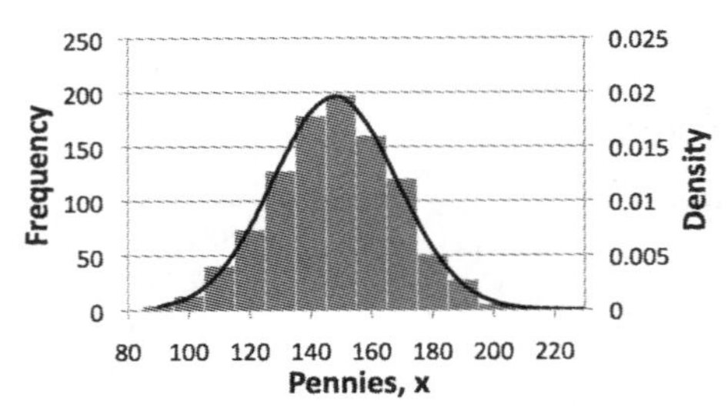

The mean and standard deviation are calculated as $\mu = 148$ and $\sigma = 20.3$ and can be used to make probability predictions. With continuous data, the probability of an individual occurrence is zero. For discrete data, this is not true. To calculate the probability of an individual value, use a continuity correction. If you want the probability that 130 pennies will be grabbed, consider $P(129.5 < X < 130.5)$ instead.

The Normal Approximation for Binomial Distributions

- For large sets of data, the normal distribution can also approximate the binomial distribution.

Example

Consider a case where 50 samples are taken from a population of fish that is known to be 30% salmon. Then, the expected number of salmon would be

$$\begin{aligned} E(x) &= np \\ &= (50)(0.3) \\ &= 15 \end{aligned}$$

So, 15 would be the mean of the corresponding normal distribution. The standard deviation for this distribution could be found by using

$$\begin{aligned} \sigma &= \sqrt{npq} \\ &= \sqrt{(50)(0.3)(0.7)} \\ &\doteq 3.24 \end{aligned}$$

By plotting both distributions on the same axes, you can see that there is a fairly good match.

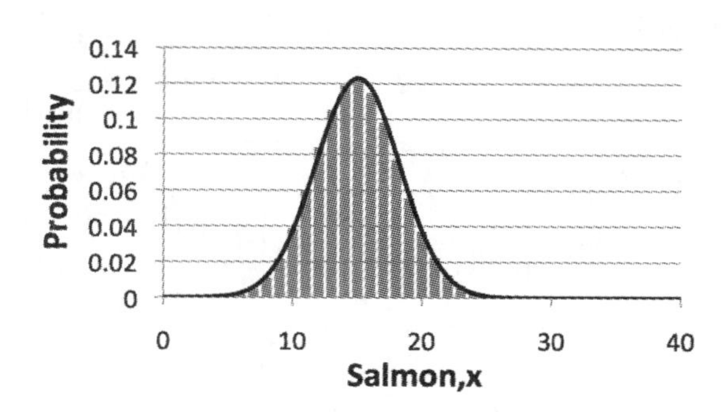

- In general, a normal distribution can be a good match to a binomial distribution, as long as $np > 5$ and $nq > 5$.

The Normal Approximation to the Hypergeometric Distribution

- For large samples, the hypergeometric distribution can be modelled by the binomial distribution. However, when the population size becomes too small (less than 5000) or the sample size is large compared to the population, the normal distribution should be used to model the hypergeometric distribution. In these situations, the normal distribution can approximate the hypergeometric situation with mean $\mu = np$ and standard deviation $\sigma = \sqrt{npq\left(\frac{N-n}{N-1}\right)}$, where N is the population size and n is the sample size. (Note that this case is not covered in the student edition; it was added for the revised curriculum.)

Example

A jar contains 50 toonies and 150 loonies. Fifteen coins are removed.

a) Create a hypergeometric distribution and a normal distribution representing the number of toonies removed.

b) Use the normal distribution to determine the probability that exactly 5 toonies are removed.

Solution

a) To create a hypergeometric distribution, use $p(x) = \frac{({}_aC_x)({}_{N-a}C_{n-x})}{{}_NC_n}$, where $a = 50$, $N = 200$, and $n = 15$.

To create a normal distribution, determine the mean:

$$\mu = np$$
$$= (15)\left(\frac{50}{200}\right)$$
$$= 3.75$$

and the standard deviation:

$$\sigma = \sqrt{npq\left(\frac{N-n}{N-1}\right)}$$
$$= \sqrt{(15)(0.25)(0.75)\left(\frac{200-15}{200-1}\right)}$$
$$\doteq 1.62$$

Number of Toonies, x	Hypergeometric $P(x)$	Normal $P(x)$
0	0.011 100 389	0.016 898 195
1	0.061 215 382	0.058 299 008
2	0.153 261 869	0.137 403 781
3	0.231 003 396	0.221 234 504
4	0.234 327 186	0.243 345 711
5	0.169 385 080	0.182 856 832
6	0.090 098 447	0.093 867 520
7	0.038 943 510	0.032 918 206
8	0.010 793 406	0.007 886 306
9	0.002 448 504	0.001 290 707
10	0.000 415 401	0.000 144 311
11	$5.173\ 12 \times 10^{-5}$	$1.102\ 26 \times 10^{-5}$
12	$4.574\ 87 \times 10^{-6}$	$5.751\ 62 \times 10^{-7}$
13	$2.710\ 68 \times 10^{-7}$	$2.050\ 27 \times 10^{-8}$
14	$9.616\ 03 \times 10^{-9}$	$4.992\ 85 \times 10^{-10}$
15	$1.538\ 56 \times 10^{-10}$	$8.306\ 20 \times 10^{-12}$

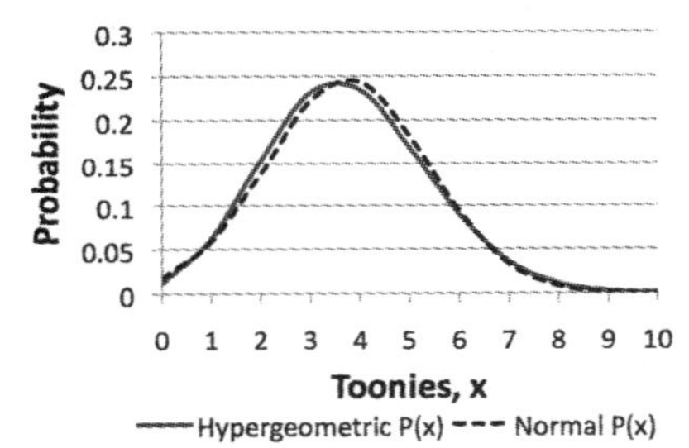

b) To find the probability of getting exactly 5 toonies with the normal distribution, a continuity correction must be used. So, $P(x = 5) \doteq P(4.5 < X < 5.5) = 0.182$, which is a little greater than the hypergeometric answer, but is a reasonable approximation.

A

1. What is the continuity correction for each situation?

a) equal to 82

b) less than 4

c) greater than or equal to 1087

d) less than or equal to 78

2. Determine the mean and standard deviation to be used to approximate each binomial distribution with the normal distribution.

a) $p = 0.45$, $n = 200$

b) $p = 0.8$, $n = 500$

c) $p = 0.1$, $n = 90$

3. Determine the mean and standard deviation to be used to approximate each hypergeometric distribution with the normal distribution.

 a) $N = 500, n = 50, p = 0.45$

 b) $N = 1500, n = 75, p = 0.15$

 c) $N = 4000, n = 200, p = 0.80$

 d) $N = 3876, n = 137, p = 0.30$

B

☆4. Penny has an Internet-based business selling hair clips. She just received a rush order for 1000 clips and enlists the help of friends to make them by morning. Through the night they each manage to make an average of 13.5 hair clips per hour with a standard deviation of 3.4. In order to meet the deadline, they must average 18 hair clips per hour each. What is the probability that they will be able to make at least 18 hair clips per hour each?

5. A student randomly chooses each answer on a 20-question multiple-choice test.

 a) Determine the probability that the student will get no more than one quarter of the answers correct using

  i) the binomial theorem

  ii) the normal approximation

 b) How do your answers in part a) compare?

 c) Repeat part a) for a 100-question test.

 d) Repeat part a) for a 1000-question test.

 e) What does this tell you about the connection between the size of the sample and the closeness of the approximation of the normal to the binomial distribution?

6. Investigate the relationship between the probability of a success and the accuracy of the approximation of the normal distribution to the binomial distribution.

 a) For a sample of 100, determine the probability of $X \leq x$ successes when

  i) $p = 0.5, x = 50$

  ii) $p = 0.4, x = 40$

  iii) $p = 0.3, x = 30$

  iv) $p = 0.2, x = 20$

  v) $p = 0.1, x = 10$

 b) What do your results say about the effect of the probability of a success on the closeness of the approximation of the normal to the binomial distribution?

7. A golfer recorded the following scores in 25 recent games:
88, 89, 102, 93, 99, 95, 94, 96, 84, 91, 95, 86, 95, 92, 102, 84, 104, 108, 92, 93, 92, 97, 96, 95, 82

 a) Determine the mean and the standard deviation.

 b) Assuming a normally distributed set of scores, what is the probability that the golfer will qualify for the club championship, which requires him to get a score under 83 on his next game?

8. In January 2010, the US-based National Safety Council stated that at least 28% of auto accidents are caused by distractions due to cell phone use. In its most recent Road Safety Report (2004), the Ontario Ministry of Transportation stated that there were more than 400 000 accidents during that year.

 a) If there were 400 000 accidents this year, what is the probability that at least a quarter of them were related to cell phone use?

 b) How do you think the probability calculated in part a) would compare to the actual number of cell-phone-related accidents in 2004?

9. A local tomato farmer's crop is used to make 4% of the ketchup a canning company produces. What is the probability that the local grocery store that has 250 bottles of ketchup has

a) at least one bottle with this farmer's tomatoes in it?

b) ten bottles with this farmer's tomatoes in them?

10. a) A bag contains 30% red marbles and the rest black marbles. Go to *www.mcgrawhill.ca/links/MDM12* and select **Study Guide**. Follow the links to obtain a copy of the file **Marbles.xls**. Use the file to investigate the connection between the normal approximations to the binomial and hypergeometric distributions when the sample chosen changes. Change the value of n using the following values.

i) 10 **ii)** 50 **iii)** 70

b) Determine the connection between the fraction of the population the sample makes up and the approximation that is closest to the actual hypergeometric distribution.

☆**11.** Statistics Canada is collecting some demographics on rural communities.

	Population	Sample Size	Percent Youth
Town A	550	60	16
Town B	1500	100	22
Town C	3500	200	18

a) Determine the probability that at least 15% of those sampled will be youths for each town, using both the normal distribution and the hypergeometric distribution.

b) Which method is easier to use?

c) Which do you think is more accurate?

d) For which of the three towns is the normal approximation closest to the hypergeometric?

12. In January 2010, Canadians were able to participate in CBC Television's *Test the Nation*, a 50-question IQ test. The scores tend to be normally distributed, and an average IQ is said to be between 90 and 109. This corresponds to a test score of 24–32 for someone under 25 years old.

a) What is the probability that someone under the age of 25 will score at least an average IQ (score of 24) just by guessing on each of the four-answer multiple-choice questions?

b) It is said that the IQ test results have a mean score of 100 with a standard deviation of 15 points. What is the probability that any person chosen at random will have an IQ greater than 130?

C

13. As stated in question 12, the score on 50 questions gives different IQs depending on the age of the person taking the test. For any population, the mean IQ is standardized at 100 with a standard deviation of 15. The table shows the mean score and standard deviation for each age group.

Age Group	Mean Score	Standard Deviation
Under 25	28	7.5
25–34	27	7.5
35–49	25	7.5
50–64	24	6
Over 65	20	5

a) Determine the probability that anyone chosen at random in each age group will have had a score of 32 or below.

b) Based on these values, determine the IQ score for each age group for a score of 32 on the 50-question test.

8.6 Confidence Intervals

KEY CONCEPTS

- The way that a sample is taken can affect the accuracy of the sample. As you learned in Chapter 2, one factor that is important with samples is their size. The larger that a sample is, the more accurately the sample will represent the population. However, making the sample too big may be unnecessary. This section looks at how the sample size affects the accuracy of the results.

- When studies or polls are conducted, the accuracy is often stated using a confidence interval. The confidence interval, P, is a measure of the likelihood of the results being correct. A confidence level of 95% means that 95% of the time the experimental results will be true.

- For a P% confidence level of a normally distributed data set, the mean, μ, will be within the range $\overline{x} - z_{\frac{\alpha}{2}}\left(\frac{\sigma}{\sqrt{n}}\right) < \mu < \overline{x} + z_{\frac{\alpha}{2}}\left(\frac{\sigma}{\sqrt{n}}\right)$, where $\overline{x}$ is the sample mean, σ is the population standard deviation, n is the size of the sample, $\alpha = 1 - P$ is the acceptable probability of error (sometimes called the tail size), and $z_{\frac{\alpha}{2}}$ is the z-score associated with it. Some z-scores for common confidence levels are shown in the table.

Confidence Level	Tail Size, $\frac{\alpha}{2}$	$z_{\frac{\alpha}{2}}$
90%	0.050	1.645
95%	0.025	1.960
99%	0.005	2.576

Example

A machine that produces control arms for a vehicle gas pedal generates pedals that have a length with standard deviation of 0.08 cm. Thirty pedals are tested to see if their lengths are acceptable. The sample has a mean of 18.2 cm. What would be the acceptable range of lengths for a 95% confidence level for the mean length?

Solution

$\sigma = 0.08$ cm, $n = 30$, $\overline{x} = 18.2$ cm, and for a 95% confidence level, $z_{\frac{\alpha}{2}} = 1.960$.
So, the range is

$$\overline{x} - z_{\frac{\alpha}{2}}\left(\frac{\sigma}{\sqrt{n}}\right) < \mu < \overline{x} + z_{\frac{\alpha}{2}}\left(\frac{\sigma}{\sqrt{n}}\right)$$

$$18.2 - (1.960)\left(\frac{0.08}{\sqrt{30}}\right) < \mu < 18.2 + (1.960)\left(\frac{0.08}{\sqrt{30}}\right)$$

Then, 18.17 cm $< \mu <$ 18.23 cm, or $\overline{x} = 18.2 \pm 0.03$ cm.

A 95% confidence level in this case means that 95% of the time (19 times out of 20), when a sample of 30 is taken, the average length of a pedal will be between 18.17 cm and 18.23 cm.

- The distance that the ends of the extremes are from the mean is called the margin of error, E. In the example, the margin of error is 0.03 cm. In general, the margin of error is given by $E = z_{\frac{\alpha}{2}}\left(\frac{\sigma}{\sqrt{n}}\right)$. Another way to state the margin of error is to use the interval width, ω. In the example, $\omega = 0.06$ cm.

- Many times, researchers want to know what an acceptable sample size would be to have a particular confidence interval. To determine this, they can use the formula $n = \left(\frac{z_{\frac{\alpha}{2}}\sigma}{E}\right)^2$ or $n = \left(\frac{2z_{\frac{\alpha}{2}}\sigma}{\omega}\right)^2$, where n is rounded up to the next whole number.

- Sometimes, you do not have any specific evidence from the population. In such cases, it is still possible to make predictions about the population based on a sample. When the proportion of the population, p, is not known, the formula used for the confidence interval is $\hat{p} - z_{\frac{\alpha}{2}}\left(\sqrt{\frac{\hat{p}\hat{q}}{n}}\right) < p < \hat{p} + z_{\frac{\alpha}{2}}\left(\sqrt{\frac{\hat{p}\hat{q}}{n}}\right)$. The corresponding sample size for a particular margin of error is $n = \hat{p}\,\hat{q}\left(\frac{z_{\frac{\alpha}{2}}}{E}\right)^2$ or $n = 4\hat{p}\hat{q}\left(\frac{z_{\frac{\alpha}{2}}}{\omega}\right)^2$. In this case, $\hat{p}$ represents the proportion found in the sample, and n is rounded up to the next whole number.

A

1. For each set of data, determine the margin of error and confidence interval for a 95% confidence level.

	n	$\bar{x}$	σ
a)	40	215	8
b)	130	35	3.4
c)	30	9.65	0.56
d)	10	0.534	0.035

2. For each set of data, determine the sample size needed.

	Confidence Level	E	σ
a)	90%	0.5	3
b)	95%	3	9.4
c)	95%	0.1	1.2
d)	99%	3.5	23.5

☆ **3.** For each of the following, where the population proportion is not known, estimate the confidence interval.

	Confidence Level	n	Number of Successes
a)	90%	50	30
b)	95%	120	45
c)	95%	190	83
d)	99%	250	115

4. For each of the following, where the population proportion is not known, determine the size of the sample needed.

	Confidence Level	$\hat{p}$	E
a)	90%	0.30	0.1
b)	95%	0.175	0.05
c)	95%	0.25	0.05
d)	99%	0.68	0.01

B

5. Investigate the effects of the margin of error on the confidence interval as the sample size increases. Go to *www.mcgrawhill.ca/links/MDM12* and select **Study Guide**. Follow the links to obtain a copy of the file **MarginofError.ftm**. This file represents a simulation of the probabilities of getting any number of heads when flipping a coin repeatedly. Follow the instructions in the file and determine what the margin of error would be to have a 95% confidence interval for flipping a coin 10, 100, 500, and 1000 times. What happens to the margin of error as the number of flips increases?

6. You used a simulation in question 5 to calculate theoretical probabilities of confidence intervals when flipping a coin. Go to *www.mcgrawhill.ca/links/MDM12* and select **Study Guide**. Follow the links to obtain a copy of the file **CoinFlips.ftm.** Use the file to confirm your results experimentally. By flipping a coin 10, 100, 500, and 1000 times, verify the margin of error needed to obtain a confidence interval of 95%.

7. A politician conducts a poll to see how the public is supporting her latest initiatives. The poll comes back with a 62% approval rating. This pleases the politician. However, the poll has a margin of error of 30%, 19 times out of 20.

   a) How should the politician read the results of her approval rating?

   b) Based on this information, how many people did she likely poll?

   c) The politician would like to have a smaller margin of error in the survey. How many people would she need to poll to ensure a margin of error not greater than 10%?

8. You have a part-time job maintaining a water-jug-refilling machine. The machine rarely fills each jug to the same volume and sometimes needs recalibrating. The manufacturer states that the standard deviation of the machine is 0.3 L. You monitor the next 20 fillings and determine that their mean volume is 18.8 L.

   a) Assuming the data are normally distributed, determine the acceptable range of volumes for a confidence level of 95%.

   b) A sign on the machine states that it dispenses 18.9 L of water. Based on this, should the machine be adjusted to ensure that at least 95% of the fills will have at least 18.9 L in them?

9. MediCaps produces a new medication using machines that fill approximately 40 000 capsules per hour. Each capsule is supposed to contain 1200 mg of medicine. An operator tests seven capsules each hour to make sure that each capsule has a mass m, in milligrams, within the range 1198 mg $< m <$ 1214 mg.

   a) Based on this information, what would the acceptable population mean be?

   b) What would the standard deviation be if the manufacturer runs its machines with a 99% confidence level?

   c) On a certain day, one of the machines is having difficulty. The operator takes 40 samples and finds that the sample mean is 1201 mg. What would the confidence interval of this machine be if the operator wants a 99% confidence level? How does this compare to the acceptable range?

10. An exit poll is done outside a voting location. People who have just voted are asked if they will state who they voted for. In a close election, an exit poll states that Larry Liberal has 48% of the vote, while Constance Conservative has 46% of the vote, with the rest split up among other candidates. The polling firm states that 500 people were polled.

a) What is the margin of error and confidence interval for each candidate if the poll is said to be correct 19 times out of 20? Is it possible that Larry might actually be losing?

b) A second poll is done later. Of 1000 people polled, 47.5% were in favour of Larry, while 46.5% were in favour of Constance. What margin of error is required in order for the confidence level to be 19 times out of 20?

c) In which situation is there a greater chance of Constance actually winning?

11. A simplified way to estimate the margin of error for a proportion for a 95% confidence level when the population is not known is to divide 98% by the square root of the sample size. For a 99% confidence level, divide 129% by the square root of the sample size. Show that these work for two different situations each.

12. A tire manufacturer is getting ready to test a new production process. The machine produces tires that have an inner diameter of 16 in. with a standard deviation of 0.25 in. and are normally distributed. How many tires would have to be tested to be sure of a 95% confidence interval with a margin of error of 0.10 in.?

C

13. Use the binomial theorem to confirm your results from question 6.

14. A newspaper prints a poll suggesting that in the latest election race, Ken Didate is leading Marge Inoverra 56% to 32%. The poll claims the data were collected from 820 respondents and is accurate to within 5 percentage points, 19 times out of 20. In a later poll of 1040 respondents, 52% stated they liked Ken while 36% said they liked Marge (with the same confidence interval and error). The *Daily Blabber* reports that Ken's support is slipping. Statistically, is the newspaper correct?

Chapter 9 Culminating Project: Integration of the Techniques of Data Management

In this course, students are required to develop a project that brings together many of the concepts of the course in a significant manner that illustrates a deep understanding of data-management concepts. The project can be done in a way that integrates both probability and statistics, or, the more common approach, as two smaller projects: one on statistics and one on probability.

This section shows some examples of what these projects might look like. You will notice that the work is not perfect but instead has parts where you can suggest improvements.

As you develop your project(s), you should keep two key things in mind:

- Think of the story you wish to tell with your data.
- The topics that have been covered in the course are the mathematical "tools" you should try to include to "build" your project.

In the coming pages, you will see an example of one project that deals primarily with statistics and one that deals with probability. Do not take these examples as the only possible ways the project(s) could be done, but instead use them as samples of things you might do.

A list of topics from the course is shown. Although it would be very difficult to incorporate all of them, you should try to incorporate as many as possible in your project.

Statistics Topics	Probability Topics
Simulations Single-variable analysis • Measures of central tendency • Measures of spread • Distributions • Graphical displays • Types of data • Outliers Double-variable analysis • Correlation • Trend discussion • Line of best fit • Graphical displays • Outliers	Simulations Independent/dependent events Mutually exclusive/non-mutually exclusive events Expected value Combinatorics Probability distributions Graphical displays

Other things that should be included are an introduction, conclusion, some discussion of the validity of your data sources, and some summary statistics.

Case Study 1: The Statistics Project

Introduction

Have you ever opened a magazine and been shocked by how many ads there are? In a recent issue of *Elle Canada*, there were 182 pages, but the table of contents did not appear until page 21. Are there some magazines that have more ads than others? Is there a relationship between the number of pages in a magazine and the number of ad pages? We will look at the answer to this and other questions in this report.

Question 1: How does the introduction set the stage for the rest of the report? Is there anything else that should have been included?

The Data

Go to *www.mcgrawhill.ca/links/MDM12*, select **Study Guide**, and follow the links to the file **MagazinesProject.ftm** to see the raw data. First, let's look at the data regarding the total number of pages and the number of ad pages.

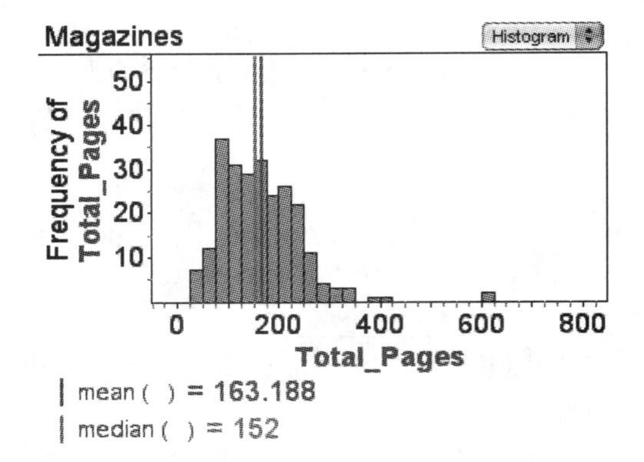

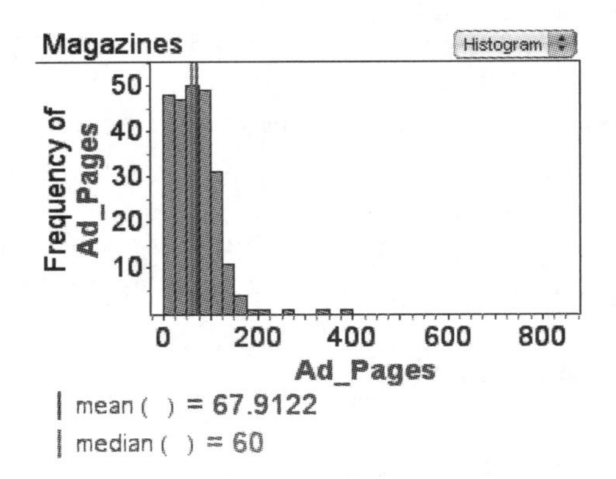

In these graphs, you can see that the average number of ad pages was approximately equal to 40% of the average number of total pages. So, it does seem as though the ad pages make up a significant portion of the actual magazine. Just looking at the averages is a relatively simple analysis. Both sets of data are relatively symmetrical (with some possible outliers), but not really normally distributed.

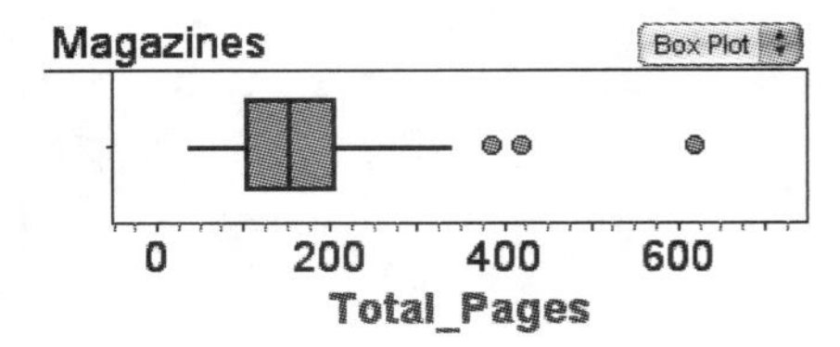

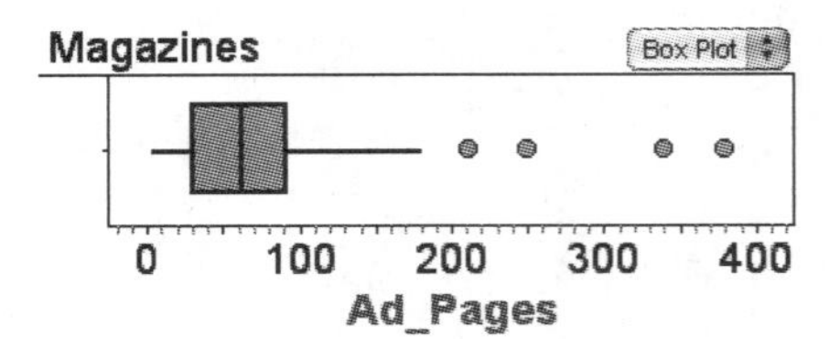

Both have the same four outliers (only three outliers are visible on the total pages graph because two are on top of each other). However, each magazine may have a different percent when it comes to advertising content.

Question 2: For clarification purposes, what else could be mentioned about the outliers?

Question 3: Why is it important to try to make the graphs with the same scales and sizes?

Another way to look at the relationship between the total number of pages and the number of ad pages is to use a double-variable graph. The scatter plot shows the relationship between the total pages and the ad pages. As you can see, there is a strong relationship (correlation is 0.91) between these two. This means as the number of pages increases, so does the number of ad pages. However, in comparison to the 40% number found in the single-variable analysis, here the ads make up 55% less 22 pages (based on the line of best fit).

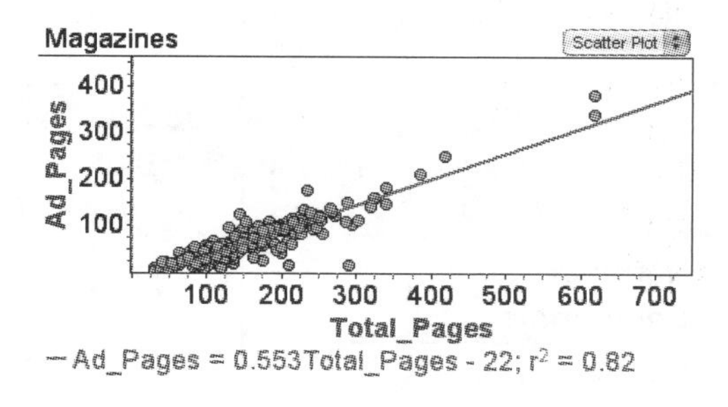

Question 4: Are there any other aspects of linear regression that should have been addressed?

The Story

Based on the double-variable graph, it is clear that ads make up a significant portion (about 55% of the total) of magazines, and as this number becomes lower, that initial value severely impacts the actual number of ads seen. One magazine type does not seem to be any different than another, since they all seem to be clustered around the line of best fit. For example, a fashion magazine like *In Style* had an issue with 620 pages and 340 ad pages, while a business magazine like *Time* had 58 pages and 13 of them were ads, yet they are both near the least-squares line. These two magazines are also on opposite ends of the line, showing that no matter what the size of the magazine, the number of ad pages follows a similar relation.

Question 5: Based on the data in the file, what kind of analysis could be done to look deeper into whether there are differences between magazine types?

Another aspect to consider is the question, what happened to the outliers from the single variable graphs? Just because a piece of data is far away from the majority of other data (making it an outlier for a single-variable analysis), that does not mean that it would be an outlier when looked at in double-variable analysis.

To determine outliers from a double-variable graph, we look for data that is far away from the line of best fit. This is most accurately done by analysing the residuals. Once the residuals are found, looking at their distribution will determine which (if any) correspond to outliers.

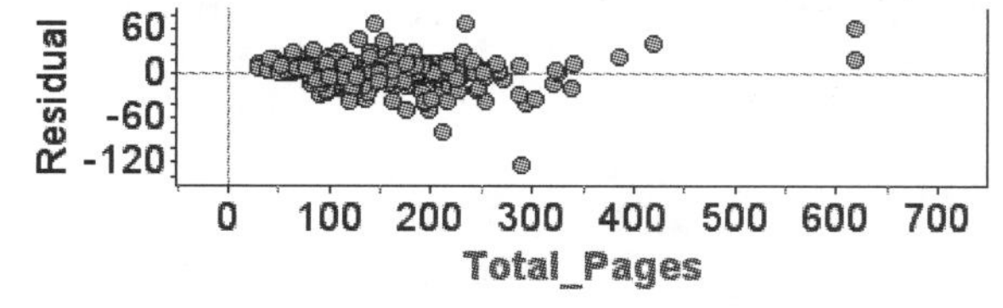

In the table below, you can see the results of the residual analysis. Of the four original outliers, only one of them is an outlier in the double-variable graph. In fact, those magazines that are outliers in the double-variable graph seem to be those that have ad percents less than 50% or greater than 60%.

Magazine	Single-Variable Outlier	Double-Variable Outlier	Total Pages	Ad Pages	Percent Ads
National Geographic		Yes	291	15	5.2%
Indoor/Outdoor Home Plans		Yes	212	15	7.1%
Baseball		Yes	177	25	14.1%
Bass Guitar		Yes	200	39	19.5%
In Style	Yes		620	340	54.8%
In Style	Yes		386	212	54.9%
Vanity Fair	Yes		420	250	59.5%
In Style	Yes	Yes	620	380	61.3%
YM		Yes	154	107	69.5%
Teen People		Yes	130	95	73.1%
Seventeen		Yes	235	174	74.0%
In Style		Yes	235	174	74.0%
House and Home		Yes	145	123	84.8%

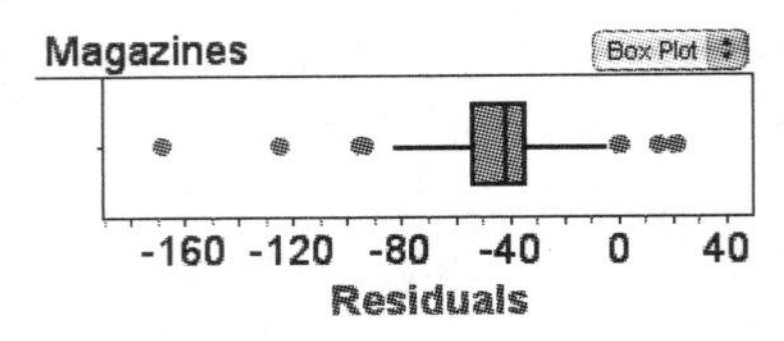

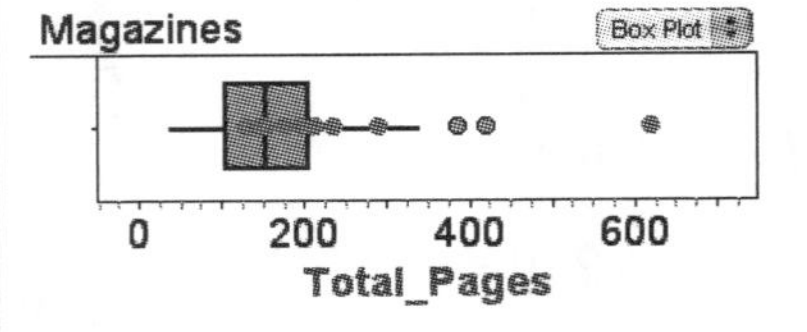

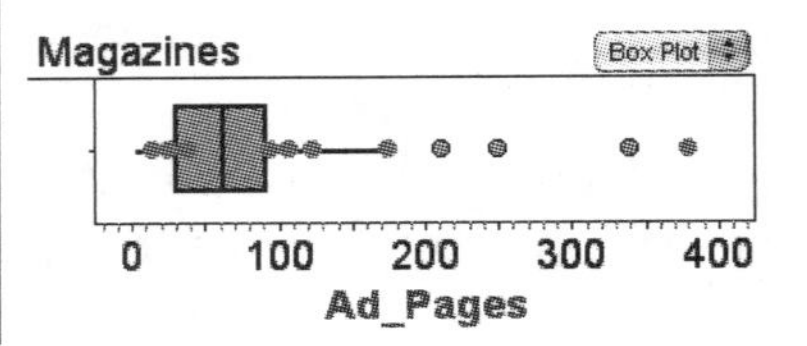

When the outliers are removed, the correlation gets stronger. So, there does seem to be a relationship between the total number of pages and the number of ad pages. Also, for magazines like those in the fashion genre, there are more ads, but since those magazines also have more pages, they have proportionally the same number of ads as those magazines with fewer pages.

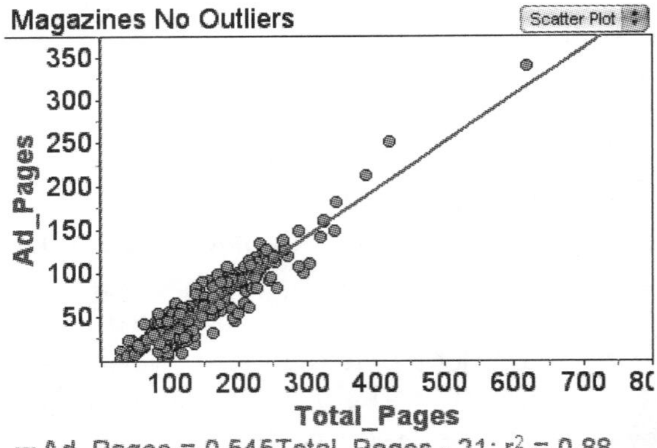

Question 6: Do the magazines at the lower end of the graph have the same proportion of ads as those at the upper right end of the graph? How could answering this question make the analysis more detailed?

Conclusion

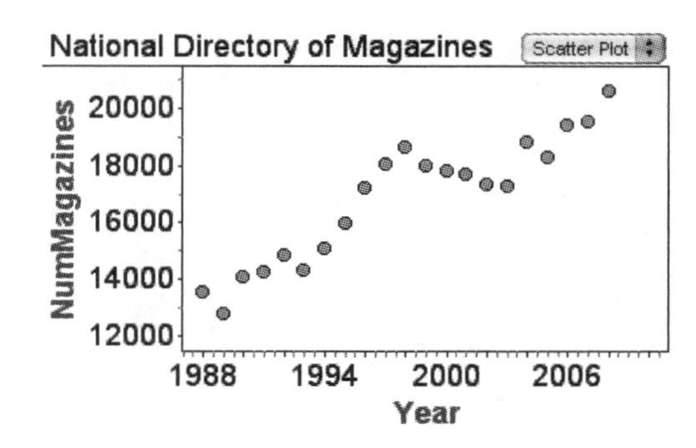

According to the American Society of Magazine Editors, there were over 20 000 different magazines published in North America in 2007. With that number seemingly on the rise, it is important for people in the industry to be aware of how much of the magazines the ads make up. The majority of the revenue that magazines get comes from these ads and not from subscriptions. This makes sense when you see that the majority of magazines have advertisements as more than 40% of their content. So, do not be surprised the next time you open a magazine and cannot find the articles because of all the ads.

Question 7: What was not in the conclusion that could be added to improve the analysis?

Case Study 2: A Probability Game Project

Introduction

For this project, we analyse a game of chance using the mathematical tools we have learned in this course. The game we are analysing is a betting game. A player predicts the outcome of rolling two dice by placing a $1 bet on his prediction. If the outcome (the sum of the two dice) matches his prediction, he wins a prize, otherwise he loses the $1.

Question 8: In this example, the introduction is a little short. What could be done to improve it?

The Game

The board is quite simple, as shown on the right, and the only other materials needed are a pair of dice.

7	2	3	4	5	6
	12	11	10	9	8

The player makes a bet on any individual number, or indicated pairs of numbers, or rows of numbers, as shown.

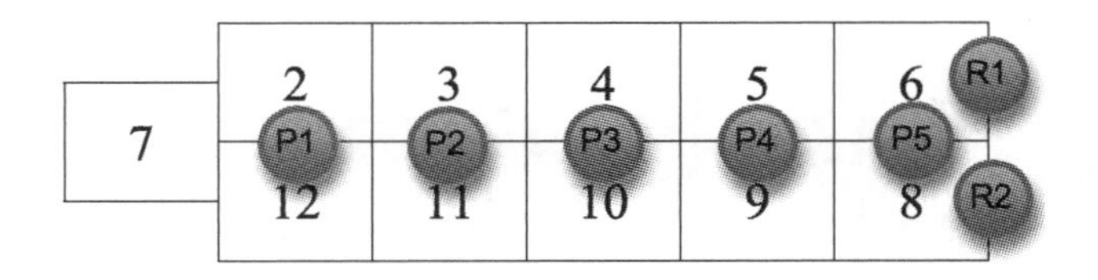

Question 9: What other bets could be included?

The Probabilities and Payouts

Below is the probability for each bet. The payouts reflect that of a $1 bet, where if the person wins, they get back their original $1 plus the payout value. As you can see, the "House" has an advantage of between 8% and 17% on each bet.

Bet	2	3	4	5	6	7	8	9	10
Probability	$\frac{1}{36}$	$\frac{2}{36}$	$\frac{3}{36}$	$\frac{4}{36}$	$\frac{5}{36}$	$\frac{6}{36}$	$\frac{3}{36}$	$\frac{4}{36}$	$\frac{3}{36}$
Payout	$30	$15	$10	$7	$5	$4	$5	$7	$10
Expected Value	−$0.14	−$0.11	−$0.08	−$0.11	−$0.17	−$0.17	−$0.17	−$0.11	−$0.08
Bet	11	12	P1	P2	P3	P4	P5	R1	R2
Probability	$\frac{2}{36}$	$\frac{1}{36}$	$\frac{2}{36}$	$\frac{4}{36}$	$\frac{6}{36}$	$\frac{8}{36}$	$\frac{10}{36}$	$\frac{15}{36}$	$\frac{15}{36}$
Payout	$15	$30	$15	$10	$4	$3	$2	$1	$1
Expected Value	−$0.11	−$0.14	−$0.11	−$0.08	−$0.17	−$0.11	−$0.17	−$0.17	−$0.17

Question 10: What could be done to clarify where these numbers came from? What could be done to give some more variation in these results?

Conclusion

This is an easy game to set up and play, and there is enough variation in the betting to give people the impression that they have a chance of winning. And much like regular casino games, the worst expected value for the player occurs on the bets with the highest probability. This is likely a bet that will be made often, and so it works out the best for the casino.

Question 11: What kinds of things could have been included in the conclusion to improve the project?

Mathematics of Data Management 12 Practice Exam

1. Use technology of your choice to complete the following.

   a) Generate a set of ten random numbers between 15 and 90 and find their mean.

   b) Generate a set of data to simulate randomly picking 10 marbles (with replacement) from a bag that contains 100 marbles: 10 yellow, 20 red, 30 blue, and 40 green.

2. For each graph, identify the type of data and tell a simple story that the data illustrate.

   a)

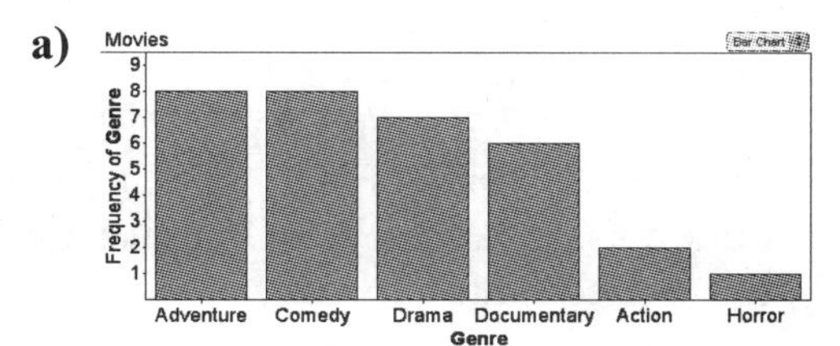

   b)

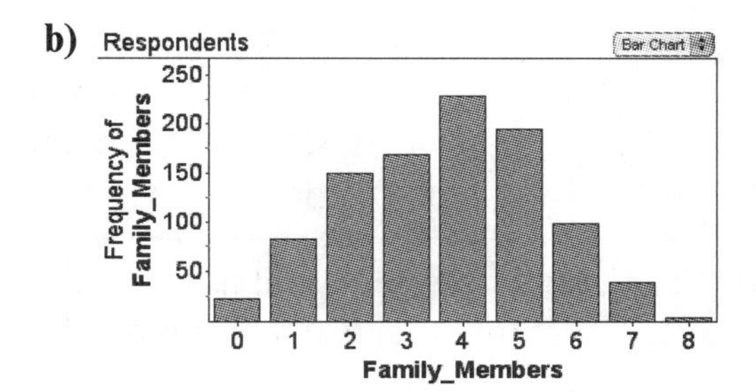

3. Complete the frequency table.

Height (cm)	Frequency	Relative Frequency	Cumulative Frequency
150–155	15		
155–160	32		
160–165	10		

4. The host of a phone-in radio talk show asked, "Who is your favourite musical artist: Bruce Springsteen or Beyoncé?" Describe the sampling method and bias.

5. Determine the mean, median, mode, standard deviation, first and third quartiles, and the interquartile range of the following data. Use technology to check your answers.

   104, 74, 84, 114, 91, 82, 95, 61, 109, 74, 69, 61, 105, 80, 97

6. Describe the type of cause-and-effect relationship in each situation.

   a) Tuition fees go up as the number of smokers worldwide decreases.

   b) The number of garbage trucks in your community increases as the number of residents increases.

   c) The more exposure you have to books, the better is your reading ability.

7. Initial disaster relief pledged to Haiti after an earthquake in that country in January 2010 totalled \$2.422 billion. Use the population of Haiti, approximately 10 million, to make sense of this amount.

8. Determine the number of possible arrangements of the letters in the word *DICE* by

   a) writing out all the arrangements

   b) creating a tree diagram

   c) using permutations

9. Evaluate.

   a) ${}_{23}P_4$ **b)** $\binom{13}{9}$ **c)** $\frac{83!}{79!}$

10. Determine the number of three-letter groups that can be made from the letters in the word *SHIFT* by

    a) writing out all the groups

    b) creating a tree diagram

    c) using combinations

11. You roll two dice. What is the probability of rolling

a) a total of 7 or two even numbers?

b) a total of 6 or two numbers the same?

12. Identify three variables that are

a) continuous

b) discrete

13. Determine the expected value for the following.

x	300	25	5	1	−5
$P(x)$	0.02	0.1	0.2	0.3	0.38

14. Give an example of a situation that could be modelled by a

a) binomial distribution

b) geometric distribution

c) hypergeometric distribution

d) normal distribution

15. Determine the expected value in each situation.

	Type of Distribution	Criteria
a)	Binomial	$n = 150, p = 0.45$
b)	Geometric	$p = 0.3$
c)	Hypergeometric	$N = 80, r = 15, a = 30$

16. Create a probability table and histogram for each distribution.

	Type of Distribution	Criteria
a)	Binomial	$n = 7, p = 0.35$
b)	Geometric	$p = 0.4$
c)	Hypergeometric	$N = 20, r = 8, a = 13$

17. For the uniform probability distribution in the range $115 < X < 195$,

a) create a distribution graph

b) determine $P(125 < X < 130)$

c) determine $P(X = 150)$

18. For each of the following, where the population proportion is not known, estimate the confidence interval.

	Confidence Level	n	Number of Successes
a)	90%	80	50
b)	99%	190	100

19. The table ranks the ten top-performing countries at the Vancouver 2010 Olympic Winter Games. Go to *www.mcgrawhill.ca/links/MDM12* and select **Study Guide**. Follow the links to obtain a copy of the file **Olympics.csv** for an electronic version of the table.

Rank	Country	Gold	Silver	Bronze
1	USA	9	15	13
2	Germany	10	13	7
3	Canada	14	7	5
4	Norway	9	8	6
5	Austria	4	6	6
6	Russia	3	5	7
7	Korea	6	6	2
8	China	5	2	4
8	Sweden	5	2	4
8	France	2	3	6

a) What is the criterion for this ranking?

b) Use another way to rank the data that takes into account the types of medals won.

c) Determine country rankings based on the number of medals won per capita.

Country	Population (millions)
USA	307.8
Germany	81.9
Canada	34.0
Norway	4.8
Austria	8.4
Russia	141.9
Korea	49.8
China	1336.1
Sweden	9.3
France	65.4

d) Another way to rank the results is by the monetary value of the medals. If each gold medal (actually 92.5% silver) is worth \$537, each silver medal is worth approximately \$300, and each bronze medal (actually made of copper) is worth only \$3.40, what would be the ranking of the countries?

20. Go to *www.mcgrawhill.ca/links/MDM12* and select **Study Guide**. Follow the links to obtain a copy of the file **Broadway2010.csv**. This file contains ticket sales data for Broadway theatre shows for the last week of February 2010.

a) Create a scatter plot using the week's gross ticket sales and the number of seats sold.

b) Determine the line of best fit and classify the relationship.

c) Explain what the slope means in the context of the data.

d) Explain what the y-intercept might mean in the context of the data.

e) Predict the weekly gross ticket sales for a theatre with 20 000 seats to sell each week. How reasonable do you think this answer is?

f) What do these data suggest about the quality of a Broadway event and its connection to ticket sales?

21. Go to *www.mcgrawhill.ca/links/MDM12* and select **Study Guide**. Follow the links to obtain a copy of the file **LongestDrive.csv**. This file contains data regarding more than 400 of the longest drives (in yards) in professional golf.

a) Which golfer's name appears most often?

b) Create a histogram of these data and determine the mean, median, and standard deviation.

c) Determine how many outliers there are in these data.

d) What type of distribution is this? Give reasons for your answer.

22. Consider choosing a password for an online account.

a) Many secure sites require you to have a password of at least six characters. There are approximately 15 000 six-letter words in the English language. Using only letters, how many more passwords are there if you do not limit yourself to actual words?

b) Most secure sites now require passwords to have at least 8 characters. How many eight-character passwords are there if letters and numbers can be used?

23. Expand and simplify $(3x - 4)^6$.

24. There are 12 girls and 9 boys in a class. How many groups of 6 students can be made if

a) there are no restrictions?

b) there must be an equal number of boys and girls?

c) there must be either one boy and five girls, or one girl and five boys?

25. Consider the situation in question 24. What is the probability that the group chosen will have twice as many girls as boys?

26. A new virus, *Algebraicuth Avoidus*, can cause people to do poorly in math. However, your teacher found a test to detect if you are carrying the virus. If you have the virus, the probability that your test will be positive is 90%. If you do not have the virus, the probability that your test will be positive is 15%. Your teacher also knows that 80% of the population has the virus.

a) Use a tree diagram to determine the probabilities that apply to this situation.

b) What is the probability that a randomly chosen person will have the virus and have a positive test?

c) What is the probability that a negative test is correct?

27. An insurance company has 1000 automobile policies in force on men who are 50 years of age. The company estimates that the probability that a 50-year-old male will have an accident is 0.01. On average, an accident claim costs $25 000.

a) Estimate the number of claims that the company can expect from this group of men within the next year.

b) What is the expected payout of each claim? Based on this, how should the company base its fee structure to cover this cost?

28. A company that makes skate-sharpening machines uses precision cutting spinners manufactured by a local supplier. The company expects that 0.1% of the spinners will be out of specification. It receives a box of 1000 spinners.

a) What is the probability that if the company checks 50 spinners, at least one defective unit will be found?

b) What is the expected number of rejected parts if 50 spinners are checked?

29. A bag contains 10 red marbles, 20 green marbles, and 30 blue marbles. Ten marbles are removed without replacement and the number of red marbles removed is counted.

a) What type of distribution is this?

b) Create a probability histogram for this situation.

c) What is the probability of removing at least three red marbles?

30. The probability of any particular egg having a double yolk is about 1%.

a) How many eggs would you expect to have to crack to find a double yolk?

b) In 2010, a woman in the United Kingdom cracked six eggs in a row with double yolks. What is the probability of this happening?

31. In the movie *Dumb & Dumber*, Jim Carrey's character, Lloyd, asked what his chances were with the beautiful Mary. Her response was "one in a million." Lloyd was relieved to hear this because, as he put it, "So you're tellin' me there's a chance."

a) Which probability is higher: the probability of Lloyd being with Mary or the probability of you getting at least a pass on a 20-question multiple-choice test (with four choices per question) by only guessing?

b) What are Lloyd's chances compared to the probability of you getting 75% or better on the test?

32. An energy company is proposing to erect wind turbines in your community. The company puts up test towers to measure normal wind conditions. The actual turbines require a wind speed of at least 5 m/s and reach optimal power output when the wind speed is 15 m/s. If the wind speed is more than 25 m/s, the turbines are shut down to prevent damage. The test towers recorded an average wind speed of 13 m/s with a standard deviation of 3.9 m/s.

a) What is the probability that the wind will blow within 3 m/s of the optimum wind speed?

b) What is the probability that the wind will blow hard enough to require the turbines to shut down?

33. There is a new water park in your neighbourhood. A rider is not allowed to get on the slide at the top until the previous rider has completely exited at the bottom. The park operators measure the ride times of 30 riders and find that the mean time is 24.5 s with a standard deviation of 6.2 s. The data are normally distributed.

a) Determine a 99% confidence interval for the mean ride time.

b) How might this information be used to create an automated timed gate system to let riders go at the appropriate time?

34. Create a simulation that would generate 100 pieces of data that

a) follow a uniform distribution

b) follow a normal distribution with mean of 28.6 and standard deviation of 3.5

35. Pi(x) is a function that determines the number of primes that are less than or equal to x.

a) What type of relationship would you expect there to be between x and Pi(x)?

b) Go to *www.mcgrawhill.ca/links/MDM12* and select **Study Guide**. Follow the links to obtain a copy of the file **Pi.csv**. This file lists the first 1000 prime numbers and gives the number of primes less than each prime. Create a scatter plot of the relationship.

c) Describe the relationship.

d) Use the relationship to predict the number of primes less than 1 000 000.

e) The actual answer for part d) is 78 498. Explain why your answer may not be accurate.

f) Determine whether the relationship is linear. Hint: Use a residual plot. How does this help explain your answer in part e)?

36. "The Nine Billion Names of God" is a short story by Arthur C. Clarke. The story is about a group of monks who believe that once they write out each of the 9 billion names of God, the universe will cease to exist. To carry out this task, the monks developed their own alphabet. A name cannot be longer than nine characters and a character may not be repeated more than three times.

a) If as many repeats as possible were allowed, how many characters would their alphabet need to have to produce at least 9 billion possible names?

b) If you could not repeat any characters, how many characters would their alphabet need to have to produce at least 9 billion possible names?

c) What does this tell you about the actual answer?

37. The Erdös Number Project studies research collaboration among mathematicians. The table shows the distribution of authors in the project's database in terms of the number of published papers each has.

Number of Papers	Number of Mathematicians
1	169 662
2	58 328
3	32 293
4	21 532
5	15 624
6	11 702
7	9 066
8	7 595
9	6 381
10	5 488

a) Create a probability-distribution graph for these data.

b) Determine the expected value of the number of papers for a randomly selected mathematician.

c) Go to *www.mcgrawhill.ca/links/MDM12* and select **Study Guide**. Follow the links to obtain a copy of the file **ErdosProject.csv**. How does the expected value change (if at all) if the entire database is used?

d) Conduct research about Paul Erdös and the meaning of an Erdös number. How is an Erdös number similar to a Bacon number? What is an Erdös-Bacon number? Name at least one famous person who has one.

38. Benford's law describes the expected distribution of first digits of numbers in any sufficiently large list.

a) Consider a list of the numbers of passenger cars exported from 110 countries. If you looked at the distribution of the first digit of each of these numbers, what pattern would you expect?

b) Go to *www.mcgrawhill.ca/links/MDM12* and select **Study Guide**. Follow the links to obtain a copy of the file **CarExports.ftm**. This file has the numbers of vehicles exported from 110 countries. Create a probability histogram of the first digits of each value. How does the distribution compare with your guess in part a)?

c) Research what Benford's law states about the relative frequency of each of the digits. How do the car export data compare?

d) Research some applications of Benford's law.

39. Go to *www.mcgrawhill.ca/links/MDM12* and select **Study Guide**. Follow the links to obtain a copy of the file **Magazines.ftm**. This file has data about 250 magazines. Create a scatter plot of Ad Pages versus Total Pages and include the least-squares line.

a) To determine the outliers on the graph, first determine the values of the residuals for each piece of data. When you have the residuals, create a box plot with them and delete the outliers.

b) Show that the residuals follow a normal distribution.

Answers
Mathematics of Data Management 12 Study Guide

Chapter 1

1.2 Data Management Software, pages 1–4

1. In Microsoft® Excel: **a)** Right-click and choose **Format Cells**, then from the **Number** menu, choose **Currency**. **b)** =average(B1:F1) **c)** =median(B1:F1) **d) Data, Sort, Decending** In Corel® Quattro Pro®: **a)** Right-click and choose **Selection Properties**, then from the **Numeric Format** menu, choose **Currency**. **b)** @avg(B1..F1) **c)** @median(B1..F1) **d) Tools, Sort**, uncheck **Ascending**
2. **a)** mean height 167 cm; mean arm span 167.571 428 6 cm **b)** mean height 167.0 cm; mean arm span 167.6 cm

c) i)

	A	B	C
1	Name	Height (cm)	Arm span (cm)
2	Britany	175.0	179.5
3	Charles	163.0	165.0
4	Clair Marie	171.0	177.0
5	Joan	165.0	163.5
6	Manuel	165.0	168.0
7	Sunita	160.0	156.0
8	Ty	170.0	164.0
9	Average	167.0	167.6

ii)

	A	B	C
1	Name	Height (cm)	Arm span (cm)
2	Sunita	160.0	156.0
3	Charles	163.0	165.0
4	Joan	165.0	163.5
5	Manuel	165.0	168.0
6	Ty	170.0	164.0
7	Clair Marie	171.0	177.0
8	Britany	175.0	179.5
9	Average	167.0	167.6

iii)

	A	B	C
1	Name	Height (cm)	Arm span (cm)
2	Sunita	160.0	156.0
3	Joan	165.0	163.5
4	Ty	170.0	164.0
5	Charles	163.0	165.0
6	Manuel	165.0	168.0
7	Clair Marie	171.0	177.0
8	Britany	175.0	179.5
9	Average	167.0	167.6

3. **a)** Open a new Fathom™ file and drag a new case table down from the shelf. Click the attribute **<new>**, type the first column heading, **Name**, then press **Enter**. Similarly, enter the remaining column headings.

Collection 1

	Name	Height_cm	Armspan_cm	<new>

Type the data into the table. To calculate the average, double-click the collection box (Collection 1) to open the **Collection Inspector**. Choose the **Measures** tab. Where it says **<new>**, type a name for the average you wish to calculate. Then, double-click the empty formula box and type the formula **mean(Height_cm)**.

Inspect Collection 1 — Cases | Measures | Comments | Display | Categories

Measure	Value	Formula
Ave_of_Height	167	mean (Height_cm)
Ave_of_Armspan	167.571	mean (Arm_span_cm)

To format for one decimal place, right-click the column (attribute) name and choose **Format Attribute**. Choose **Fixed Decimals** and type **1**. To sort by any column, right-click the column name and choose **Sort Ascending** or **Sort Descending.**

b) Collection 1

	Name	Height	Arm_span
units		centimeters	centimeters
1	Sunita	160.0 cm	156.0 cm
2	Joan	165.0 cm	163.5 cm
3	Ty	170.0 cm	164.0 cm
4	Charles	163.0 cm	165.0 cm
5	Manuel	165.0 cm	168.0 cm
6	Clair Marie	171.0 cm	177.0 cm
7	Britany	175.0 cm	179.5 cm

4. **a)** Division 1: Economy, 4.9; Four Door, 24; Mini Van, 23; SUV, 8; Division 2: Economy, 62; Four Door, 34; Mini Van, 35; SUV, 4

b)–c) AutoSales

	A	B	C	D
1		Economy	Four Door	Mini Van
2	Division 1	='Division 1'!B8	='Division 1'!C8	='Division 1'!D8
3	Division 2	='Division 2'!B9	='Division 2'!C9	='Division 2'!D9
4	Totals	=SUM(B2:B3)	=SUM(C2:C3)	=SUM(D2:D3)

Division 1 / Division 2 / Totals

d)

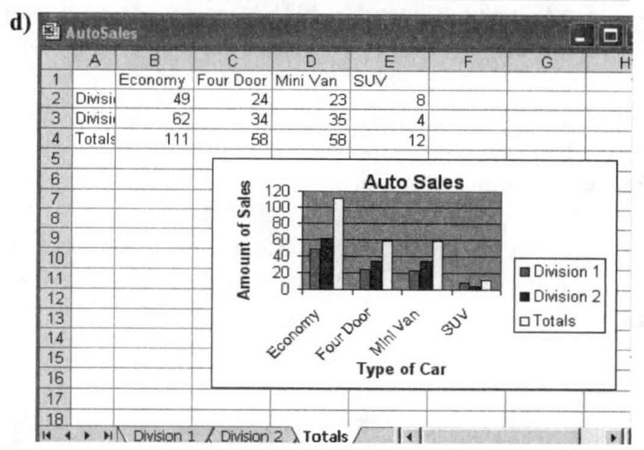

AutoSales

	A	B	C	D	E
1		Economy	Four Door	Mini Van	SUV
2	Divisi	49	24	23	8
3	Divisi	62	34	35	4
4	Totals	111	58	58	12

5. **a)** \$2697 **b)** \$404.55 **c)** cameras: \$1500, \$225; laptops: \$2400, \$360 **d)** Total Value: \$6597, Total Commission: \$989.55. Make the formula for the Total Value with relative referencing and then copy and paste it over to the Commission.
6. **a)** Click **Data**, **Autofilter** and choose **BN** from the **Type** menu. b) From the **Total Pages** menu, choose **Custom AutoFilter** and type **is greater than, 100**. Click **OK**. Then, 189 magazines have more than 100 pages. **c)** Use a **Custom AutoFilter** to show **is greater than or equal to 4**; select **And** and type **is less than or equal to 5**.
7. To see the data, click the **Magazines** collection to select it. Then, drag a table to the Fathom™ workspace.
a) Right-click the table and choose **Add Filter**, then enter **Type = "BN"** in the editor.
Double-click the filter (at the bottom of the table) to edit it:

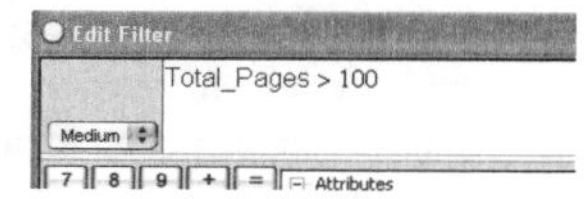

To get the greater-than-or -equal-to symbol, hold down the **Ctrl** key when clicking.

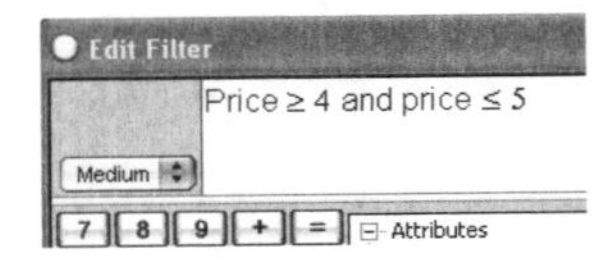

b) Drag a blank summary table Summary to the workspace. Drag the **Type** attribute name from the table to where it says **Drop an attribute here**.

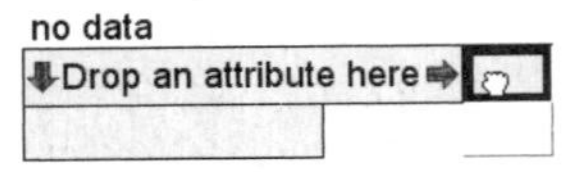

There are 109 heath/fashion magazines.
To get the averages, right-click the summary table and choose **Add Formula**. Type **mean(price)**. This shows that the average price of science/technology magazines is $6.10.

Magazines	Type					Row Summary
	BN	HF	HG	SE	ST	
	20	109	26	78	17	250
	4.82222	4.35606	4.9484	5.43513	6.102	4.90118

S1 = count ( )
S2 = mean (Price)

8. **b)** You may have to delete the last two cases.
c) 1224.4 cinemas per country

9. **a)** 526 males, 520 females **b)** People who like both English and math a lot. **c)** People who like math a lot but hate English are all males.

10. If you rent the guitar at this rate for more than 17 weeks, it would cost less to buy one instead.

11. Enter formulas: In B6
=D1 + D2*$A6/100 + D3*B$5/100;
in C5 change the last letter to C and so on. Then, change the charges in D1, D2, and D3 to see the change in all values.

12.

	A	B
1	Fibonacci	Ratio
2	1	
3	1	1
4	2	2
5	3	1.5
6	5	1.666667
7	8	1.6
8	13	1.625
9	21	1.615385
10	34	1.619048
11	55	1.617647
12	89	1.618182
13	144	1.617978
14	233	1.618056
15	377	1.618026

	A	
1	Fibonacci	Ratio
2	1	
3	1	=A3/A2
4	=A2+ A3	=A4/A3
5	=A3+ A4	=A5/A4
6	=A4+ A5	=A6/A5
7	=A5+ A6	=A7/A6
8	=A6+ A7	=A8/A7
9	=A7+ A8	=A9/A8
10	=A8+ A9	=A10/A9
11	=A9+ A10	=A11/A10
12	=A10+ A11	=A12/A11
13	=A11+ A12	=A13/A12
14	=A12+ A13	=A14/A13
15	=A13+ A14	=A15/A14

1.3 Databases, pages 5–8

1. **a)** This is a database of answers to a survey that youths took and is about their feelings on school. **b)** 1046
c) 19 attributes per person **d)** Some are demographic information, such as the sex of the person. Other items are opinions on how they like various aspects of school.

2. things often seen: name, time, size, artist, album, rating, date added, composer; things sometimes seen: bit rate, description, last played, sample rate, disk number

3. a library reference system, a baseball card with a player's game statistics and personal background information, a CD cover with data about lengths of songs

4. type, location, size, name, date created, date modified, title, subject, author, manager, company, keywords, comments

5. **a)** Name, Artist, Composer, Album, Genre, Size, Time, TrackNumber, Year, Bitrate, SampleRate **b)** Alternative
c) Talking Heads

6. The Internet is an electronic network connecting stores of information. It can be searched and the data can be organised and categorized.

7. **a)** Based on this graph, they were making a profit for a brief period in 1999 and from about 2002 on.

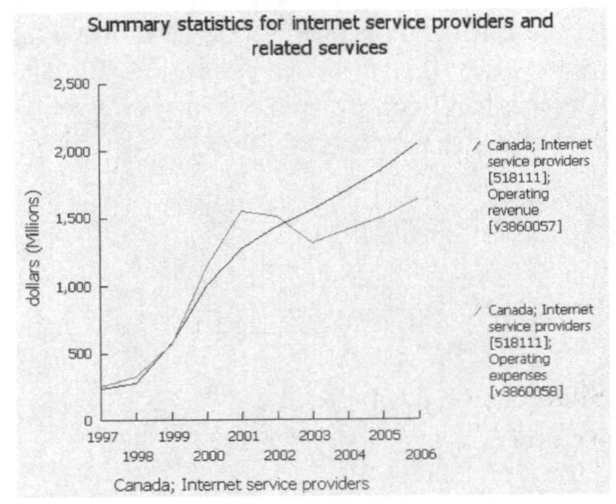

b) vertical bar graph
c) When importing the data into a spreadsheet, you may have to use the **Text to columns** feature.

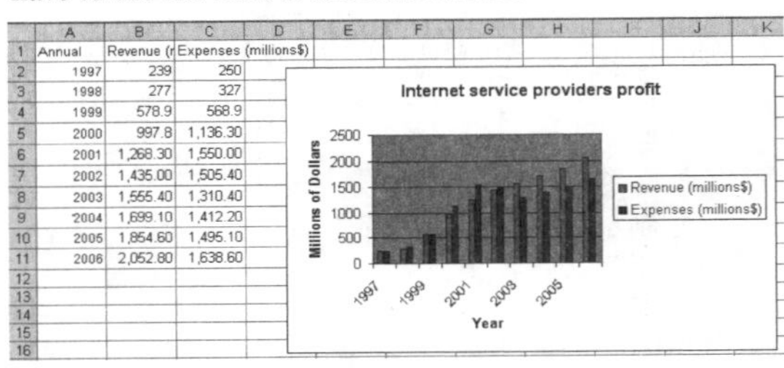

	A	B	C	D
1	Annual	Revenue (r	Expenses (millions$)	
2	1997	239	250	
3	1998	277	327	
4	1999	578.9	568.9	
5	2000	997.8	1,136.30	
6	2001	1,268.30	1,550.00	
7	2002	1,435.00	1,505.40	
8	2003	1,555.40	1,310.40	
9	2004	1,699.10	1,412.20	
10	2005	1,854.60	1,495.10	
11	2006	2,052.80	1,638.60	

d) The tabular data lets you see a more accurate record of the numbers, while a graph can quickly give you a sense of relative size.
e) Select **Operating Profit margin** instead of **Operating revenue** and **Operating expenses**.
f) The metadata includes the variable reference numbers, the table numbers, and the labels on the axes.
g) According to *The Daily*, "In recent years, revenues from broadband access have increased most rapidly as a result of Internet users shifting from narrowband to broadband services." Narrowband service primarily transmits data via voice, whereas broadband service has the capability of transmitting voice, data and video.

8. **a)** The incidence of cancer increased until the mid 1980s and then started to decline.
b) The maximum was 97.1 cases per 100 000, and this occurred in 1984.

c) 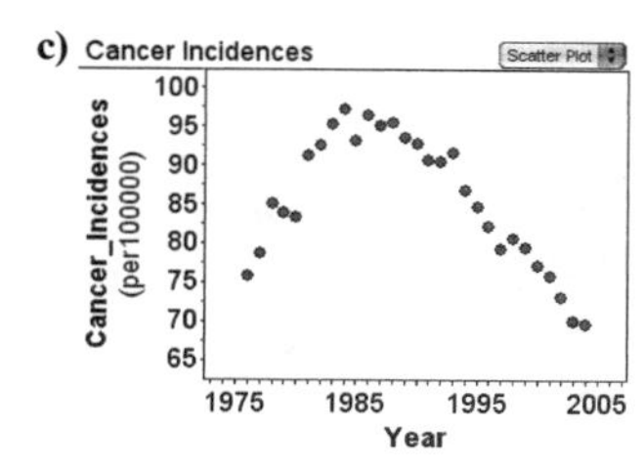

9. Using 2006 census data: **a)** London population: 457 720, private dwellings: 198 144 **b)** 16–19%; Ingersoll is the closest community that has a lower percent of allophones. Toronto has a greater percent than London. **c)** politicians, community groups, advertising agencies
10. **a)** The population centres are noticeable since there will be a greater concentration of fire hydrants in those areas. **b)** four parks **c)** Harrow District High School, Harrow Jr. Elementary, Harrow Sr. Elementary, St. Anthony's Elementary **d)** sometime between 2001 and 2004 **e)** approximately 5.3 ha **f)** Land owners could use it to measure areas of their property. The local public works might use it to track where drainage ditches are, where things like bridges are, or what type of road surfaces exist in various places.

1.4 Simulations, pages 9–12

1. **a)** use **rand*26** **b)** use **randInt(25,50,20)**
2. **a)** Use the formula **=RAND()*20.**
   b) Use the formula **=INT(RAND()*(20-11)) + 11**. Note that because the INT() function always rounds down, to guarantee the numbers are between 10 and 20, start at 11.
   c) You would expect all of the values to occur about the same number of times. This will only occur, however, when you have done enough experiments. In this case 100 is not enough.

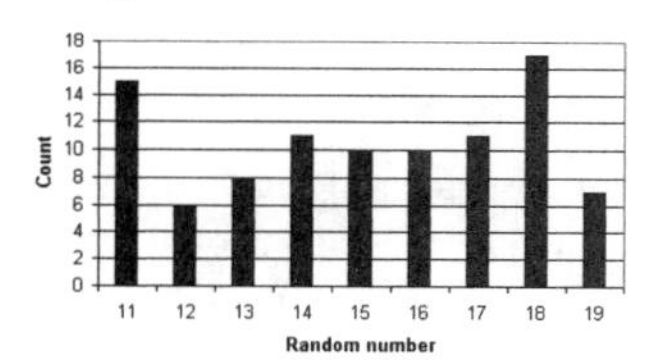

3. **a)** Open a new Fathom™ file and drag a table down to the workspace. Click **<new>** and create an attribute called **Random1**. Right-click the table and choose **New Cases**. Enter the value **30**. Right-click the attribute name and choose **Edit Formula**. In the formula window, type **random(–10,10)**.

Collection 1

	Random1
=	random (−10, 10)
1	-8.47903
2	0.0676926
3	4.25412
4	-1.01676

b) Change the formula from part a) to read **randomInteger(–9, 9)** to generate integer values between (but not including) –10 and 10.

c) Change the formula as shown. For this question, 100 cases are needed. Right-click the table and choose **New Cases**. Add 70 more cases.

Collection 1

	Random1
=	randomPick ("red", "green", "blue")
1	blue
2	green
3	blue
4	blue

d) Drag a graph to the workspace. Drag the **Random1** attribute to the horizontal axis.

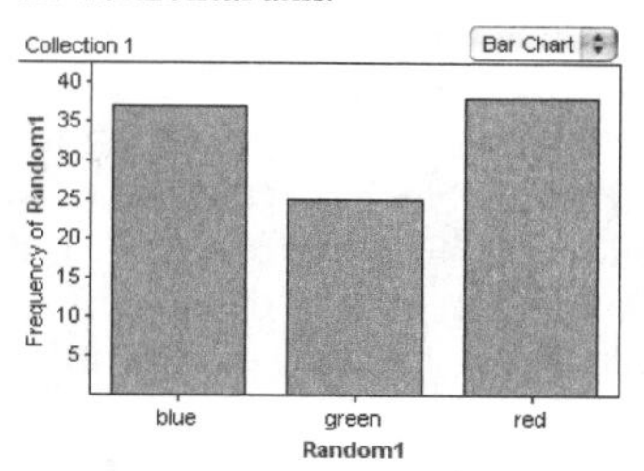

4. The more realistic the simulation games are, the harder they are to play. Turning some of the simulation features off allows players to learn the controls and slowly move to the more realistic situations.
5. **a)** Use playing cards to represent the days of the week (Sunday = ace, Monday = 2 and so on).
   b) and **c)** Use playing cards, but this time the cards represent the digits (ace = 1, 2 = 2 … 10 = 0). One card can be drawn for each digit of the day of the month or day of the year.
   d) Cards can be adapted to represent the days, months, and years.
6. **a)** Since the year was limited to between 1950 and 2010 and the day was between 1 and 31, there may be some unrealistic dates in February and months with only 30 days. In column A, for Day, use **=INT(RAND()*31+1**; in column B, for Month, use **=INT(RAND()*12+1**; in column C, for Year, use **=INT(RAND()*59+1950**.

b)

	Month	Day	year
=	randomPick ("January", "Febr	randomInteger (1, 31)	randomInteger (1950, 2010)
1	July	3	1980
2	September	6	1967
3	January	15	1952

7. Any situation that requires either 30% or 70% will do. For example, a student guesses on tests and gets answers correct 30% of the time, or the batting average of a baseball player is 0.300.
8. One method could have been to use a deck of cards and to randomly draw cards, red = true and black = false.
9. Assign letters to playing cards. Shuffle, randomly pick a card, and replace it. Do that 5 times.
   Another way would be to use Fathom™ with the **randomPick("a","b","c" … "y","z")** function and 5 cases.
10. **a)** Although it is not likely, you may only have to pull out as few as 3 letters. However, it would not be uncommon to find a word before you pull out as few as 10 letters. To be certain that your average is relatively accurate, the more trials the better.
    b) This is not likely to happen right away, and since you are looking for a specific word, it will probably take longer, perhaps over 100 letters. The longer it takes to get the word to show up, the more likely the average is to be realistic. It may take up to 100 trials.
    c) If the word *too* was used, then it likely would have a lower average, since "o" and "t" are the two most abundant letters.
    d) Since a four-letter word is more complex, you should expect it to take longer.

11. Use five cards: one with an H on it for a hit and four with an N for no hit. Randomly pick a card 20 times in a row (replace the card each time) and record the result. A quicker way, with Fathom™, is to use **randomPick ("H","N","N","N","N")** with 20 cases. To rapidly do 20 trials in a row, press **CTRL-Y** to re-randomize. For example, doing this 5 times yielded 15%, 10%, 25%, 10%, 75%. Notice none of the trials gave 20%. Some were close, but that is the nature of this type of experiment.

12. **a)** Let the numbers 1 to 6 represent each prize. Then, roll a six-sided die and keep a tally. Once you have a tally mark in each number (1 to 6), the experiment is done and you record the total number of rolls, or "buys." Doing this many times will be time consuming, but you will likely get an average of over 12 before you get one of each item.
b) With Fathom™, after 20 trials, you have an average of 13.5. Each dot represents one experiment and the number of rolls it took to get at least one of each number.

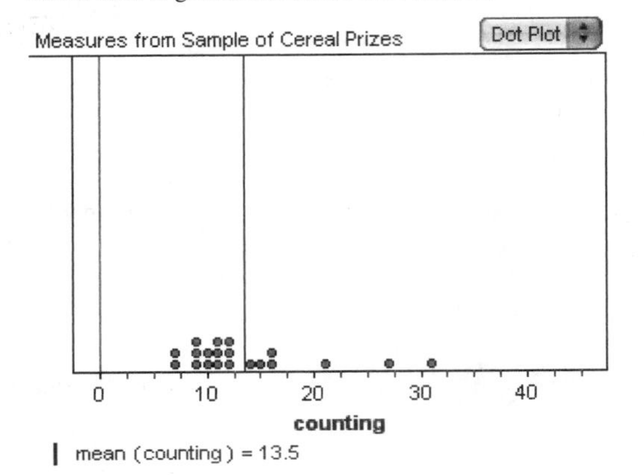

c) With Fathom™, you can do multiple sets of 20 and have them all contribute to the overall average to get a better picture of the information and a more accurate average.

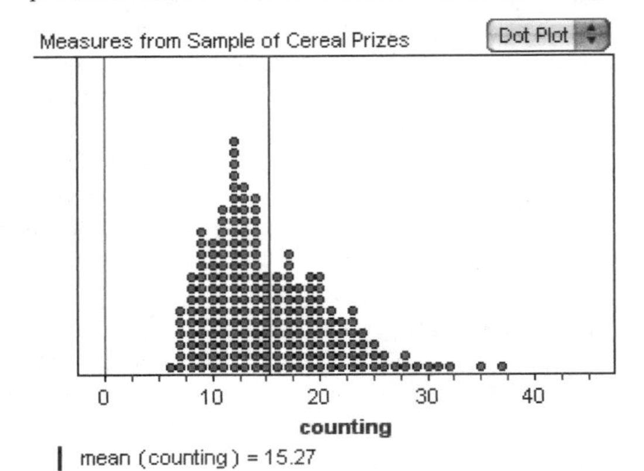

13. **a)–e)**

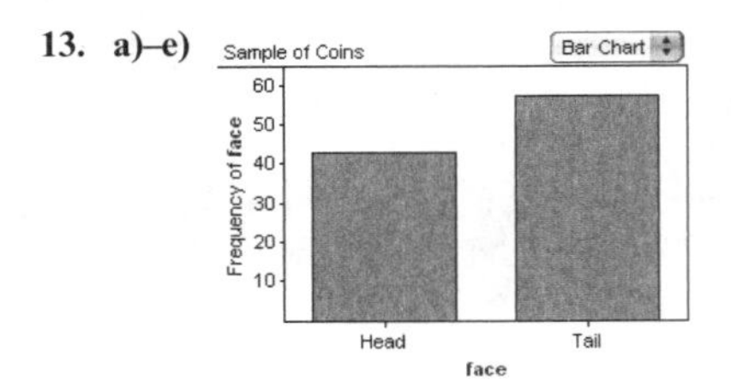

f) When sampling more cases, it seems to be rare that Heads and Tails are equal.
The **Measures** tab should look like this:

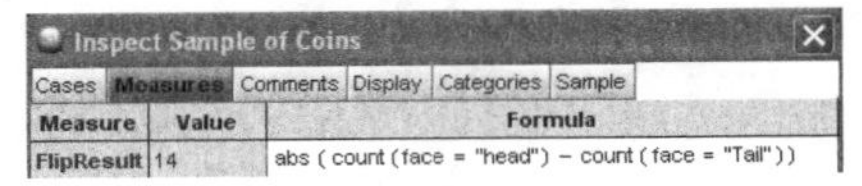

g) Each dot represents one trial of flipping the coin 100 times and the difference between the numbers of heads and tails. Any point that is on zero represents an instance when the number of heads is equal to the number of tails. In one trial there was a large difference of 26, while in most trials the difference was 10 or fewer.

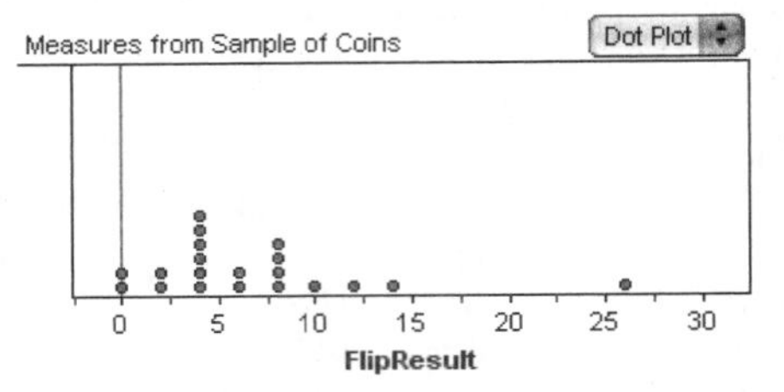

14. **a)** First, right-click the collection and choose **Sample Cases**, then change the number of cases to 100 and click **Sample More Cases**.
Then, select the **Sample of Single Die** collection and drag a table down to the workspace. Drag a graph onto the workspace. Then, drag the **Face** attribute to the horizontal axis of the graph.

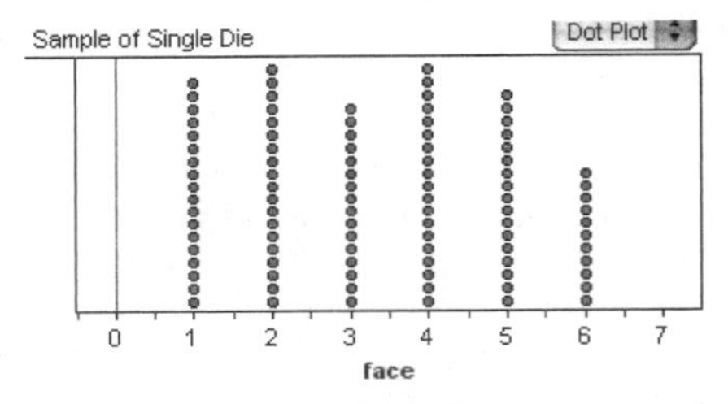

b) Once you create the new **Sample of Single Die** and change the number of samples to 2, you should have a roll of two dice showing in the collection. When you click the **Measures** tab and create the **Total Measure**, it should match the sum of your two dice.

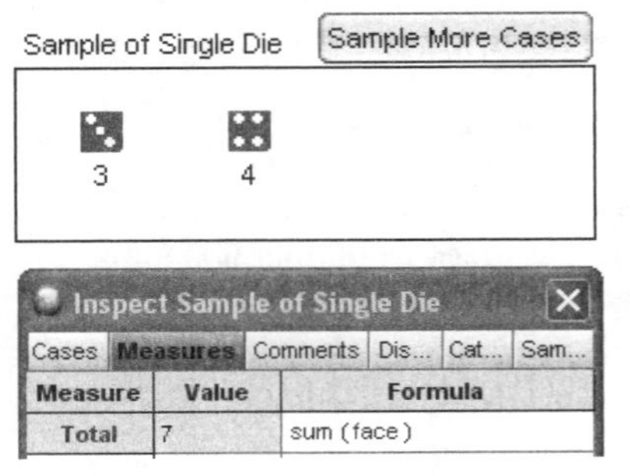

Once the measure is created, right-click **Sample of Single Die** and choose **Collect Measures**.
Once the measures are collected, create the table and graph in the same way as before. Each of the dots represents

the sum of rolling two dice once. There will be a different graph for anyone doing this; in some cases, all possibilities will show up, and in others, they will not.
After 100 trials, a pattern emerges.

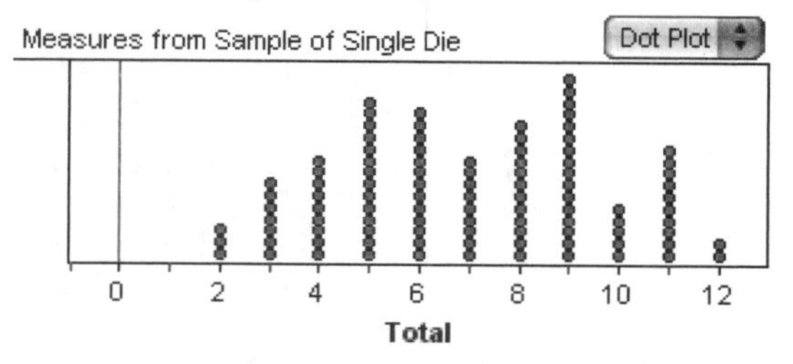

c) It seems that there is a pattern to the rolls. Getting 2 or 12 is not as likely, and that is why they win the larger prize. The numbers in the middle, 5, 6, 7, 8, and 9, show up more often, so people are more likely going to roll those values and win the lower amount.

15. A simulation method would be to use 3 red cards and 7 black ones. Draw a card at random to simulate whether it will rain or not. Doing the experiment several times will give a better picture of the possibilities. A more efficient method would be to use software like Fathom™ with **randomPick("R", "R", "R", "N", "N", "N", "N", "N", "N", "N")** to simulate as many days as possible.

16. Pseudo-random numbers are the random numbers that calculators and computers generate. Generating a truly random number electronically is not easy. What is typically done is a large bank of random numbers is generated (likely by hand) and is stored on the computer or calculator. If the calculator just used this set of numbers, then all sets of random numbers would be the same. So, whenever a computer or calculator needs a random number, it moves to a random number on the list and starts there. Many computers use the current time as a way to define the starting point. With that random starting point, an "almost" random number, or pseudo-random number, is born.

17. Start by right-clicking the collection and choosing **Sample Cases**. From the **Sample of Deck** collection inspector, choose 2 cases. Next, choose the **Measures** tab and create a new measure called **Total**, which will have the formula **sum(blackjackvalue)**. Next, right-click the **Sampled** collection and choose **Collect Measures**. Choose 100 measures to collect and click the button to collect more measures (check **Replace existing cases**). The graph shows that there are only 4 out of 100 that are actually guaranteed a win by getting 21.

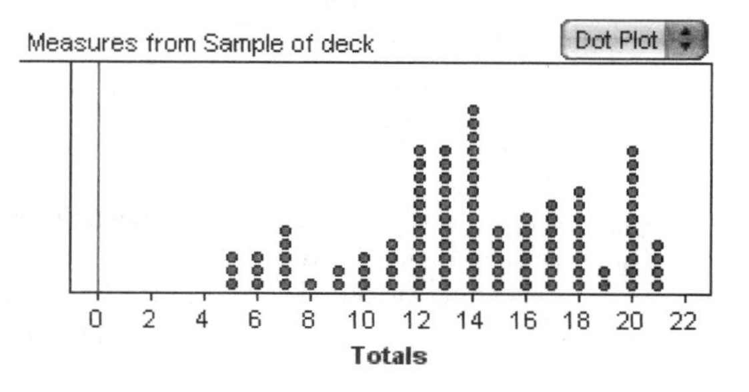

18. a) Select **Toss Coins** from the menu. On the next screen, choose **Set** and then change the **Trial set** to 999 (the greatest number possible).

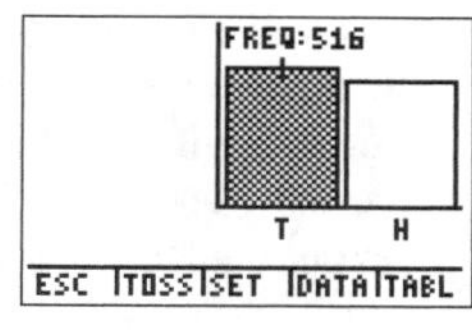

b) Choose **Esc**, then **Yes** to get back to the menu. Choose **Spin Spinner**. Choose **Set**. Under **Trial Set**, type **100**. Now choose **Adv**. This lets you set the weight of each section: 1 = 50%, 2 = 25%, 3 = 12.5% and 4 = 12.5%. Then, click **OK** and **Spin** to get the experiment to start. In this case, the number of 4s was 13, which is almost a perfect number of hits to represent 12.5%.

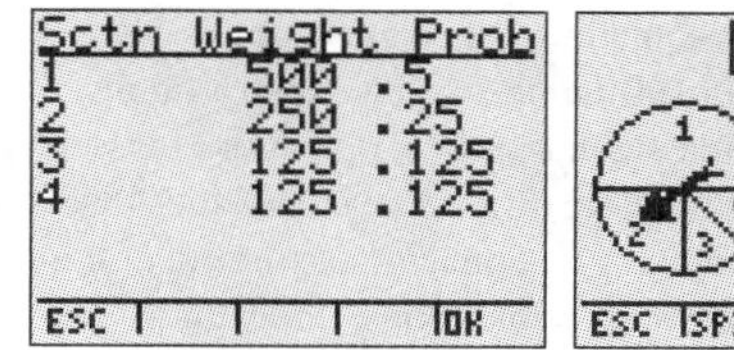

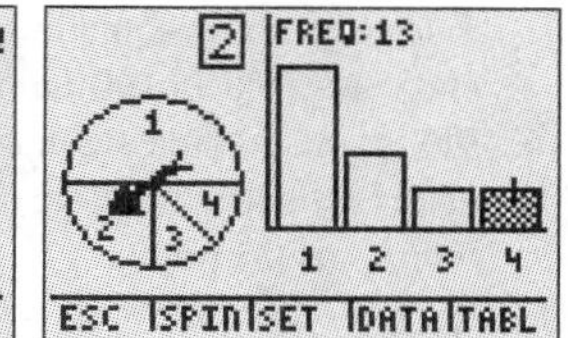

Chapter 2

2.1 Data Analysis With Graphs, pages 13–18

1. **a)** categorical, nominal, non-binary
b) numerical, continuous
c) categorical, nominal, binary
d) categorical, ordinal, binary

2. **a)** Midfielders and defenders make up almost two thirds of all soccer players.
b) Most Ironman champions finish in under 9.5 h.
c) Almost 80% of the respondents were from urban areas.
d) Most people do not drink milk.

3. Histograms and bar graphs are both for numerical data. Bar graphs are used for discrete data, and so are drawn with separate bars to indicate there is no data between any two bars, as in question 1c). Histograms are for continuous data, which means the data can have any value, so the bars are touching, as in question 1b).

4. A dot plot is similar to graphing points on a number line. It shows the relative position of data on the number line, and if there is more than one piece of data with the same value, they are stacked on top of each other.

5. **a)** This is about NASCAR drivers and how much money they have won during a race.
b) Each bar tells how many people have won a certain range of prize money. For example, the first bar shows that 5 people have won between $50 000 and $75 000.
c) The interval $75 000–$100 000 has the most, with 16 people.
d) Use the midrange value for each bar and multiply that by the frequency of each bar. The total prize many is $4 370 000.
e) There is over $4 million won in each NASCAR race, and even those who finish at the end of the pack win over $60 000.

6. **a)** bin width 10; 20–30, 30–40, 40–50, ...
b) bin width 50; 100–150, 150–200, 200–250, ...
c) bin width 500; 21 000–21 500, 21 500–22 000, 22 500–23 000, ...
d) bin width 0.2; 1.2–1.4, 1.4–1.6, 1.6–1.8, ...
7. **a)** 20–30 **b)** 30–40 **c)** 30–40
8. **a)** Pictographs are used to pique people's interest in whatever topic the data is about, whereas a simple bar graph may not.
c) Pie graphs can be used only for data that show how some sort of total is shared. The costs of a movie are independent of each other.
9. Sample answers: **a)** heads or tails **b)** eye colour **c)** yes or no **d)** feelings about something e.g. happy, indifferent, sad **e)** heights of people **f)** number of people in a family
10. This seems to indicate that there is no pattern in the areas of the provinces and territories. The size of the circles represents the population. Nunavut has the greatest area but has one of the least populations, while Ontario has the fourth largest area but has the most population.
11. **a)** Since the data are about obesity, the information is incorporated into a scale.
b) just past the 30 mark **c)** Between 1990 and 2006, the rate went from 11.6% to 30.6%, an increase of 21%, or about 1.3% per year. From 2006 to 2018, the rate is predicted to go from 30.6% to 43%, an increase of 12.6%, or about 1.05% per year. So, obesity appears to be increasing at a slower rate.
12. Pictographs will vary to illustrate the following data.

	Number of Countries
Illegal	7
No Law Against Drinking	25
Drinking Age 14	1
Drinking Age 16	11
Drinking Age 17	1
Drinking Age 18	32
Drinking Age 19	1
Drinking Age 20	2
Drinking Age 21	3

One story is that there are more countries that have no law against drinking than there are countries where it is illegal for anyone to drink (mostly Middle Eastern countries).

13. **a)**

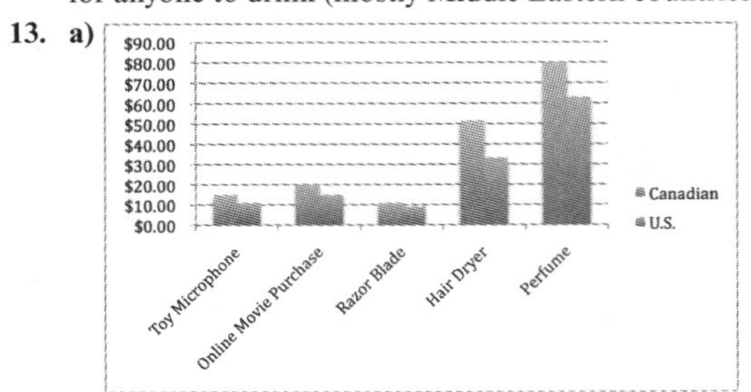

b) The article was probably about how Canadians pay more for the same thing than people in the US do.

14. **a)** This seems to be telling us that health-care spending is really increasing.
b) Although there are no mistakes on the graph, there are two main misleading things on this graph. The scale does not start at 0, so the 1995 bar looks several times smaller than the last one. There is a time period missing between the second-last bar and the last bar (three years of data), so that last bar looks like a significant jump.
c)

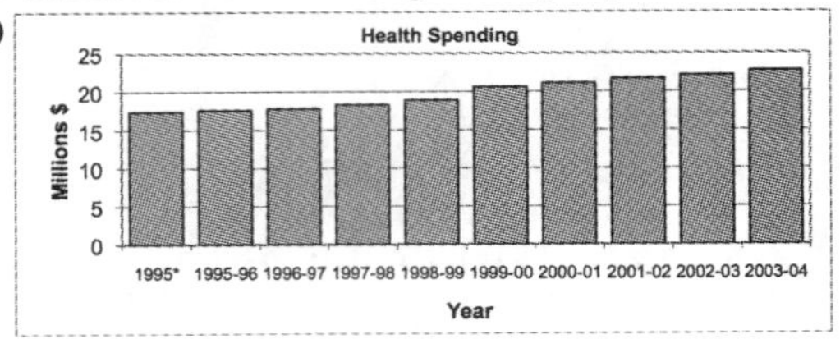

d) This graph seems to indicate that health-care spending is rising steadily (mostly on par with inflation), with the exception of a little jump in the 1999–2000 year (when the pamphlet was published).

15. **a)**

Interval	Frequency	Relative Frequency
6.0–6.2	14	45%
6.2–6.4	5	16%
6.4–6.6	2	6%
6.6–6.8	3	10%
6.8–7.0	4	13%
7.0–7.2	0	0%
7.2–7.4	0	0%
7.4–7.6	1	3%
7.6–7.8	1	3%
7.8–8.0	1	3%

b) and **f)**

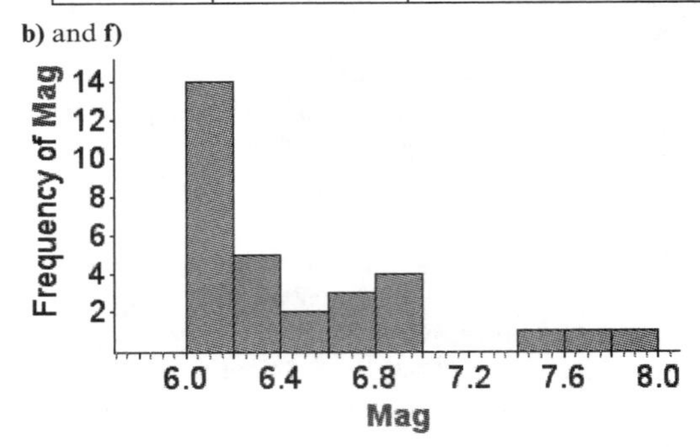

c) 6.0–6.2
d) 10%
e) Santa Cruz Islands, Vanuatu Islands, Celebes Sea **g)** date, time, latitude, longitude, and depth
h) This creates a map of Earth with the magnitude of the earthquakes colour coded. The earthquakes that should be red are not seen, since there are actually 10 in the same area. In fact, 6 of them happened within 24 h.

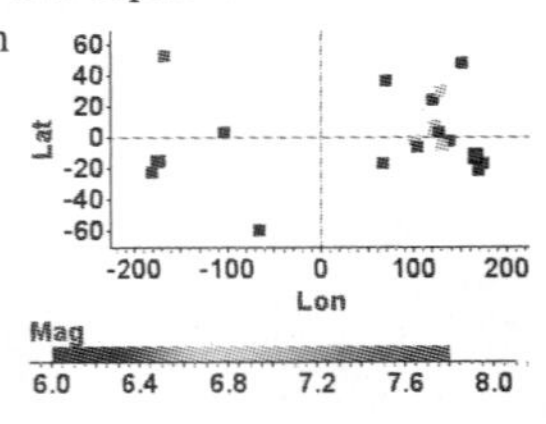

16. a) 30 s or 60 s (minute) intervals

b)

Interval (s)	Frequency	Relative Frequency
120–180	6	12%
180–240	7	14%
240–300	16	32%
300–360	12	24%
360–420	4	8%
420–480	2	4%
480–540	3	6%

c)–e)

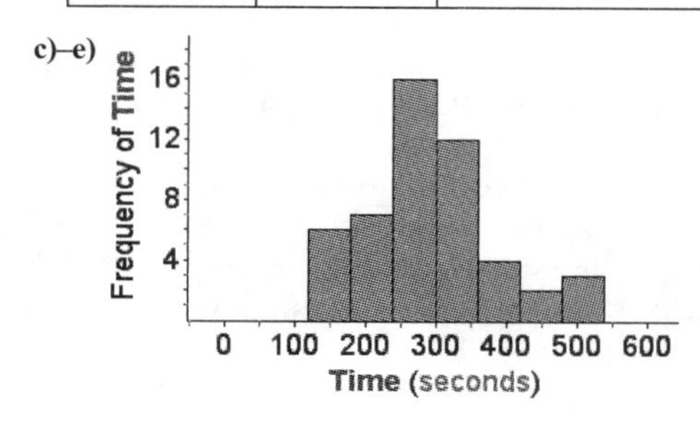

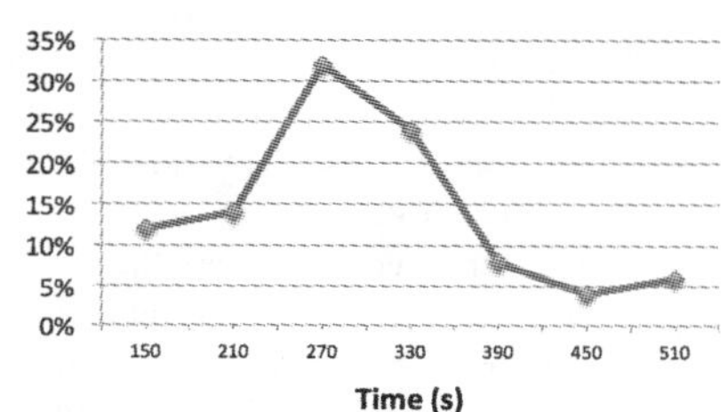

f) 88%

17. Sample answers: **a)** and **b)** numerical, continuous: This histogram shows that the majority of people surveyed (50.1%) earn $25 000 or less:

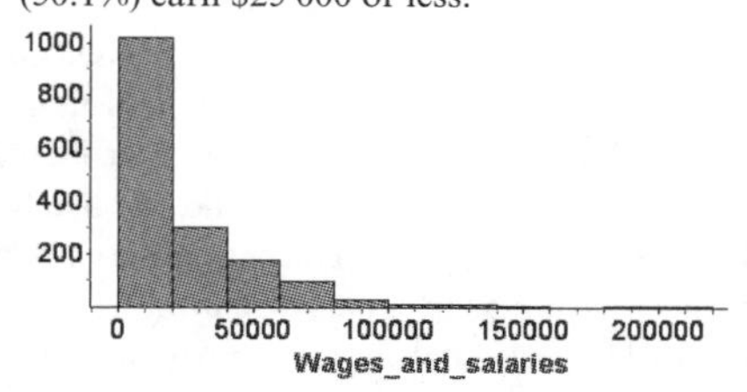

categorical, nominal, non-binary: A bar graph of religious affiliation shows that 16.1% of those surveyed have no religious affiliation.

numerical, discrete: A bar graph of household size seems to show that the most prevalent household size is 2 people (27.2%), and over 70.5% have between 2 and 4 people.

categorical, ordinal, non-binary: A bar graph of education level shows that almost 10% of those surveyed had less than a grade 8 education.

categorical, nominal, binary: A bar graph shows that 71.3% of the people surveyed own their own home, while 28.7% rented or lived in band housing.

There are no data that are categorical, ordinal, and binary.

c)

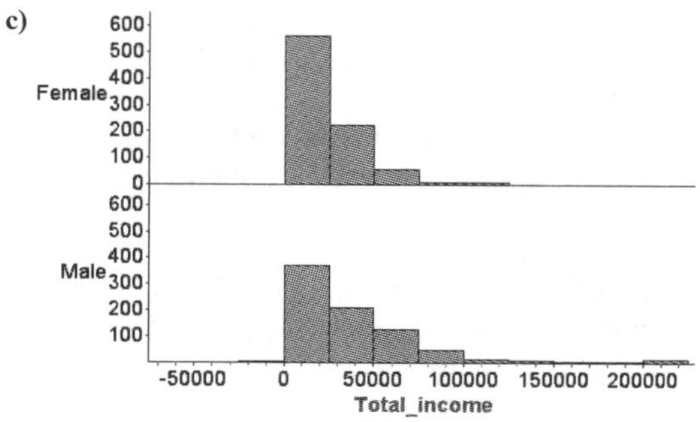

This graph shows that more females seem to have jobs with salaries below $50 000, and that more men have jobs that pay more than $50 000 a year.

18. a) The dot plot seems to suggest that there are four separate groups of pennies. It seems to suggest that there is more to sorting this data than just by mass.

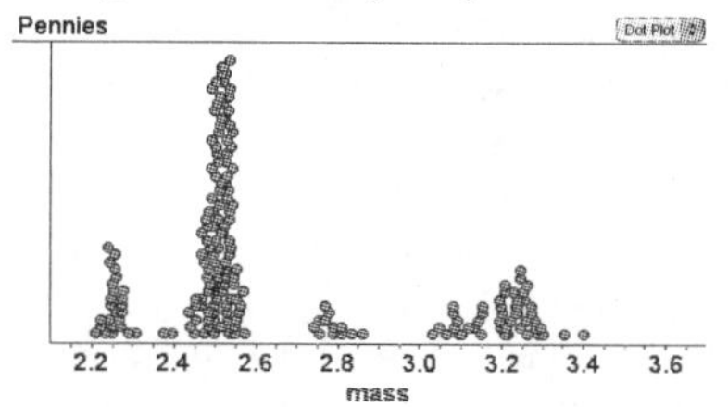

b) This seems to indicate that Canadian pennies have four different masses while American pennies seem to have two different masses.

c) This seems to suggest that different groups were made in different years. For example, the greatest number of pennies seems to have been made between 1982 and 1998. The heaviest group of pennies were all made before 1980.

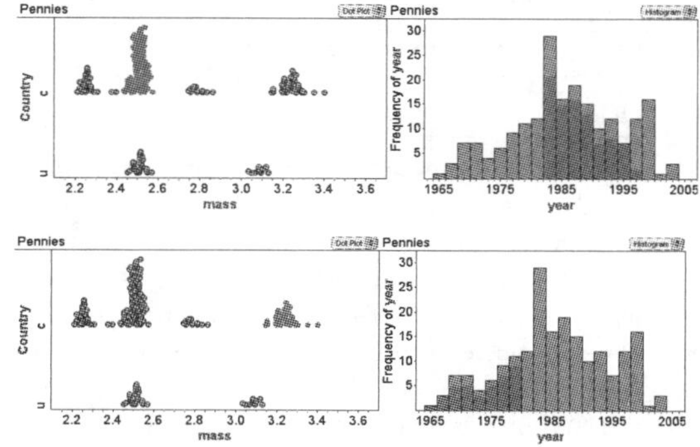

d) This adds a colour coding to the graph, which seems to indicate that the pennies get lighter over time. An Internet search of "minting Canadian Pennies" confirms that these dates correspond to times when the Canadian mint changed the composition of the metals used in making pennies.

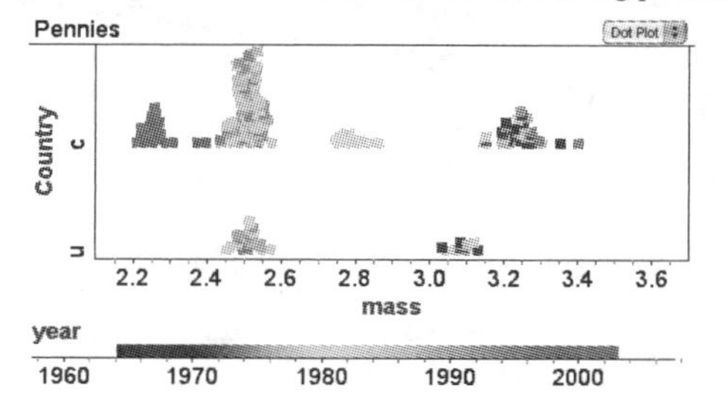

19. As the years of experience increases, there are fewer workers. This makes sense, since as people gain years of experience, they get closer to retirement age. Also it seems that there are groups of people with similar years of experience. This could indicate that there were time periods when many people were hired. This could indicate a period when the company was expanding. Conversely, there were periods when very few were hired. This could be when times were lean. In general, the graph seems to suggest that hiring was cyclical.

2.2 Indices, pages 19–22

1. **a)** 2002 **b)** 50 points **c)** \$28.18 **d)** \$39.13
2. body mass index, Dow Jones Industrial Average, humidex
3. **a)** 30, 30% **b)** −33, −33% **c)** 55, 55% **d)** −50, −50%
4. **a)** From 1950 to 1974, the price of bananas remained relatively the same, even going down a bit at the end of that time. Between about 1975 and 1995, the price varied but in general rose. Then, until about 2008, the price fluctuated but did not have an overall increase. After that the price had a big increase.
   b) A sharp increase: the index was about 100 in 2008 but rose to about 140.
   c) The lowest index value occurred in about 1971, when it was about 15. That is a difference of about 125 points. So, bananas in 1971 would have cost about 14 ¢/kg.
5. **a)** and **b)**

Team	Single Adult (with parking, drink & hot dog)	Family of 4 (with parking, drinks & hot dogs)
Bears	\$22.50	\$55.00
Tigers	\$22.75	\$59.00
Flash	\$23.75	\$51.00
Riders	\$24.50	\$54.00

   c) In both cases, the cheapest ticket price did not mean the lowest overall cost. This is due to the extra costs for drinks and hot dogs. Note also that although the most expensive ticket cost was almost the most expensive single adult cost too, it actually turned out to be the cheapest family cost.
6. **a) i)** 36.3, obese **ii)** 16.1, underweight **iii)** 25.3, slightly overweight
   b) The BMI oversimplifies the situation. For example, someone who has a very muscular build will likely rate as overweight since their muscles are quite heavy.
7. **a)** 2005 **b)** Canada: 2.13, US: 2.68, Japan: −0.09
   c)

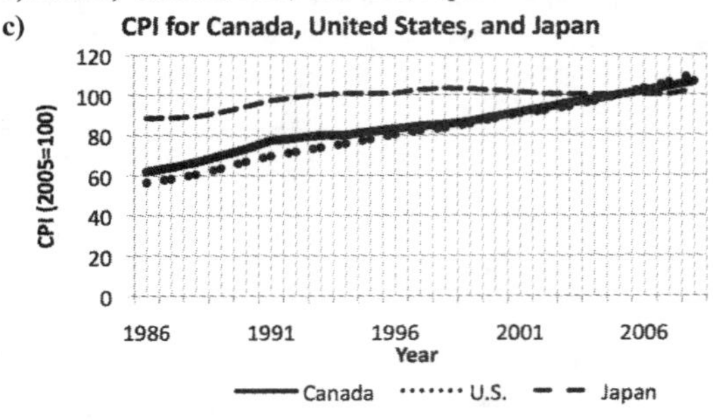

   Japan's CPI seems to have been fairly steady, whereas the Canadian and US CPIs are on a steady rise. So, while prices seemed to be going up in Canada and the US, in Japan they remained constant.
8. **a)** The index went from 80.2 to 115, so the CPI went up 34.8 points. That is a percent increase of $\frac{115 - 80.2}{80.2} \times 100$, or 43.4%.
   b) In 1990 the price of gas was 55.7 ¢/L in 1990 dollars. Inflation has increased the cost of things by 43.4%, so the price (in 1990 dollars) is 55.7 × 1.434 or 79.9 ¢/L.
   c) 1998
   d) Gas prices do not seem to follow inflation. In earlier years, gas was actually cheaper than expected. Although there was a big jump in 2000, it was still cheaper than in 2008. If the price of gas followed inflation, then each year the price in today's dollars would be about the same.
9. The most expensive was the newest, Yankee Stadium, and the least expensive was the oldest, Wrigley Field.
10. **a)** The DJIA is an average of 30 large companies' stock prices at any one time.
    b) Between about 1991 and 2000 the DJIA increased very rapidly, then there was a rapid drop. Between about 2003 and 2008 the DJIA again increased very rapidly—a correction was likely.
    c) It may not be realistic to use the trend from 1991 to 2008 to extrapolate into the future.
11. In both graphs, all three countries have the same CPI at the point where the base year is 100. This emphasizes the fact that these are indices. That means they are only relative to themselves. Putting data from different sources on the same graph does not make them related. It also reminds us that since an index is a relative value, it could be anything. If the base value were 1000, the index would still have the same meaning.
12. The volume controls on any listening device are based on an index related to sound intensity. The electronics inside determine what the maximum volume can be. Usually, that is marked as a 10 on the dial and then the rest of the numbers are filled in at equal intervals. So, in the movie, the amps were not "one louder." The manufacturer just marked 11 at the top end and divided the rest of the scale into more intervals.

2.3 Sampling Techniques, pages 23–26

1. **a)** primary, micro
   b) summary, aggregate
   c) primary, micro
2. A population is the entire group that you are studying. A sample is a small portion of the population that typically has many of the same characteristics of the population itself. For example, you might want to find out what everyone in your school thought of the latest vampire movie. Everyone in the school would be the population, and a smaller group of students you chose to survey would be the sample.
3. **a)** voluntary response **b)** simple random
   c) multi-stage **d)** stratified
   e) cluster **f)** convenience
   g) systematic

4. a) There are 800 students in the school. List them in alphabetical order and randomly choose one of the first 8 students (800 ÷ 100 = 8) and then choose every 8th student after that.
b) Of the 800 students, 425 are male. Males are 53%. So, sample 53 males and 47 females.
c) Choose the sample by grade:

Grade	Number	Percent	Sample
9	300	38%	38
10	250	31%	31
11	150	19%	19
12	100	12%	12

5. a) Experimental data are collected when you have a hypothesis that needs to be tested where the characteristics of what is being tested are not factored in.
b) Observational data are collected when you want to determine how various groups behave under specific circumstances.
6. a) Answers will vary.
b) The more pennies you pull out the more likely you will get the same proportion of Canadian pennies as is in the population. More of them were way off when the number sampled was lower. In the case of $n = 16$, most of those results were around 75%.
7. a) about 40 cases b) about 150 c) about 250
d) Generally, a larger sample is better, but not always. The fewer the number of possible answers, the smaller the sample can be. Since there are fewer choices, there is a better chance of the sample being representative of the population sooner. Sometimes, there are so many choices that it is not realistically possible for a sample to be accurate.
8. a) By taking several samples they are getting a better picture of the overall makeup of the load.
b) 12 734 kg of pods
9. a)

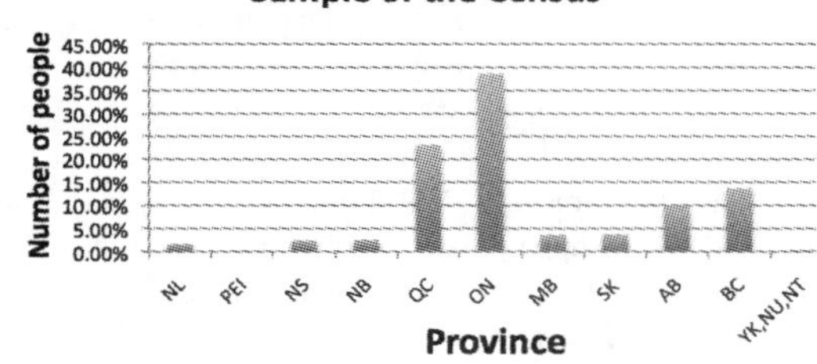

b)

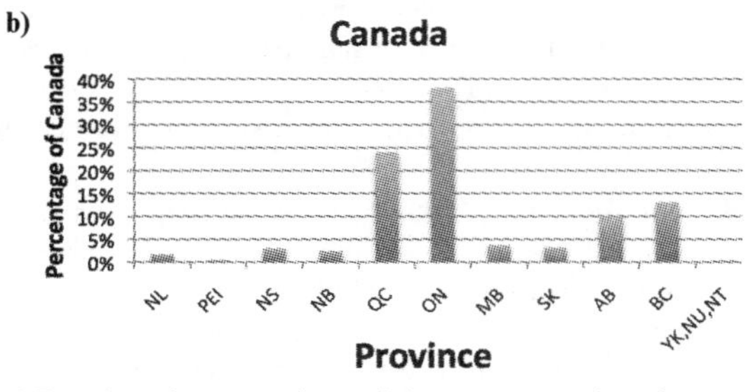

c) Based on the comparison of the two proportions, it seems that this small sample (relative to the population) actually is stratified fairly closely to the population, even though it is less than one hundredth of a percent of its size.

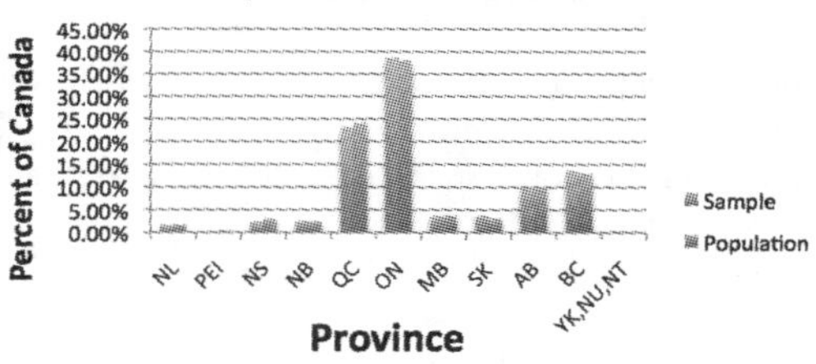

10. This is a cluster sample, since they randomly chose 33 locations, then randomly chose a neighbourhood in each of those locations, then sampled the nearest 30 homes. Due to the two levels of randomness, everyone in the area had an equal opportunity to be chosen, so the sample is valid.
11. The privacy officer is referring to the fact that when people supply personal information via the Internet, they may not be truthful, since no one would know and it does not harm them to give false information.
12. Since the 1950s, the Neilson Corporation has kept track of what people watch on television. Originally, they gave a sample of households diaries to record what they watched. More recently, they have installed devices to measure what people are watching in real time. Approximately 25 000 people are in the sample of people who participate in the rating system.

2.4 Bias in Surveys, pages 27–28

1. a) response b) sampling c) non-response d) measurement e) measurement f) sampling g) measurement
2. a) Do the survey at a concert. b) Ask a question such as "Do you like this cheesy band The Nothings or would you rather listen to a sophisticated band like Majesty?"
c) Send a mass e-mail out to all the people you know.
d) Ask a question like "How many cool concerts have you been to in the last year?"
3. The group that gets the actual drug ensures that someone gets the medication to be tested. The purpose of the placebo group is to make sure people are not getting better just because they think they are getting medication. The response to the real drug must be significantly better than that of the placebo. The control group is used to make sure that if some of people are getting better on their own, that group is factored in. For example, if 10% of the people in the control group get better and 50% of the people in the group that gets the drug get better, then it is more likely that only 40% got better because of the drug.
4. In a double-blind study, neither the people receiving the medication nor the people who administer the medication know which is the real thing and which is the placebo.
5. a) This person wants to learn about how different age and gender groups feel about eating healthy.
b) Question 5 is biased, as it is a loaded question. A more appropriate question would be "Should people eat healthier?"

6. **a)** Because the company is showing data it gathered, it might be thought that it might not include information it found that could show wind turbines in a negative light.
b) They could have an independent research firm do the study.
7. **a)** People who do not have cars, or access to a ride, will not have a chance to vote. Since these tend to be people of lower socio-economic status, the vote may underestimate what this demographic wants.
b) This question has measurement bias. It is written in a way to sway people to vote for the increase in funding.
c) Restricting when the voting takes place limits who can actually get out and vote. In Canada, you are allowed to take time off work to vote.
8. They might have asked loaded questions like "Would you recommend chewing on the end of a pencil or chewing Superbubble gum?" Also, they could just keep asking dentists until they get the desired results.
9. **a)** At a reunion, people might state they are making more money than they actually are, since it would make them appear more successful. This would be an example of response bias.
b) In this case, it would overestimate the average salary of the group of people.
10. **a)** One of the problems occurred in a small town where 100% of both the control group and the vaccinated group survived an epidemic. The incidence rate was so low in general that given their small numbers, they should have expected only one or two cases. It is possible that the village had no cases simply by chance. Another incident occurred when the groups in an experiment were differentiated by socio-economic status: people with higher incomes were given the drug, and many poorer people were in the control group. The researchers found that the control group actually had a smaller incidence rate than the other group. Researchers thought this might have been because the squalid conditions that poorer people lived in built up their natural resistance to some viruses.
b) Canada helped in the discovery of the epidemiology of the virus and also helped develop a process that sped up the production of the vaccine. Go to *http://www.healthheritageresearch.com/Polio-Conntact9606.html* for more details.

2.5 Measures of Central Tendency, pages 29–33

1. and **2. a)** 51.3, 53, 53 **b)** 668.75, 667, 665 **c)** 1.264, 1.28, 1.28
3. **a)** symmetrical; mean and median are too close to identify
b) skewed to the right; mean is to the right of the median **c)** skewed to the left; mean is to the left of the median
d) symmetrical; mean and median are too close to identify
e) skewed to the right; mean is to the right of the median
4. **a)** mean: 3.2, median: 2, mode: 2
b) Both the median and mode are 2. The mean is supposed to represent what is common, but it does not do this well when 5 of the 6 pieces of data are less than 3.2.
5. 8.67 kg
6. **a)** The median will be the 34th piece of data. Since there are 15 pieces of data in the first interval and about 27 in the 2nd, the 34th piece will be in the 2nd interval. So, the median is between 100 and 200, perhaps approximately 150. **b)** 100–200
7. **a)** Using the midrange value for each interval, the mean is \$45 769.
b) Using the least value in each interval, the mean is \$40 769; using the greatest value in each interval, the mean is \$50 769.
c) The least, midrange, and greatest values in each interval differed by multiples of \$5000, and the means differed by \$5000. So, if all the values in the mean calculation change by a certain amount, the mean will change by that same amount.
8. **a)** mean: \$46 000, median: \$30 000, mode: \$20 000
b) When the term *average salary* is used, it could represent the mean, median, or mode. In this case, all three are significantly different, so this tells us that any measure of average will not be very informative.
c) The company might state the mean to suggest the average salary is quite high, although only 8 of the 25 employees have salaries above that. The union might use the mode, since it is the lowest, to suggest that most workers get paid a low amount. This is closer to the truth than the company's using the mean, but it does not represent more than half of the workers (the upper 13) very well.
d) The mean goes up to \$57 000, while the median and mode stay the same.
e) The top salary is definitely an outlier. These values tend to pull the mean toward the outlier and away from the median.
9. Each measurement should have been 2 s less. So, the mean and median should go down by 2 s, to 3.7 s and 3.9 s, respectively.
10. **a)** Each mark is worth a different part of the total, because they are not equally weighted. Using Stuart's method assumes that they are all worth the same.
b) 73% **c)** Knowledge; It has the greatest category weight.
11. 90%
12. The student is incorrect. The data are sorted by order of height, but the arm span is not. So, choosing the middle value is not appropriate until the data are sorted by arm span.
13. Machine A: mean = 72.997 mm, median = 72.997 mm; both are within 3 μm of 73 mm but 72.982 mm is 15 μm away from the mean, so this machine fails.
Machine B: mean = 72.999 mm, median = 72.998 mm; both are within 3 μm of 73 mm and all measurements are within 12 μm of the mean, so this machine passes.
Machine C: mean = 72.997 mm, median = 72.996 mm; The median is more than 3 μm away from 73 mm, so this machine fails.
14. **a)** It would be half only if the average she was reporting was the median.
b) He uses the median instead of the mean to find the total number of strokes.
c) The company is incorrect, since this can be true, especially since some very high salaries could skew the mean above what is most common.

15. **a)** 545 **b)** mean: 7.336 59, median: 7.2, mode: 7.0
c) The data are skewed to the right.
d) The median is probably most appropriate, since close to half of the data are at or below 7.2.
e) The strongest earthquake was 9.0, and it happened on Dec. 26, 2004. This was a huge earthquake that caused a tsunami in the Indian Ocean.

16. **a)** A histogram is chosen, because the bars give a good approximation of the number of data files in each range.

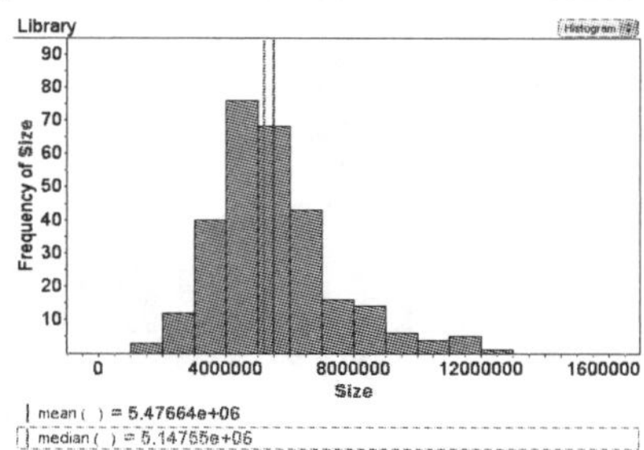

b)

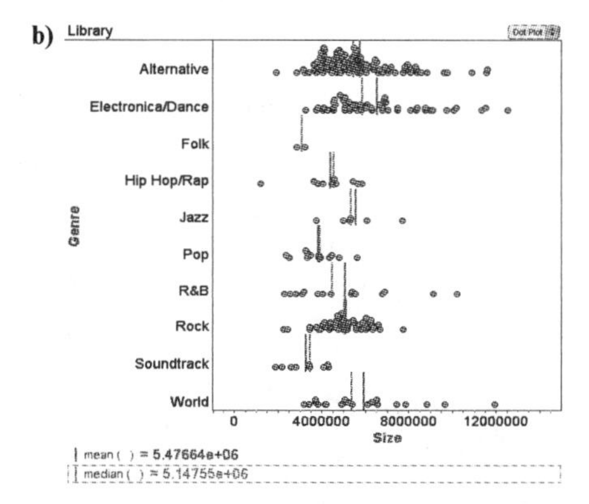

c) Based on the songs here, folk has the least average song size (mean or median), so that would be the best. However, there are only two folk songs, which is not a very good sample. Of the genres with a lot of songs, rock seems to have the lowest averages and thus would probably be the best bet for getting more songs on.

17. **a)**

Year	Mean Salary	Median Salary
2006	\$75 608	\$120 874
2009	\$45 000	\$65 000

b)

Year	Least Salary	Greatest Salary
2006	\$11 700	\$900 000
2009	\$15 300	\$5 500 000

c) By creating both graphs with the same scales and the same intervals, you can create a better comparison. In 2009, there were a few more players making over \$1 000 000, but also many more players getting paid at the lowest end.
d) In 2006, the outliers would probably be the four highest paid players: Landon Donovan, Eddie Johnson, Juan Francisco Palencia, and Juan Pablo Garcia. In 2009, the outliers would probably be the eight who are the furthest away from the group: David Beckham, Cuauhtemoc Blanco, Juan Pablo Angel, Fredrik Ljungberg, Julian de Guzman, Landon Donovan, Luciano Emilio, and Guillermo Barros Schelotto.
e) There is a large gap between the salaries of the highest-paid players and the lowest-paid players. Also, the overall pay has gone up since 2006, though not by much for the lowest-paid players, and by a huge amount for the top players.

18. For an average to exist, some of the data have to be above the average and some below. So, all of the data cannot be above the average

19. From the department data, it appears that women have higher admission rates in each of the departments. However, the number of women compared to men is quite small in the two departments where the women had really high admission rates. So, even though individual department admission rates were higher for women, the much larger total number of male applicants results in higher admission rates for men overall.

20. 7, 7, 21, 22, 23

2.6 Measures of Spread, pages 34–38

1. and **2.** **a)** 57, 555.2, 23.56 **b)** 1.43, 0.0757, 0.275
c) 5.605, 0.104, 0.322

3. and **4.** **a)** 24, 40, 87, 65, 47, 91
b) 127, 142, 174, 145, 32, 180
c) 12.4, 12.6, 13.3, 12.95, 0.7, 13.4

5. **a)** 35 **b)** 50 **c)** 82.5 **d)** 32.5

6. **a)** 26, 39, 50, 68, 72

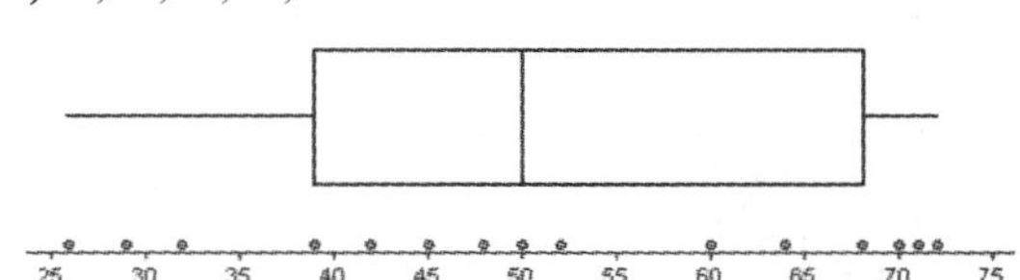

b) 39, 46.5, 59.5, 72.5, 78

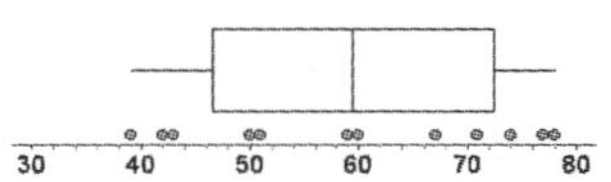

7. **a)** 0.80 **b)** −1.66 **c)** −1.14 **d)** 1.47

8. If it is in the 5th percentile for weight, that means that 95% of all babies were heavier. If it is in the 95th percentile for height, that means 5% were taller. So, this is a tall thin baby.

9. **a)** By looking only at the averages, you are ignoring any possible spread of the data. Two sets of data could have identical averages but one could be more spread out. The one that is more spread out has a less-reliable average.
b) Without doing any calculations, it can be seen that Machine B is producing pistons with less variation in the diameter, so they should be more cylindrical.

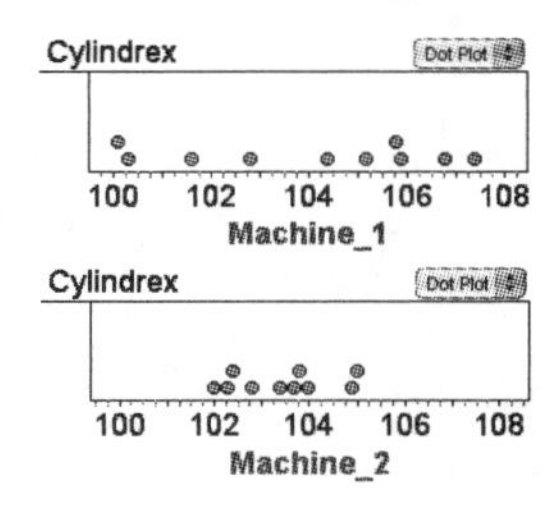

c) Machine A: 104.0 ± 0.8 mm, Machine B: 103.4 ± 0.3 mm

10. **a)** The Ad Pages attribute seems more consistent, since its interquartile range is noticeably smaller.

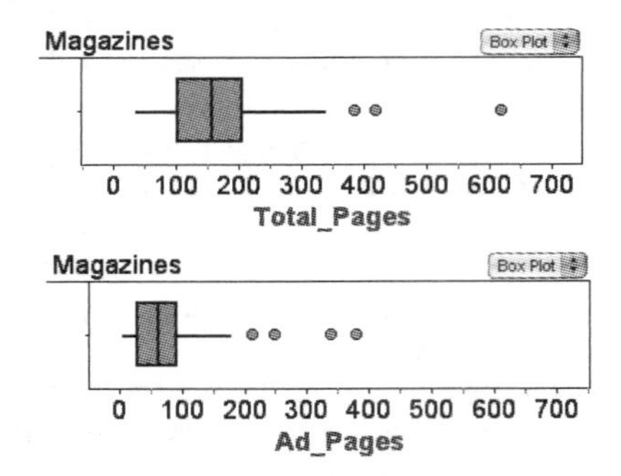

b) This shows the box plot for each type of magazine, with a visual cue for the mean and median of each set of data and the overall mean of each attribute. The overall mean for Total Pages is 163.592, and for Ad Pages it is 67.828.
c) There are five outliers in the Ad Pages graph. Three are *In Style*, one is *Vanity Fair*, and the last is *Flex Magazine*. In the Total Pages graph there is one less *In Style* and there is also a *National Geographic*. So, there are four that are in both.
d) **i)** health/fashion **ii)** business/news
e) The highest mean for the Total Pages attribute is 198.4 for HF, and for Ad Pages it is also HF, with 87.7.
f) This confirms that in both cases BN is the most consistent, since it has the lowest standard deviation.
g) There is not a lot of variation in size for business and news magazines, while health and fashion magazines have the largest variation, with some only 50 pages and some as many as 600 pages. Health and fashion magazines are the most popular, since there are far more of them than of the other magazine types.

11. **a)** For each city, the mean and median equal 15.
b) Based on the averages, one might think that these cities have fairly similar climates.
c) Mathland: $s_x = 8.1$, Geometown: $s_x = 3.0$, Algeville: $s_x = 20.1$. These values indicate that the only reasonably reliable average is that of Geometown.
d) The dot plots agree with the standard deviations, as the points for Geometown are the most localized.

12. **a)** Team A: 16.30 km/h, Team B: 16.21 km/h, Team C: 15.30 km/h
b) Since each leg was a different distance, a weighted mean is used: Team A: 16.37 km/h, Team B: 16.26 km/h, Team C: 15.31 km/h. These values are relatively close to the mean speeds calculated using the total distance and time in part a). Weighted means should be close to those calculated using total distance and time because the weighted means take into account the distance travelled at each speed.
c) Team A: $s_x = 1.37$, Team B: $s_x = 1.21$, Team C: $s_x = 0.90$. Based on these weighted standard deviation values, Team C was the most consistent. Further evidence can be seen in the fact that its speeds in parts a) and b) are almost identical.
d) The fact that the Team C box (IQR) is almost half as wide as the others was an indication that these data are some of the most consistent.

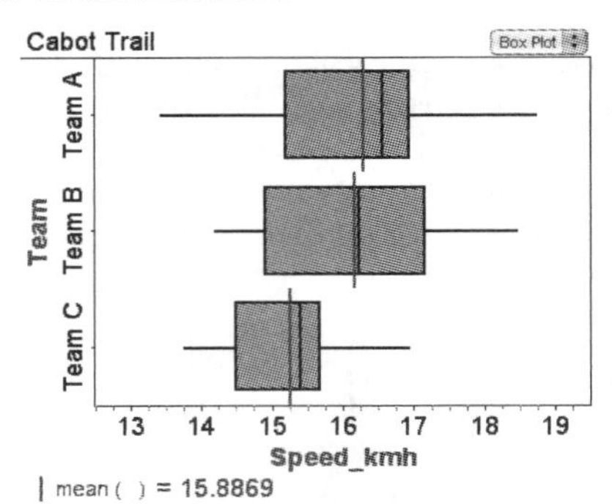

13. **a)** 88 336 287, 81 309 612, $s_x = 33\ 215\ 287$, $Q_1 = 67\ 101\ 667$, $Q_2 = 102\ 996\ 415$
b) The New York Yankees are the only outlier. This is due to the significant difference between its z-score and that of the next nearest team. The Yankees' payroll is over $65 million more than the next nearest team's. That is more than the entire payroll of each of the 6 lowest teams. This also shows up in the z-score, as it is a full 2 standard deviations greater than that of the next closest team.
c) Since 1999, the Yankees have won the eastern division championship every year but 2007 and 2008. They have won the American League championship 5 times and the World Series 3 times. So, they have had all but two winning seasons. One could argue, however, that their payroll is so much higher that every season should be a winning season.

14. **a)** Footlong Sweet Onion Teriyaki. It is a sub with sweet onions and teriyaki sauce, so you would expect it to have more sugar.
b) The two with the least IQR are Sugars and Saturated Fat, but it is hard to tell which is lower. However, when the standard deviation is calculated, Saturated Fat has a lower value. Even if you delete the outlier, Sugars just matches Saturated Fat. Also, the first quartile is quite tiny, indicating that they are all close together. So, any average will be more accurate with this set of data.

15. **a)** These two plots are mirror images of each other. This makes sense, because the time is how long it takes for them to drive around the track once. The speed is related to that time, hence the similar shape.

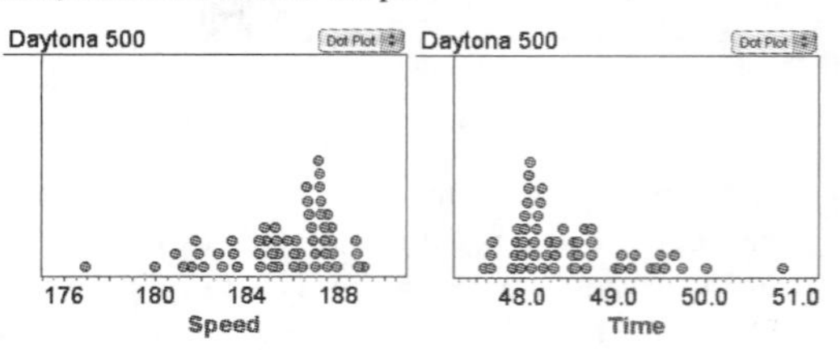

b) With the exception of the two outliers, all of the cars had lap times within about 2.5 s of each other. There are a few standouts though. If you make a percentile plot, you can see that the top four drivers were distinct from the rest of the pack. Then, the next 38 cars would have finished within 1 s of each other.

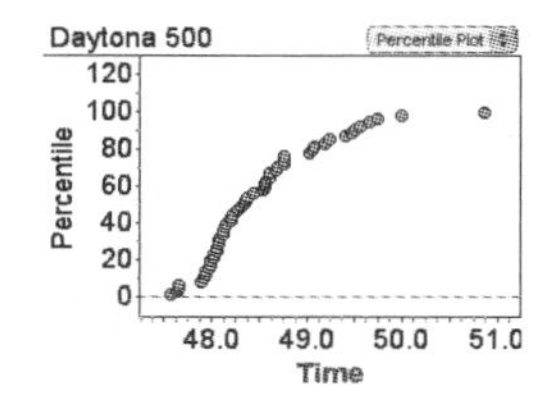

16. In general, the movies that play in the most theatres make the most money. There are, however, exceptions to that rule, for example, *Me and Orson* and Disney's *The Princess and the Frog*, which played in only a few theatres but made over $100 000 at each.
17. **a)** The weighted average, using the mid-point of each interval, is $32 518.
b) Based on the number of people who are working, one story could be that approximately 3% of the total workforce make $100 000 or more. When you tally up their wealth, it is actually about 15% of the total value. Conversely, there are about 21% of the people who make less than $10 000, and they account for only about 3% of the wealth.

Chapter 3

3.1 Scatter Plots and Linear Correlation, pages 39–43

1. **a)** strong, positive **b)** weak or none **c)** strong, negative **d)** strong, positive **e)** strong, positive **f)** strong, negative **g)** strong, positive **h)** moderate, positive
2. **a)** number of cigarettes smoked per day
b) hours of yoga practice
c) amount of candy consumed
d) number of concerts attended
e) years of schooling
f) temperature
g) speed
h) air pressure
3. **a)**

	A	B	C	D	E	F
1						
2		x	y	x^2	y^2	xy
3		1.1	2.5	1.21	6.25	2.75
4		2.8	8.2	7.84	67.24	22.96
5		1.3	7.3	1.69	53.29	9.49
6		3.5	13.4	12.25	179.56	46.9
7		5.4	9.5	29.16	90.25	51.3
8		5.8	21.5	33.64	462.25	124.7
9		7	19.5	49	380.25	136.5
10		7.1	29.1	50.41	846.81	206.61
11	Sums:	34	111	185.2	2085.9	601.21

b)–c) $r = 0.868$; strong
4. **a)** strong, positive **b)** strong, negative **c)** moderate, positive **d)** strong, positive
5. **a)** single-variable; There is only one variable.
b) two-variable; There are two variables: time and temperature.
c) single-variable; There is only one variable being measured repeatedly.
d) both; Different attributes are measured repeatedly for several sandwiches.
6. **a)** $0 \le r^2 < 0.1089$ **b)** $0.1089 < r^2 < 0.4489$ **c)** $0.4489 < r^2 \le 1$
7. **a)** Average ticket price increases each year.
b) $r = 0.9950$; strong, positive **c)** Sample answer: Since the average price generally increases by about 20 cents each year, the price could be about $7.38 in 2010.
8. **a)** at bats
b) New York; $r_{\text{Yankees}} = 0.8993$, $r_{\text{Orioles}} = 0.8390$
c) These high correlation values indicate a very strong positive relationship between times at bat and home runs hit.
d)

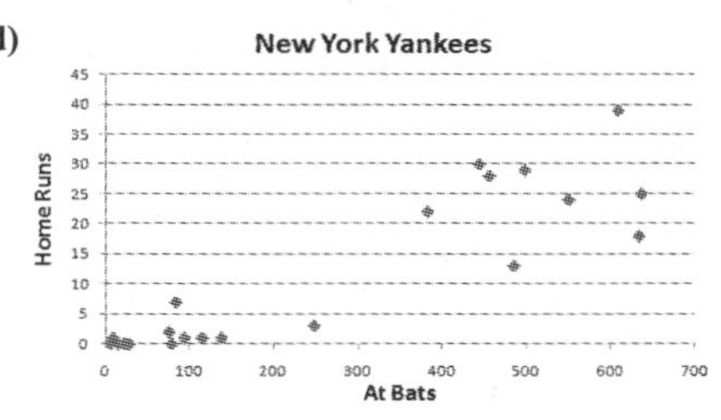

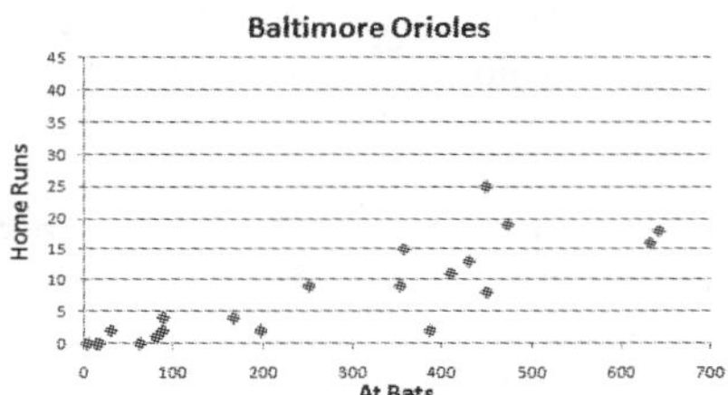

The data appear to suggest that a team having a winning season will likely have more home runs than a team that is not winning as many games.
9. **a)** Property costs rise as you advance around board.
b) Rent increases as cost increases.
c) Baltic Avenue, Park Place, Boardwalk
d) Rent goes up by $2 for each $20 increase in cost. A new property's rent would be $46.
10. Aaron's. The correlation between height and arm span ($r = 0.850$) appears to be stronger than the correlation between forearm and foot lengths ($r = 0.596$).
11. **a)** The number of games played increases with the number of years active, as shown in a scatter plot:

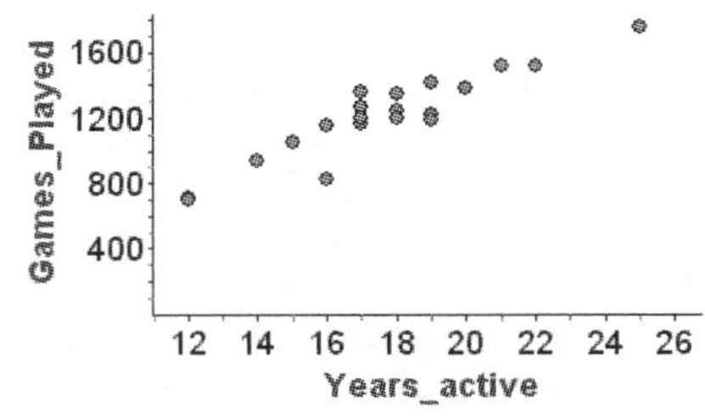

b) The number of points should increase with the number of games played, as suggested in the scatter plot. The correlation coefficient is $r = 0.766$ (strong, positive).

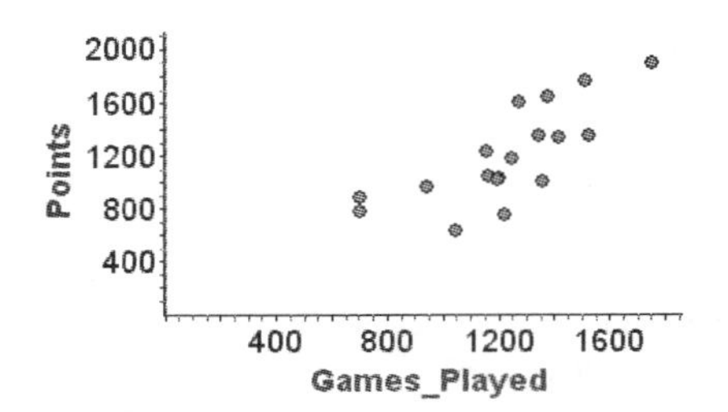

c) Each goal adds to the point total, so more goals results in more points.

d) The total salary might be expected to increase with years active, but a scatter plot suggests there is no relationship between total salary and years active. The correlation coefficient is weak and negative ($r = -0.013$).

e) One would expect that the more points a player gets, the higher would be the player's salary. However, a scatter plot shows differently, and the correlation is $r = 0.245$.

f) A cost-effective player scores a lot of points but does not get paid the highest salary. Based on an index of salary paid per point scored, Mark Messier appears to be most cost-effective at just over $31 000 per point. Chris Pronger appears to be least cost-effective at more than $100 000 per point, but he is on defense and might not be expected to score a lot of points.

12. a) The better the finish (i.e., closer to first), the higher the winnings. The graph and correlation coefficient ($r = -0.678$) suggest this is true for certain top drivers, but less so for most racers.

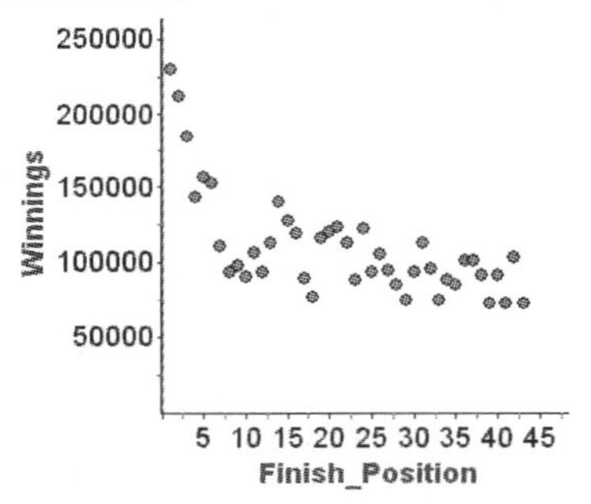

b) The data suggest that, with a few exceptions, there is no relationship between a driver's start and finish positions. The scatter plot has no pattern.

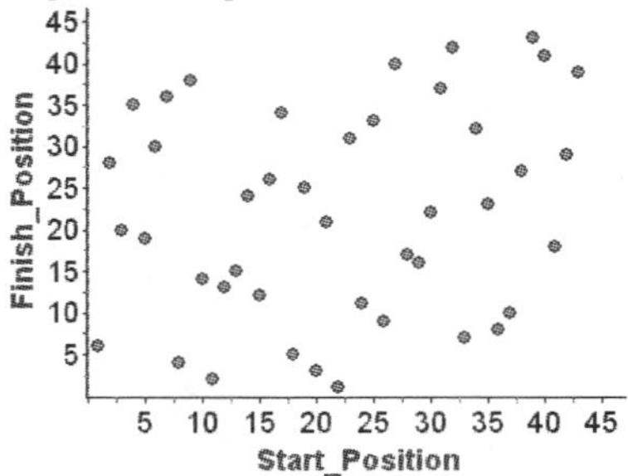

13. a) You probably would expect a strong and positive correlation; however, $r = 0.516$, so the correlation is only moderate and positive.

b) It appears there are actually two relationships, one that goes up more steeply than the other. This would explain why the overall correlation is not strong.

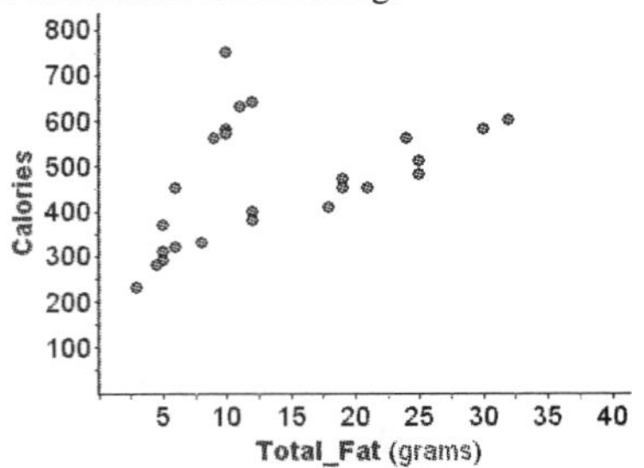

c) This shows there is one relationship for "healthy" (i.e., less total fat) sandwiches and one for "regular" sandwiches.

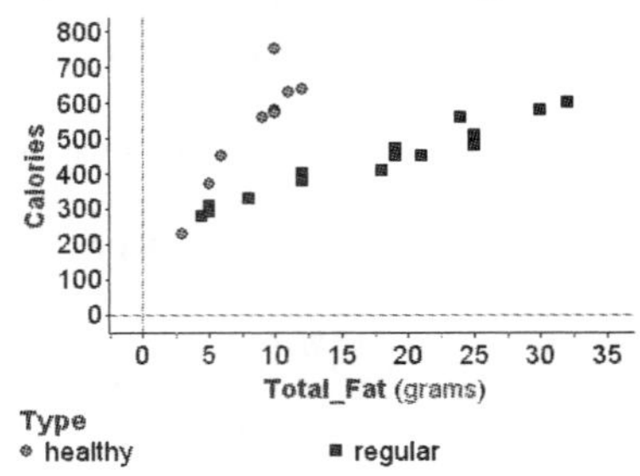

d) 6" Subway Club® is low fat but relatively small compared to most healthy "foot longs," so it likely belongs in the regular category.

14. Answers will vary.

3.2 Linear Regression, pages 44–48

1. a)

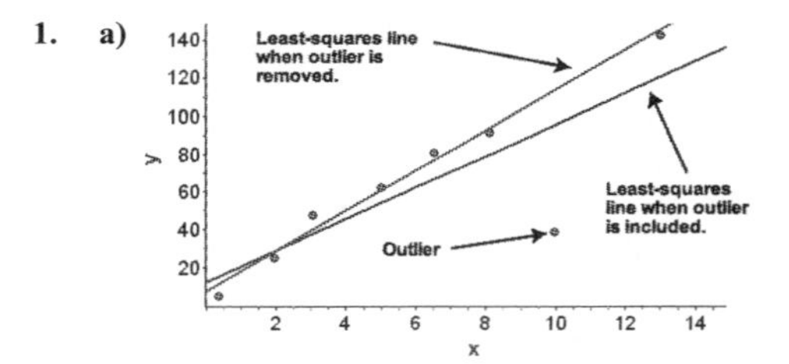

b)

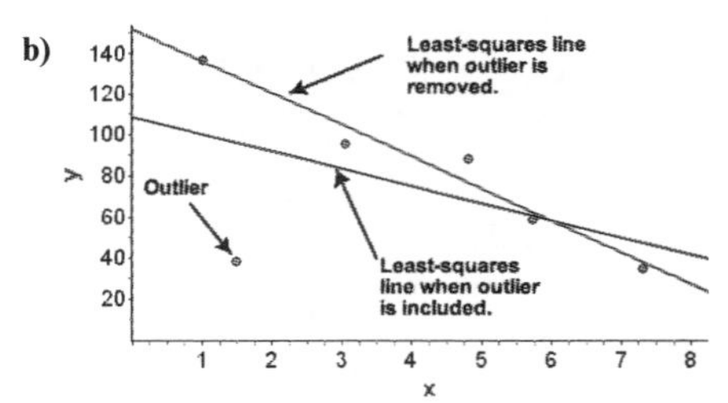

2. a) Select all the data. From the **Charts** menu, choose **Scatter Plot**. Then, select the actual data points on the graph and right-click them. Choose **Add Trendline** (you can choose to show the equation and the coefficient of determination as well). The correlation can be found either by using the **CORREL** formula or by taking the square root of the coefficient of determination. In this case

$r = 0.636$, so the relationship is moderate, positive, and linear. The line of best fit is $y = 0.2648x + 14.666$.

b) Enter the data into **L1** and **L2**. Turn diagnostics on by pressing [2nd] **0**, scrolling down to **DiagnosticsOn** and pressing [ENTER] twice. To get the regression, press [STAT] and select **CALC 4: LinReg (ax + b)**. Then, press [VARS] and select **Y-VARS 1: Function**. Finally, press [Y1] and [ENTER] twice. This will do the regression and will enter the formula into Y1 so you can graph it easily. To see the graph, press [Y=], make sure **Plot1** is selected, and use **ZOOM:9** to display the data. Since $r = -0.925$, the relationship is strong, negative, and linear. The equation of the least-squares line is $y = -8.2x + 43.3$.

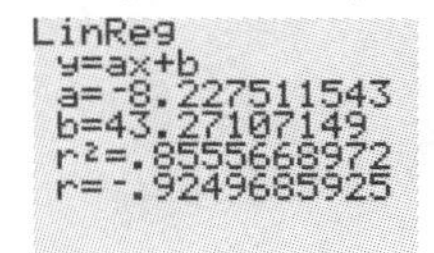

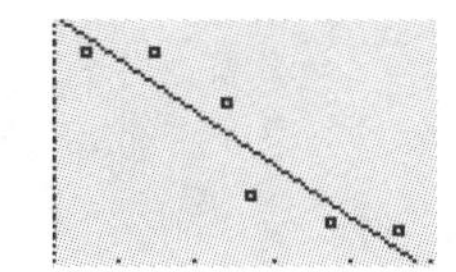

c) Enter the data into a new case table. Create a scatter plot and right-click the plot. Select **Least Squares Line**. The equation of the line of best fit and the coefficient of determination will be shown below the graph. Note that the r^2 value will be accurate to two decimal places and thus, so will the correlation. For the purposes of this question, take the square root to get $r = 0.95$. The correlation is strong, positive, and linear, and the equation is $y = 2.03x + 37$.

3. **a)** As temperature decreases, amount of gas consumed increases (relationship is strong, negative).
b) $y = -36.4x + 695$; The relationship is strong, negative, and linear, with temperature as the independent variable.
c) The slope, –36.4, represents change in gas consumed for each degree change in temperature. Since the slope is negative, gas consumption decreases by 36.4 m^3 for each degree increase in temperature.
d) 1241 m^3
e) −215 m^3, which is not realistic, because you cannot use a negative amount of gas. The model is good only until the amount of gas needed is zero. This makes sense, because when the temperature is warm enough, gas heat is not needed.

4. **a)**

	Set I	Set II	Set III	Set IV
i)	9.00	9.00	9.00	9.00
ii)	7.50	7.50	7.50	7.50
iii)	3.32	3.32	3.32	3.32
iv)	2.03	2.03	2.03	2.03
v)	0.82	0.82	0.82	0.82
vi)	$y = 0.5x + 3$	$y = 0.5x + 3$	$y = 0.5x + 3$	$y = 0.5x + 3$

b) It would appear that each set represents the same relationship.
c) The fact that each set of data and its scatter plot are completely different says several things: 1) Graphical representation of data helps the reader understand the numbers. 2) Even if a correlation is strong, the model may not be correct. 3) Outliers can have a great effect on the line of best fit.

5. **a)** strong, positive for both
b) For both, the scatter plot, equation of line of best fit, and correlation coefficient show a strong, positive relationship: volume versus beads correlation is 0.998, mass versus beads correlation is 1.000. (Note: Fathom™ rounds r^2 to two decimal places, so it will indicate that these are both perfect correlations when, in fact, they are not.)

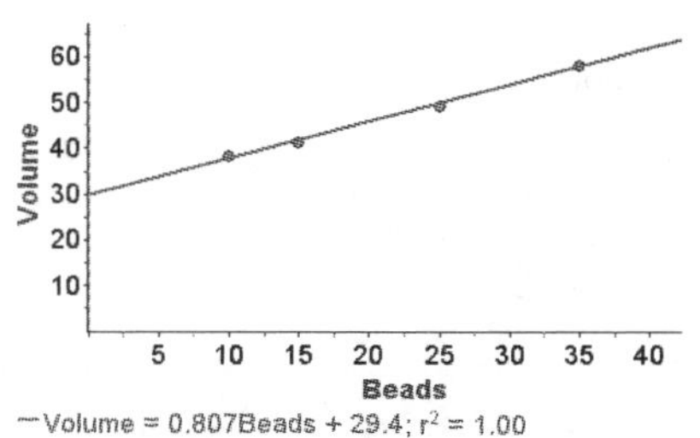

—Volume = 0.807Beads + 29.4; r^2 = 1.00

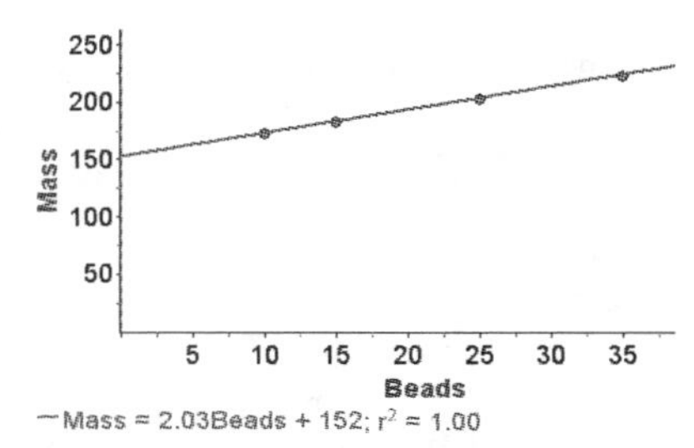

—Mass = 2.03Beads + 152; r^2 = 1.00

c) The volume of one bead is approximately 0.8 mL with mass approximately 2 g
d) y-intercepts represent initial volume (29 mL) and initial mass (152 g)
e) Yes, there is a strong, positive, and linear relationship (almost perfect: $r = 0.998$). Make volume the independent variable, because when volume is zero, the cylinder has a mass.

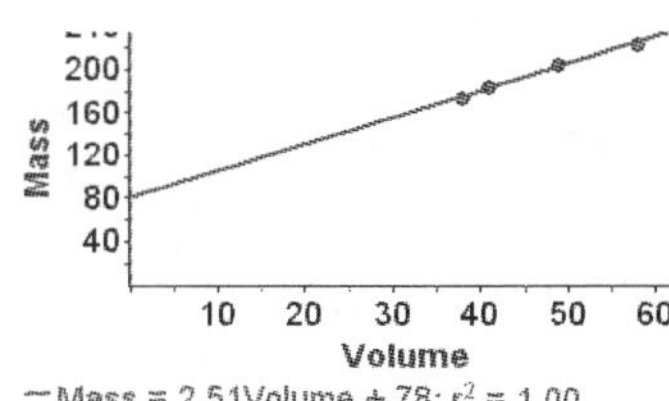

—Mass = 2.51Volume + 78; r^2 = 1.00

6. **a)** $r = 0.9997$, line of best fit is $y = 0.171x + 65$, where x is the day of the year and y is the number of videos, in millions
b) the number of videos, in millions, added each day
c) 189.8 million
d) in 5467 days, or about 15 years
e) Since the predictions are made from only three data points and extrapolated well beyond the range of values, they are probably not very accurate.

7. **a)**

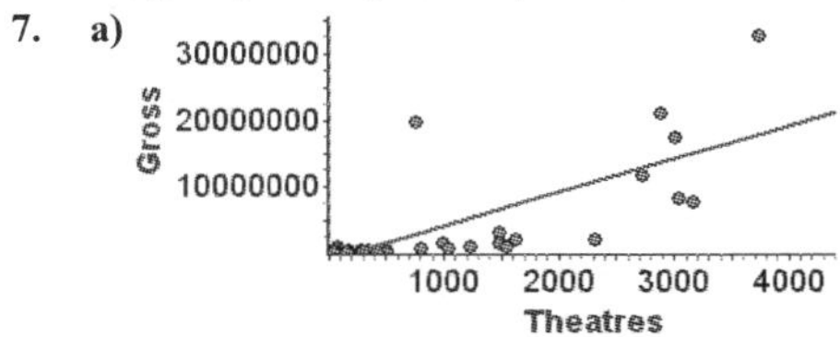

—Gross = 5020Theatres - 1300000; r^2 = 0.55

b) strong and positive, $r = 0.74$, $y = 5020x - 1\ 300\ 000$
c) Outliers include *Paranormal Activity* and *Where the Wild Things Are*. When they are removed, the correlation coefficient increases to $r = 0.81$. It could be argued that any of the top five movies are outliers. Removing them raises the correlation to $r = 0.88$.

8. a) Though the correlation is strong and positive with $r = 0.9$, the relationship could be described as non-linear, or as two separate linear functions. One least-squares line does not model the correlation very well, two linear correlations are better.

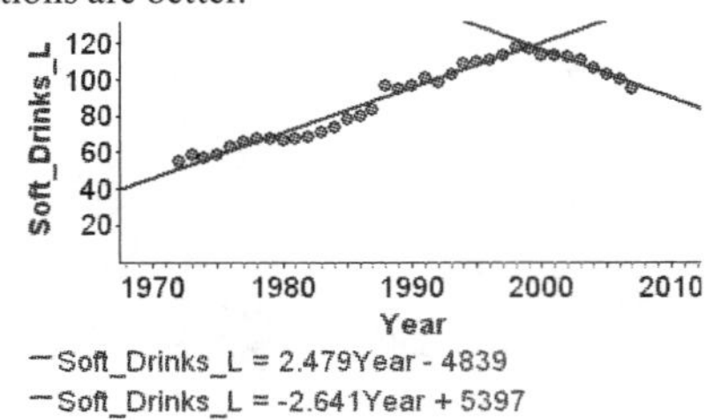

b) The line of best fit for time, x, and bottled water, y, is strong: $y = 1.8868x - 3746.4$, with $r = 0.99$. Soft drink consumption has had a fairly steady rise since the 1970s, but since about 2000 there has been a noticeable decline. Data begin in the mid 1990s for bottled water and show consumption increasing over time while soft drink consumption begins to fall.
c) For soft drinks, $y = 2.5695x + 5254.6$. Consumption of bottled water will surpass consumption of soft drinks in about 2020.
d) These data seem to show that soft drink consumption has fallen with the rise in bottled water consumption.

9. a) Though the regression is strong and positive ($r = 0.97$), a single linear model is not correct, since the data seem to be organized into two distinct lines.
b) Adding the BitRate attribute creates two separate regression models, one for each bit rate.

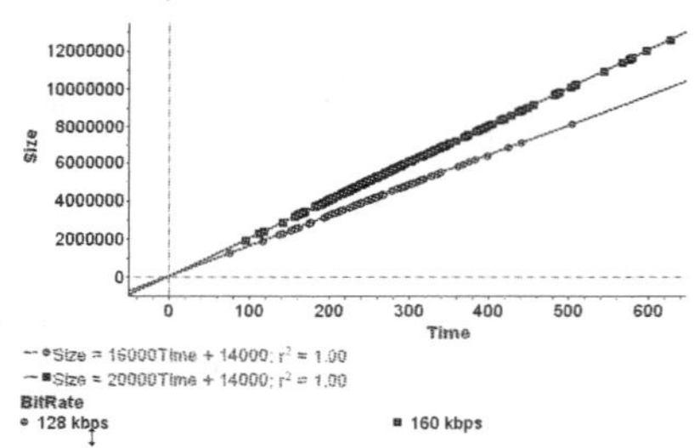

Bit rate is the compression rate of a music file. A CD normally has a bit rate of about 1440 kbps (kilobits per second). The average MP3 file has a bit rate of about 128 kbps. A higher bit rate means the file will be larger. So, it makes sense that the bit rate used would affect the storage space required for a song of a given time length.
c) The y-intercept is the size needed for a file that has no music but has metadata about the music (e.g. song name, length, etc). Intercepts are the same for both models, since most music files will use the same amount of space to store metadata.

10. a) The regression is strong and positive but is not really linear.

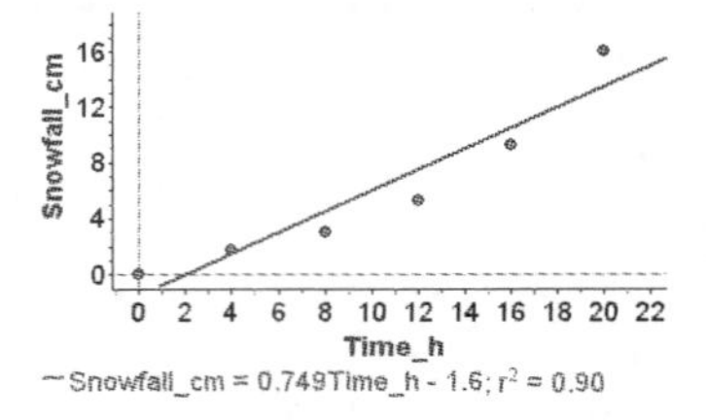

b) The pattern in the residual plot confirms that the set of data is non-linear.

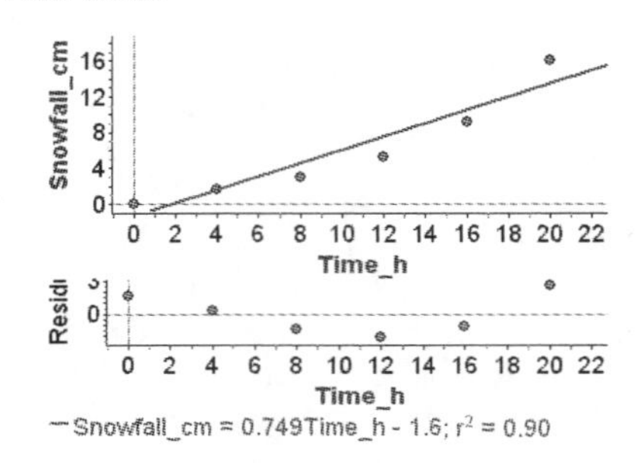

c) i) 17.21
ii)

Snowfall_cm
Time_h
—Snowfall_cm = 0.749Time_h - 1.6; r² = 0.90
Sum of squares = 17.21
—Snowfall_cm = a time_h²
Sum of squares = 270.4

iii)

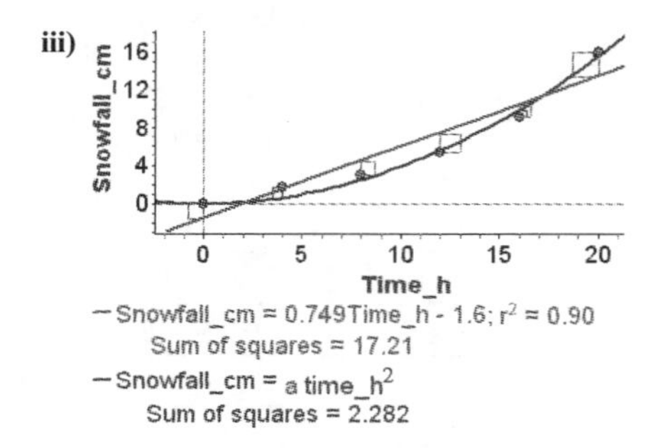

Equation of the curve of best fit is $y = 0.0388x^2$.
d) Since the non-linear sum (2.282) is less than the linear sum (17.21), the non-linear model fits the data better.
e) linear model: $y = 0.749(24) - 1.6$ or 16.4 cm, non-linear model: $y = 0.0388(24)^2$ or 22.3 cm. The non-linear model is more likely correct, because it fits the data better.

11. a)

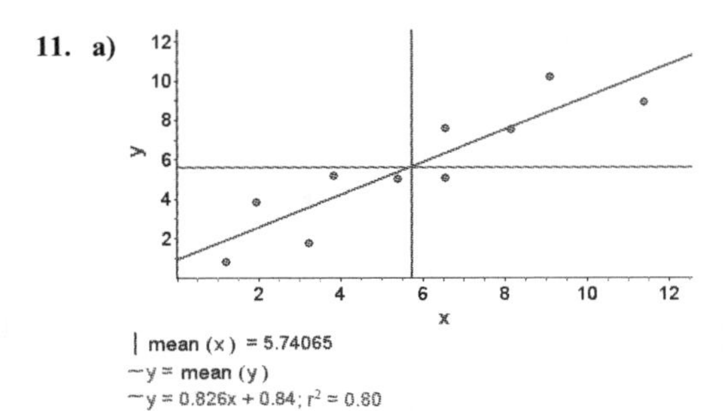

b) The vertical line marks the mean of the x-values, and the horizontal line marks the mean of the y-values. Their intersection is a new point, $(\bar{x}, \bar{y})$. **c)** The point represents the mean of x and the mean of y and therefore will always lie on the line of best fit.

3.4 Cause and Effect, pages 49–51

1. **a)** presumed **b)** cause-and-effect **c)** cause-and-effect **d)** accidental **e)** common-cause **f)** presumed **g)** reverse **h)** cause-and-effect
2. **a)** characteristic of normal human growth pattern **b)** environment in which reading is valued and encouraged **c)** inflation
3. **a)** The graph implies that as the number of pirates increases, average global temperature decreases. **b)** accidental
4. **a)** reverse **b)** There is a strong, positive correlation; however, it appears that the number of police was starting to level off and that the relationship is not actually linear. The line of best fit is $y = 0.008\ 75x + 29\ 279$, where x is the number of crimes and y the number of police. **c)** Beginning in 1991, the number of crimes and the number of police start to decline. As crime continues to fall, the number of police starts to rise. Surprisingly, the number of crimes begins to climb again along with increase in police. Eventually, the trend reverses to a point where the number of crimes in 2007 is below the number in 1986.

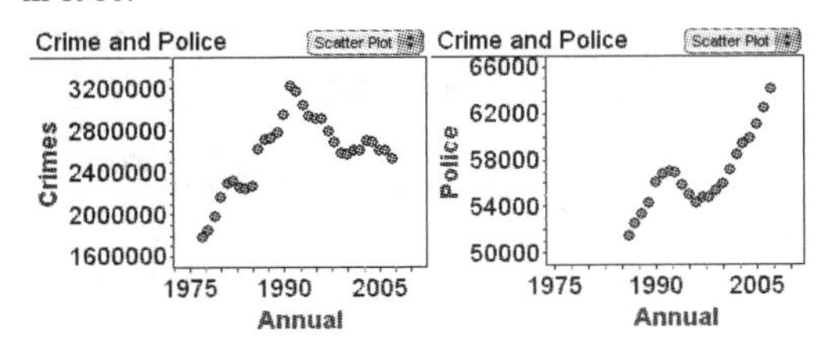

5. **a)** accidental **b)** The relationship is more likely common cause rather than accidental, based on the population increasing.
6. **a)** common cause (wealth of country) **b)** Denmark, with disproportionately more coasters (correlation goes from $r = 0.65$ to 0.70 when Denmark is removed) **c)** Canada sits close to the line of best fit and has the fifth greatest number of broadband connections.
7. **a)** The line of best fit is $y = 1212x + 89\ 500$. With correlation of $r = 0.72$, this is a strong positive relationship. The slope represents 1212 pregnancies per 100 000 cancer cases. **b)** Increase in pregnancies does not appear linked to an increase in incidence of cancer. Though both increase over time, there is a different pattern for pregnancies. The female population is very strongly correlated with the year. Cancer incidence is correlated exactly the same way with both year and female population, which suggests that its increase is more attributed to population growth. Pregnancy numbers do not seem to coincide well with either population or year. Thus, pregnancy versus cancer correlation is likely accidental.
8. **a)** cause-and-effect **b)** Fatality rate when a child safety seat is used properly is one quarter of the rate when none is used, just over one fifth of the rate when a child safety seat is used improperly, and almost one third of the rate when a belt is used improperly. So, it seems the original claim is correct. However, note that the fatality rate when a lap and shoulder belt is used is almost identical to the rate when a child safety seat is used. Although a child safety seat is effective, using lap and shoulder belts properly is equally effective.

Child Restraint Used	Fatalities	Total Accidents	Percent Fatalities
None Used/Not Applicable	1151	2680	42.9
Shoulder Belt	3	10	30.0
Lap Belt	79	553	14.3
Lap and Shoulder Belt	192	1754	10.9
Child Safety Seat	413	3858	10.7
Motorcycle Helmet	4	9	44.4
Restraint Used - Type Unknown	7	69	10.1
Safety Belt Used Improperly	14	39	35.9
Child Safety Seat Used Improperly	90	191	47.1
Unknown	105	521	20.2
Totals	2080	9852	21.1

c) Seat manufacturers might not want data disclosed because the information appears to suggest that a car seat is not needed as long as you correctly use the lap and shoulder belt. In fact, improper use of a car seat accounts for the highest fatality rate.

9. One example is that as the number of hours of radio listening goes up, life expectancy goes down. This is a strong, negative, linear correlation, but it is likely that both are more correlated with population.

3.5 Critical Analysis, pages 52–55

1. **a)** 17.6% **b)** 4.5% **c)** –11.1%
2. **a)** Assuming you work for 40 years, this is an extra \$50 000 per year. **b)** This is about \$636.66 per person. **c)** This is about \$5.56 per viewer. **d)** This is \$1169.59 per new cancer patient.

e) Assuming 25 years of actual riding, this is about 2.7 races a day with a win about every 2 days.
f) This is about $154 321 per game or $17 147 per inning.

3. **a)** CO_2 levels are definitely rising.
b) Beginning just after the time period represented in part a), there actually was a drop in emissions.
c) She will likely suggest that although the CO_2 level fluctuates on a yearly basis, in general levels have gone up during the last 45 years.
d) A sample that is too small has a large potential to misrepresent the data.
e) The least-squares line is $y = 1.4228x - 2477.1$, with a correlation of $r = 0.989$. So, in 2015, $y = 1.4228(2015) - 2477.1$ or 389.8. It might be that a curve is a better model for the data. This would mean that the linear estimate would underestimate the expected CO_2 level.

4. **a)** people who run these web sites and their advertisers
b) The biggest problem is that the entire graph is based on estimates. No real measurements are taken.
c) With a big jump in April 2009, Facebook surpassed Myspace as the most popular site and by the end of 2009 is on pace to have increased its membership by more than four times the 2007 level. In the same period, the number of Myspace users fell by more than a third.

5. **a)** There is only a weak correlation between the number of pole positions and points, and only a moderate correlation between the finish position and the points. Both top 5s and top 10s have strong relationships to points, but the top 10s have the stronger correlation ($r = 0.89$ versus $r = 0.80$).

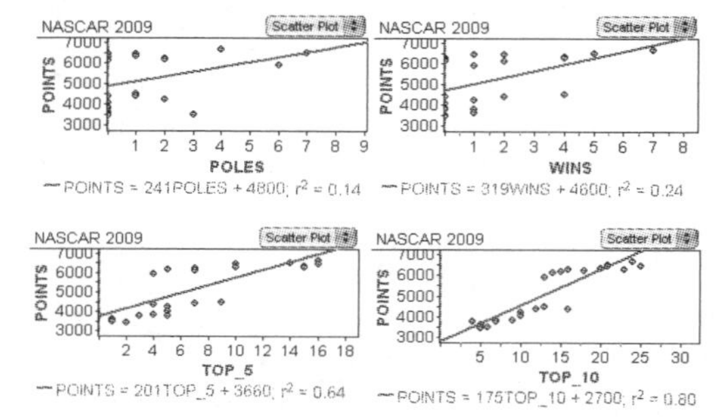

b) A single linear model is not correct, because there are clearly two sets of data. After 26 races of the 36-race season, any driver in the top 10 or within 400 points of the leader is placed in the "Chase for the Cup" series with a new point structure. However, when the data are separated into the two groups, a top-10 finish still has a higher correlation than a top-5 finish.

6. **a)** $r = -0.40$, which is barely moderate. The model is not the most appropriate, because there are two gender-based sets of data.

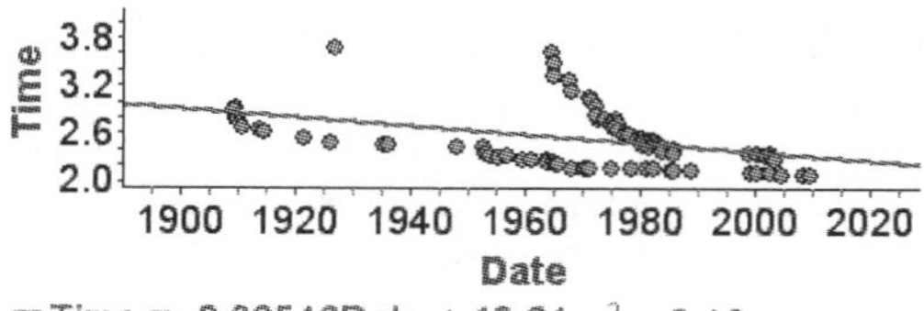

b)

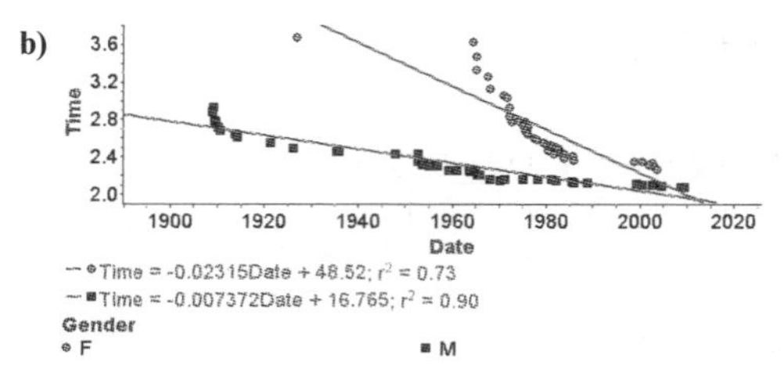

Based on these models, women and men will be running at the same speed by about 2018. However, the two linear models may not be adequate due to the rapid decline at the beginning of each set. A better approach might be to model only the more recent data. This gives an intersection date of approximately 2078 with a time of about 1:45. This is a large extrapolation and the time would mean running at a pace of 1 mi in less than 2 min, which is not realistic either!

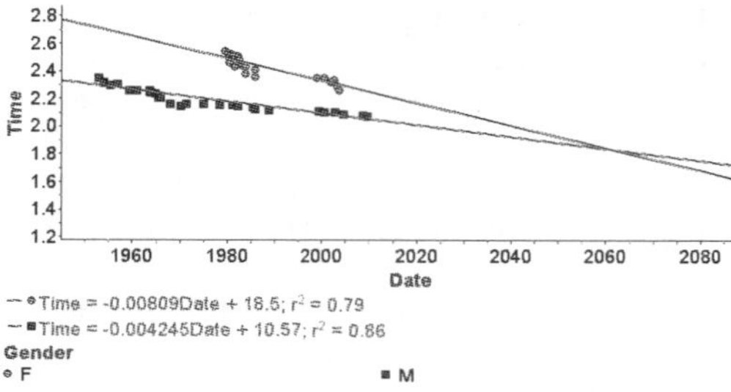

7. A better headline might be "British Columbians: The Most Active Canadians"

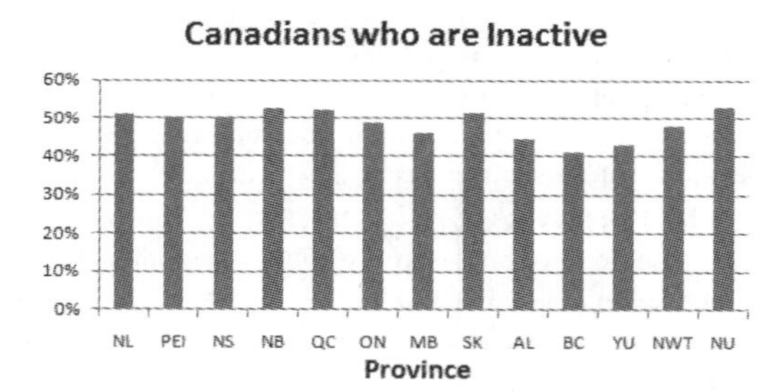

8. Answers will vary.

9. **a)** Answers will vary. **b)** In this case the model is strong, positive, and linear with a correlation of $r = 0.79$ and an equation for the line of best fit is $y = 0.501x + 14$.
c) A good guesser would be expected to have a correlation close to 1. Here, the correlation is $r = 1$ and the line of best fit is $y = x + 10$. A person consistently guessing over is not a good guesser but would have a perfect correlation, so correlation alone cannot determine the best guesser.
d) The slope represents the average relationship between the guessed and actual prices. For double the actual price, this would also be a perfect positive correlation with a line of best fit of $y = 2x$.
e) A person who is a good guesser would have a very strong correlation, with a slope close to 1 in the equation of the line of best fit and a y-intercept close to 0.

Chapter 4

4.1 Organized Counting, pages 56–60

1\.

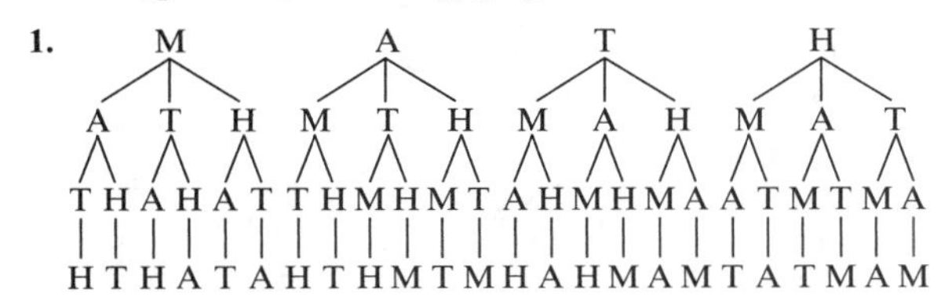

2\. 24

3\. a)

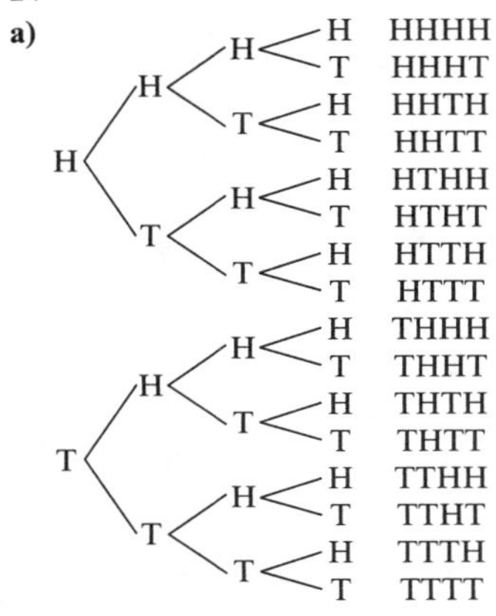

b) 6

4\.

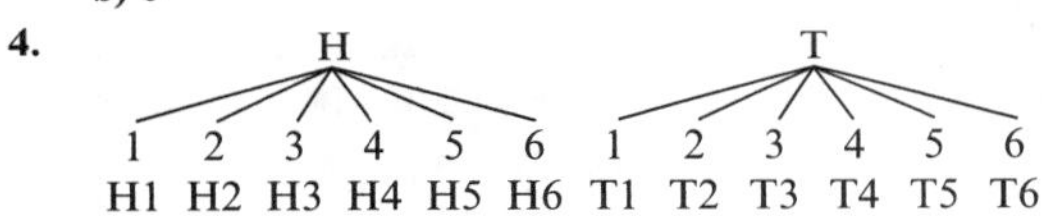

5\. 26^5 or 11 881 376

6\. 120

7\. 360

8\. a) 7 741 440 000 000
b) This gives users a unique gaming experience that encourages them to play more often and purchase new games by the same company.
c) Since the manufacturing process is automated, it is more cost effective to create packages and make larger numbers of each package. Customers can still order options individually but this will delay delivery and will be more expensive.

9\. 11 100

10\. 64 000

11\. a)

25 — 25 — 25: 75
25 — 25 — 10: 60
25 — 10 — 25: 60
25 — 10 — 10: 45
10 — 25 — 25: 60
10 — 25 — 10: 45
10 — 10 — 25: 45
10 — 10 — 10: 30

The possible point totals are 30, 45, 60, and 75.
b) The possible point totals are 0, 10, 20, 25, 30, 35, 45, 50, 60, and 75.

12\. a) 216 b) 384

13\. a) The total number of possible codes is 36^4 or 1 679 616. The site could have 679 616 more users.
b) The total number of possible codes would be 36^6 or 2 176 782 336. The site could have 2 175 782 336 more users.
c) The third letter is B, P, or R, so it could take up to 3×36 or 108 possible tries.

14\. a) 10 000 000 b) 1000 c) 2 100 000 d) 23 100 000

15\. 124 416

16\. Determine the total number of codes possible. Then, subtract the number of codes with three letters the same.
Total number of codes
$= 26 \times 10 \times 26 \times 10 \times 26 \times 10$
$= 17\ 576\ 000$
Number of codes with the same letter three times
$= 26 \times 10 \times 1 \times 10 \times 1 \times 10$
$= 26\ 000$
Number of codes without three repeated letters
$= 17\ 576\ 000 - 26\ 000$
$= 17\ 550\ 000$

17\. 20 736

18\. a) Only one finger can be on each of the six strings, so, there are $6 \times 5 \times 4 \times 3$ or 360 chords. The four fingers can move up the guitar neck in 18 positions. Each position will have 360 chords, so the total number of possible chords is 18×360 or 6480 chords.
b) At 2 s per chord, the time is $2 \times 6480 = 12\ 960$ s or 3.6 h.
c) This may underestimate the number of chords, since a guitarist is not restricted to using only one finger per fret, using only four adjacent frets, or always using each finger. On the other hand, not all combinations of strings and frets make musically valid chords.

19\. a) The total number of serial numbers is 36^{24} or 2.2×10^{37}. Each serial number requires 24 bytes, for a total of $24(2.2 \times 10^{37})$ or 5.4×10^{38} bytes or 4.9×10^{26} terabytes (1 terabyte = 2^{40} bytes) of information. This is more information than exists on all the computers in the world.
b) The characters bolded and underlined are the same in each serial number, so assume that only the other characters are variable. Wherever a letter appears in a serial number, only a letter appears in the same position in the other serial number. The same pattern appears to be true for digits.
VMT4-57XK-9**BBD**-2B3Y-VQ34-**BB46**
VMG3-52XT-6**BBD**-2P2Y-YQ31-**BB46**
VMR1-58XQ-8**BBD**-2R6Y-RQ39-**BB46**
Applying these conditions, the total number of serial numbers is $26 \times 10 \times 10 \times 26 \times 10 \times 26 \times 10 \times 26 \times 10$, or 4.6×10^{10}. At 24 bytes per serial number, this is 1.1×10^{12} bytes or 1 terabyte. This is about the size of a large computer hard drive, so clearly not all these serial numbers will be valid.

20\. Case 1: One scoop: $3 \times 36 = 108$,
Case 2: two scoops: $3 \times 36 \times 36 = 3888$,
Case 3: two scoops: $3 \times 36 \times 36 \times 36 = 139\ 968$ for a total of 143 964 cones.
Each of those can have up to 2 of 10 toppings:

Case 1: no topping = 1,
Case 2: one topping = 10,
Case 3: two toppings = 10 × 10, which is 111 possible toppings.
The total number of ice cream cones is 143 964 × 111 = 15 980 004, which is large but not quite 20 million.

4.2 Factorials and Permutations, pages 61–63

1. **a)** 720 **b)** 479 001 600 **c)** 7.11×10^{74}
2. **a)** False **b)** False **c)** True **d)** True
   e) False **f)** True **g)** True **h)** False
3. Most calculators will not calculate any value greater than $69! = 1.71 \times 10^{98}$.
4. **a)** $\frac{8!}{3!}$ **b)** $\frac{78!}{75!}$ **c)** $\frac{n!}{(n-4)!}$
5. **a)** $32 \times 31 \times 30 \times 29 \times 28 \times 27 \times 26 = 1.70 \times 10^{10}$
   b) $89 \times 88 \times 87 \times 86 \times 85 = 4\,980\,917\,040$
   c) $120 \times 119 = 14\,280$
   d) $10! = 3\,628\,800$
   e) $17! = 3.56 \times 10^{14}$
6. **a)** ${}_{12}P_3$ **b)** ${}_{32}P_9$ **c)** ${}_kP_z$ **d)** ${}_{52}P_8$
   e) ${}_{19}P_4$ **f)** ${}_7P_7$ **g)** ${}_{101}P_3$
7. **a) i)** 24 **ii)** 24 **iii)** 24
   b) The situations in part a) are all the arrangement of four distinct items.
8. **a) i)** AB, AC, AD, AE, BA, BC, BD, BE, CA, CB, CD, CE, DA, DB, DC, DE, EA, EB, EC, ED
   ii)

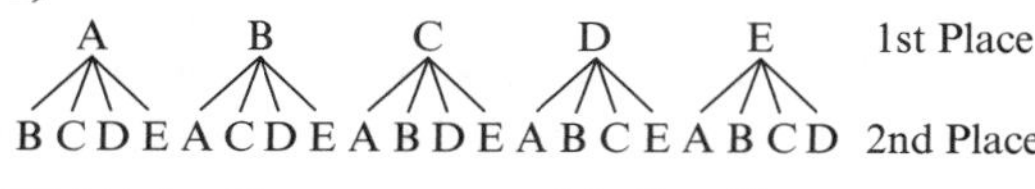

   iii) $5 \times 4 = 20$ **iv)** ${}_5P_2 = 20$
   b) permutations
   c) a tree diagram
9. $16!$ or 2.09×10^{13}
10. **a)** 479 001 600
    b) 7 257 600
    c) 7 257 600
11. **a)** When people are seated around a table, some of the arrangements that would exist if they were in a line are simply the same arrangement rotated.
    i) For example, there are 3! or 6 ways to arrange three people in a line. However, at a table, there are two sets of 3 that are just the same arrangement, rotated. So, in this case there are two distinct ways.

ABC BCA CAB	A, B, C around circles
ACB CBA BAC	A, C, B around circles

    ii) Similarly, when there are four people, there are 4! or 24 arrangements in a straight line but only 6 distinct ways to position them around a table. If you are not sure if these sets are the same, ask yourself the question "Is the same person on my right, on my left, and across from me?" If the answer is yes, then these are all the same arrangement.

ABCD BCDA CDAB DABC
ACBD CBDA BDAC DACB
ADBC DBCA BCAD CADB
ABDC BDCA DCAB CABD
ACDB CDBA DBAC BACD
ADCB DCBA CBAD BADC

    iii) For five people, there are 5! or 120 arrangements in a line, but 24 possible arrangements in a circle.
    b) Compare circular arrangements to arrangements in a line.

n	Arrangements in a Line	Arrangements Around a Circle
3	$3! = 6$	2
4	$4! = 24$	6
5	$5! = 120$	24

    In general, if n people are sitting around a table, the number of arrangements is $(n-1)!$
12. 24
13. **a)** 24 **b)** 9
14. 720
15. **a)** 5040 **b)** 151 200 **c)** 112
16. $0! = 1$. Sample explanation: Use the fact that ${}_nP_n = n!$ This represents the number of ways to arrange n items taken n at a time. So, $0! = {}_0P_0$ which is the number of ways to arrange 0 items if you do not take any of them. Even though you do not actually have any items, mathematically there is one way to not arrange any of them.
17. **a)** 215! **b)** 1.71×10^{23} **c)** 3.34×10^{20}
18. **a)** 57 120 **b)** 1260 **c)** 2520
19. **a)** $6! = 720$ **b)** $2 \times 5! = 240$ **c)** ${}_6P_5 = 720$
    d) $1 \times 2 \times {}_4P_3 = 48$
20. **a)** 9000 **b)** 4536 **c)** 154
21. $n < 7$
22. 8

23. **a)** $10 \times 8 \times 6 \times 4 \times 2 = 3840$
b) $17 \times 15 \times 13 \times 11 \times 9 \times 7 \times 5 \times 3 \times 1 = 34\,459\,425$
c) $9 \times 7 \times 5 \times 3 \times 1 \times 8 \times 6 \times 4 \times 2 = 9!$
$= 362\,880$

24. $70! = 70 \times 69!$
$= 70 \times 1.722 \times 10^{98}$
$= 1.198 \times 10^{2} \times 10^{98}$
$= 1.198 \times 10^{100}$

25. **a)** $\frac{(n+2)!}{n!} = \frac{(n+2)(n+1)n!}{n!}$
$= (n+2)(n+1)$

b) $\frac{(n+2)!}{n} = \frac{(n+2)(n+1)n(n-1)!}{n}$
$= (n+2)(n+1)(n-1)!$

c) $\frac{(2n)!}{(2n-2)!} = \frac{2n(2n-1)(2n-2)!}{(2n-2)!}$
$= 2n(2n-1)$

d) $\frac{(n+2)!}{(n-2)!} = \frac{(n+2)(n+1)n(n-1)(n-2)!}{(n-2)!}$
$= (n+2)(n+1)n(n-1)$

26. **a)** 30 **b)** 30 030 **c)** 510 510

4.3 Permutations With Some Identical Elements, pages 64–65

1. **a)** 360 **b)** 420 **c)** 4 989 600 **d)** 180 **e)** 831 600

2. **a)** **i)** LGOO, LOGO, LOOG, OLOG, OLGO, OOLG, OOGL, OGLO, OGOL, GLOO, GOLO, GOOL
ii)

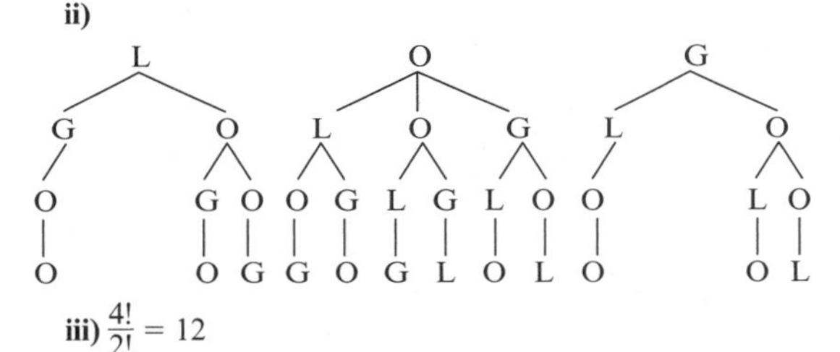

iii) $\frac{4!}{2!} = 12$
b) factorials **c)** tree diagram

3. **a)** 907 200 **b)** 30 **c)** 6

4. **a)** 362 880 **b)** 27.5 times more **c)** 45 times more

5. **a)** **i)** ${}_6P_4 = 360$
ii) Once you have decided which three colours to use, there are $\frac{4!}{2!}$ or 12 ways to arrange them. You need to know how many different ways to have three colours. Write them out, using B for blue and B_K for black: RWB, RWB_K, RWY, RWG, RBB_L, RBY, RBG, RB_KY, RBG, RYG, WBB_K, WBY, WBG, WB_KY, WB_KG, WYG, BB_KY, BB_KG, BYG, and B_KYG. There are 20 ways and each one can be arranged 12 ways. Also there are 3 possibilities for which of the colours will be the repeated one. Hence, there are $20 \times 12 \times 3 = 720$ possible codes with three colours.
iii) There are 15 ways to select the two colours, so there are $15 \times 6 = 90$ possible codes with two pairs of colours. However, if there are three of one colour and one of the other, there will be $\frac{4!}{3!}$ or 4 ways to arrange them. There are 2 choices for which of the two colours will be the triplet. So, there are $15 \times 4 \times 2 = 120$ possible codes with a triple and a single.
Overall, there are $90 + 120 = 210$ possible codes if two different colours are used.
b) Four pegs with 6 possible colours for each gives $6^4 = 1296$. This is 6 more than the sum of the answers in part a), since there are also 6 codes using only one colour.
c) **i)** Regardless of the type of peg, since there are four correct colours, there are 4! or 24 possible moves.
ii) Regardless of the pegs, there are $\frac{4!}{2!}$ or 12 possible next moves.

6. **a)** 24 **b)** 2520 **c)** 90

7. You need a total of 36 codes for the 26 letters and 10 digits. Using one or two blocks, there are a total of 11 codes, 8 of which are listed in the sample.

# of Characters	Reasoning	# of Codes
1	Since there are only three distinct colours, there are only 3 one-letter codes.	3
2	Case 1: two different colours, ${}_3P_2 = 6$ codes Case 2: two colours the same, 2 codes	8
3	Case 1: three different colours, ${}_3P_3 = 3! = 6$ Case 2: one colour repeated, one unique. Four different sets of colours are possible in this case: RRB, RRG, BBR, BBG and each can be arranged $\frac{3!}{2!}$ or 3 ways, for a total of $3 \times 4 = 12$	18
4	Case 1: two pairs of colours, $\frac{4!}{2!2!} = 6$ Case 2: one pair and two unique. Two sets of colours are possible in this case: RRBG and BBRG. Each can be arranged $\frac{4!}{2!} = 12$ ways, for a total of $12 \times 2 = 24$	30

With up to three blocks, there would be only 29 characters possible. Using four blocks, there would be 59 characters possible.

4.4 Pascal's Triangle/4.5 Applying Pascal's Method, pages 66–68

1. **a)** 1 9 36 84 126 126 84 36 9 1; row 9
b) row 4, element 3 **c)** $t_{8,6}$
d) 21; The two values above add to 21, the row is symmetrical and 21 appears in the matching place elsewhere in the row, and 21 adds with the 7 before it to give the 28 in the row below.
e) Each row is symmetrical. When there is an odd number of elements in the row, the centre element is unique; otherwise there are an even number with the same elements on each side.

2. a) The sum of a selection of values down a diagonal is the same number as the value just below and to the right or left, to form a "hockey stick."
b) $1 + 3 + 6 + 10 + 15 + 21 = 56$
3. In a row in which the first element is prime, each element, except for the first and last elements, is a multiple of the row number.
4. a) 32 b) 72
5. a) 53 b) 450
6. a)

Flip	Possible Outcomes	Number of Each Outcome
1	1H, 1T	1, 1
2	2H, 1H1T, 2T	1, 2, 1
3	3H, 2H1T, 2T1H, 3T	1, 3, 3, 1
4	4H, 3H1T, 2H2T, 1H3T, 4T	1, 4, 6, 4, 1
5	5H, 4H1T, 3H2T, 3T2H, 4T1H, 5T	1, 5, 10, 10, 5, 1

b) The numbers of each different possible outcome correspond to the values in Pascal's triangle.
7. a)

Circles	Points	Line Segments	Triangles	Quadrilaterals	Pentagons
1	1				
1	2	1			
1	3	3	1		
1	4	6	4	1	
1	5	10	10	5	1

b) The entries correspond to the values in Pascal's triangle.
c) 28
8.–9. Answers will vary.

Chapter 5

5.1 Organized Counting With Venn Diagrams, pages 69–71

1. a) b)

c) d)
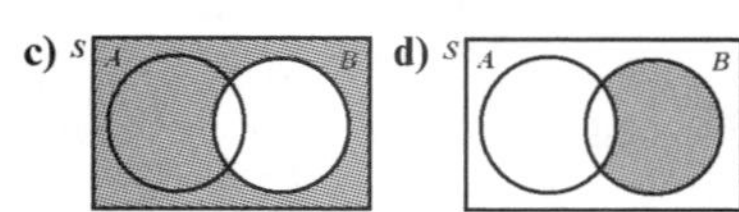

2. a) {10} b) {1}
c) {2, 3, 4, 6, 8, 9, 10, 12, 14, 15, 16, 18, 20}
d) {12}
e) {6, 12, 18}
f) {Ø}
3. a) Start with the most specific information, that there are 4 shoes that have JumpGoo soles and are basketball shoes. So, 4 is in the intersection of the JumpGoo and basketball circles. Next, because 12 shoes have JumpGoo soles but 4 of those are basketball shoes, then 8 are shoes that just have JumpGoo soles. Likewise, there are 14 basketball shoes but 4 of them have JumpGoo soles, so there are 10 that are just basketball shoes. Finally, since there are 60 styles of shoes altogether, but 8 + 4 + 10 or 22 of them are either basketball shoes or have JumpGoo soles, then the number of remaining shoes is 60 − 22 or 38.

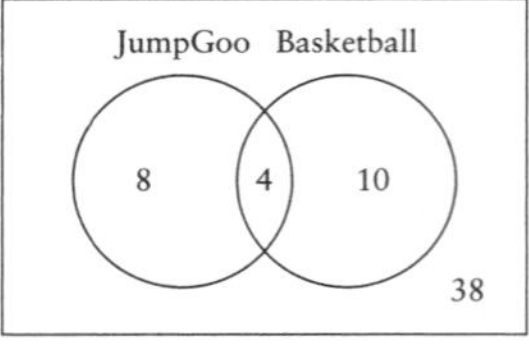

b) There are 38 models that are not basketball shoes, nor do they have JumpGoo soles.
4. a)

b) Since there are only five people who either like or do not mind working midnights and there are seven days in the week, then those employees likely will get more hours.
5.

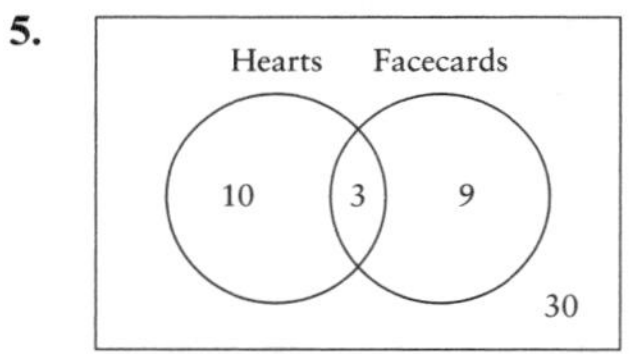

6. a)

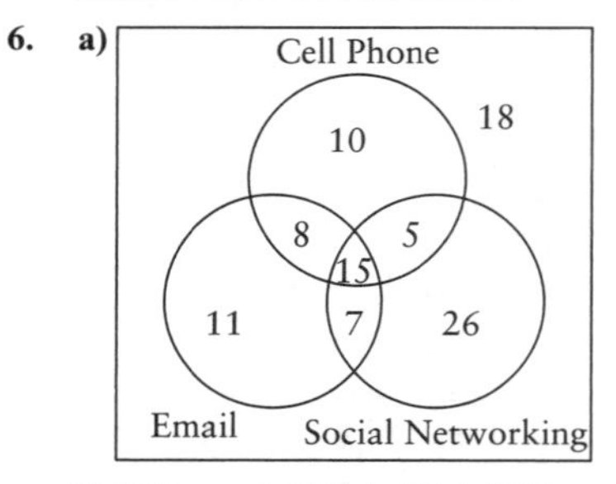

b) 10% c) 10% + 5% + 26% or 41%
d) 10% + 5% + 15% + 8% + 7% + 11% or 56%
7. When producing the Venn diagram for this situation, it might seem that there should be four subsets. Only two are needed. Make males one subset. Then females are anything outside that subset.

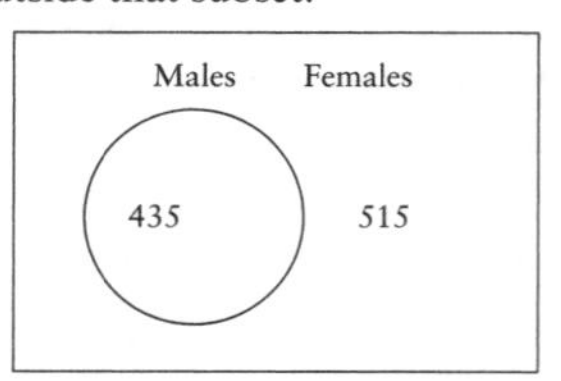

Similarly, you could use participants as one subset, and non-participants would be whatever is left.

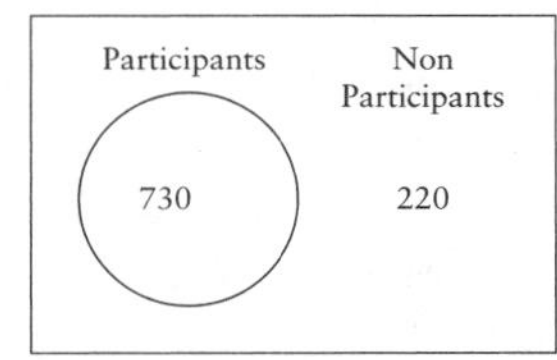

In both of the previous cases, there is the same total of 950 students. Now the two diagrams are put together.

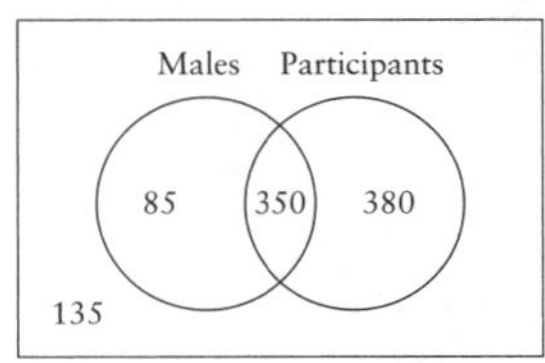

8. In terms of technology, females use the Internet and cell phones more. The company should target males.

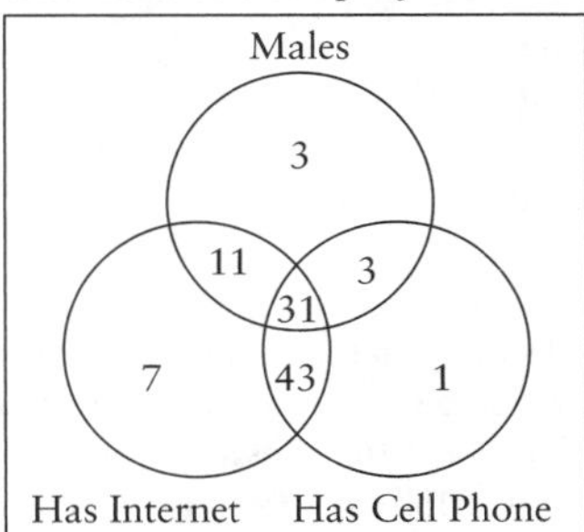

5.2 Combinations, pages 72–75

1. and **2. a)** 120 **b)** 1287 **c)** 635 013 559 600

3. a) $\binom{10}{4}$ **b)** $\binom{12}{8}$ **c)** $\binom{20}{6}$ **d)** $\binom{25}{18}$ **e)** $\binom{15}{6}$ **f)** $\binom{32}{8}$

4. For a combination lock, the order of the numbers matters. However, order does not matter when dealing with combinations in mathematics.

5. a) 2 **b)** 5 **c)** 9 **d)** 17 **e)** 32 **f)** 49

6. ${}_nC_r$ means choose r things from n of them. There is no way to choose more things than there are, so r cannot be greater than n.

7. a) i) RGB, RGY, RGO, RBY, RBO, RYO, GBY, GBO, GYO, BYO

ii)

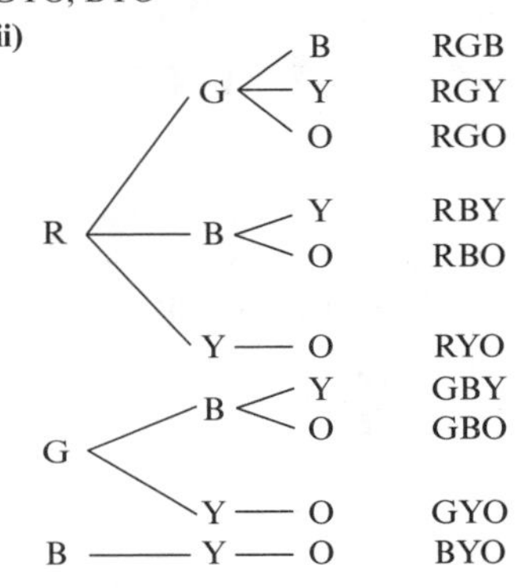

iii) ${}_5C_3 = 10$

b) tree diagram **c)** combinations

8. a) i) 45 **ii)** 45 **iii)** 45

b) Though the contexts are different, the questions ask the same thing: how many ways can a pair be made from ten items?

9. a) L.S. $= C(8, 5) = \frac{8!}{(8-5)!5!} = \frac{8!}{3!5!}$

R.S. $= C(8, 3) = \frac{8!}{(8-3)!3!} = \frac{8!}{5!3!}$

b) L.S. $= C(13, 7) = \frac{13!}{(13-7)!7!} = \frac{13!}{6!7!}$

R.S. $= C(13, 6) = \frac{13!}{(13-6)!6!} = \frac{13!}{7!6!}$

c) L.S. $= C(19, 2) = \frac{19!}{(19-2)!2!} = \frac{19!}{17!2!}$

R.S. $= C(19, 17) = \frac{19!}{(19-17)!17!} = \frac{19!}{2!17!}$

d) L.S. $= C(n, r) = \frac{n!}{(n-r)!r!}$

R.S. $= C(n, n-r) = \frac{n!}{(n-[n-r])![n-r]!} = \frac{n!}{r!(n-r)!}$

10. a) ${}_{52}C_{13} = 635\ 013\ 559\ 600$

b) For the first player there are 52 cards to choose from, of which the person gets 13; that is, ${}_{52}C_{13}$ ways. For the second player there are now only 39 cards, of which the person gets 13; that is, ${}_{39}C_{13}$ ways. For the third player there are now only 26 cards, of which the person gets 13; that is, ${}_{26}C_{13}$ ways. The fourth player gets the remainder of the cards; that is, ${}_{13}C_{13}$ or 1 way.

So, the number of different 13-card hands that can be dealt to four players is ${}_{52}C_{13} \times {}_{39}C_{13} \times {}_{26}C_{13} \times 1$ or 5.36×10^{28}. This is true whether the cards are handed out one at a time to each player or all at once to each player.

c) Part a) looks only at how many groups of 13 can be made. In each case, for every set of 13 cards, there is a remaining set of 39 cards. The process goes further in part b) in that the remaining 39 cards for each original set of 13 are then used to make another set of 13, with 26 cards remaining. These 26 are used to create another set of 13, leaving the last 13 cards for the last player.

11. a) 35 **b)** 15 **c)** 127

12. a) 20 475 **b)** 4410
c) 9310 **d)** 20 440

13. a) $({}_{24}C_4 \times {}_{20}C_4 \times {}_{16}C_4 \times {}_{12}C_4 \times {}_8C_4 \times {}_4C_4) \div 6!$
$= 4.51 \times 10^{12}$ (Divide by 6!, since here, for the groups, order does not matter.)

b) 1.37×10^{11}

14. a) ${}_{50}C_{10} = 10\ 272\ 278\ 170$

b) Ways of choosing the groups of 10:

${}_{20}C_5$	$\times$	${}_{30}C_5$
Five of the ten come from the athletes.		The last five come from the non-athletes.

$= 15\ 504 \times 142\ 506$
$= 2\ 209\ 413\ 024$

c) There are two ways to do this. Both use cases. One way is to figure out each of the cases explicitly for 3, 4, 5, ..., 10 athletes. The other way is to figure out the cases for 0, 1, and 2 athletes (the opposite of what is needed) and subtract them from the unrestricted case. The second method requires less work.

Case 1: no athletes

$_{30}C_{10}$ = 30 045 015
All ten come from the non- athletes

Case 2: one athlete

$_{20}C_1$ × $_{30}C_9$
One of the ten comes from the athletes. / The last nine come from the non-athletes

= 20 × 14 307 150
= 286 143 000

Case 3: two athletes

$_{20}C_2$ × $_{30}C_8$
Two of the ten come from the athletes. / The last eight come from the non-athletes

= 190 × 5 852 925
= 1 112 055 750

Thus, the total number of ways to have at least three athletes is 10 272 278 170 − (30 045 015 + 286 143 000 + 1 112 055 750) = 8 844 034 405.

15. **a)** 4368 **b)** 9100 **c)** 5940
16. 3160 days
17. **a)** 252 **b)** 140 **c)** 126
18. **a)** $({}_{16}C_2 \times {}_{14}C_2 \times {}_{12}C_2 \times {}_{10}C_2 \times {}_{8}C_2 \times {}_{6}C_2 \times {}_{4}C_2 \times {}_{2}C_2) \div 8! = 2\,027\,025$
b) First you need to choose which of the 15 teams gets a bye. Then proceed as in part a) with the remaining 14 teams, to get 2 027 025.
19. **a)** 455, **b)** 198 **c)** 180
20. **a)** 20 358 520 **b)** 1.91×10^{14}
c) 15 **d)** 6 points **e)** 12 points
21. **a)** 360 **b)** 20 **c)** 18 **d)** 758

5.3 Problem Solving With Combinations, pages 76–77

1. 63
2. 59
3. **a)** 2, 3, 5, 7
b)

# of Factors	Factors
1	1, 2, 3, 5, 7, 210
2	2 × 3 = 6, 2 × 5 = 10 2 × 7 = 14, 3 × 5 = 15 3 × 7 = 21, 5 × 7 = 35
3	2 × 3 × 5 = 30, 2 × 3 × 7 = 42, 2 × 5 × 7 = 70, 3 × 5 × 7 = 105
4	2 × 3 × 5 × 7 = 210

The number of different factors is 16.
c) $2^4 = 16$
4. **a)** 216 ways **b)** 384 days
5. **a)** 2 × 2 × 2 × 3 × 3 × 5 = 360 **b)** 24
6. **a)** Case 1: 1 boy, 3 girls

$_7C_1$ × $_5C_3$ = 7 × 10
One of the 4 comes from the 7 boys. / The other 3 come from the 5 girls.

Case 2: 2 boys, 2 girls

$_7C_2$ × $_5C_2$ = 21 × 10
Two of the 4 come from the 7 boys. / The other two come from the 5 girls

Case 3: 3 boys, 1 girl

$_7C_3$ × $_5C_1$ = 35 × 5
Three of the 4 come from the 7 boys. / The other one comes from the 5 girls

Total = 70 + 210 + 175 or 455
b) Case 1: unrestricted, $_{12}C_4 = 495$
Case 2: 4 boys, $_7C_4 = 35$
Case 3: 4 girls, $_5C_4 = 5$
Total = 495 − 35 − 5 or 455
7. 12 people
8. 22 548 578 304
9. **a)** 74 **b)** 5476
10. **a)** 96 **b)** 7 toppings
11. **a)** There are 361 places to put stones and three possibilities for each place (white stone, black stone, no stone), so there are a total of 3^{361} or 1.74×10^{172} possible boards.
b) $1.74 \times 10^{172} - 4.63 \times 10^{170} = 1.69 \times 10^{172}$
12. **a)** 60 **b)** 50 **c)** 44
13. 12 perfect square factors
14. Consider ten apples in a row: AAAAAAAAAA. To divide them into five piles you need four dividers (e.g., A|AA|AAA|AAA|A). There are nine possible places for dividers, so there are $_9C_4 = 126$ ways to divide the apples up.
15. 270

5.4 The Binomial Theorem, pages 78–80

1. **a)** $_{11}C_6 + {}_{11}C_7$ **b)** $_{15}C_{12}$ **c)** $C(20, 7) + C(20, 8)$
d) $C(9, 4)$ **e)** $\binom{9}{6} + \binom{9}{7}$ **f)** $\binom{17}{6}$

2. **a)**

$$\text{L.S.} = {}_{18}C_{15} = \frac{18!}{(18-15)!15!} = \frac{18!}{3!15!}$$

$$\text{R.S.} = {}_{17}C_{14} + {}_{17}C_{15} = \frac{17!}{(17-14)!14!} + \frac{17!}{(17-15)!15!} = \frac{17!}{3!14!} + \frac{17!}{2!15!} = \frac{17!}{3!14!}\left(\frac{15}{15}\right) + \frac{17!}{2!15!}\left(\frac{3}{3}\right) = \frac{15(17!) + 3(17!)}{3!15!} = \frac{18(17!)}{3!15!} = \frac{18!}{3!15!}$$

b)

$$\text{L.S.} = C(23, 10) \qquad \text{R.S.} = C(22, 9) + C(22, 10)$$

$$= \frac{23!}{(23-10)!10!} \qquad = \frac{22!}{(22-9)!9!} + \frac{22!}{(22-10)!10!}$$

$$= \frac{23!}{13!10!} \qquad = \frac{22!}{13!9!} + \frac{22!}{12!10!}$$

$$= \frac{22!}{13!9!}\left(\frac{10}{10}\right) + \frac{22!}{12!10!}\left(\frac{13}{13}\right)$$

$$= \frac{10(22!) + 13(22!)}{13!10!}$$

$$= \frac{23(22!)}{13!10!}$$

$$= \frac{23!}{13!10!}$$

c)

$$\text{L.S.} = \binom{32}{25} - \binom{31}{24} \qquad \text{R.S.} = \binom{31}{25}$$

$$= \frac{32!}{(32-25)!25!} - \frac{31!}{(31-24)!24!} \qquad = \frac{31!}{(31-25)!25!}$$

$$= \frac{32!}{7!25!} - \frac{31!}{7!24!} \qquad = \frac{31!}{6!25!}$$

$$= \frac{32(31!)}{7!25!} - \frac{31!}{7!24!}\left(\frac{25}{25}\right)$$

$$= \frac{32(31!) - 25(31!)}{7!25!}$$

$$= \frac{7(31!)}{7(6)!25!}$$

$$= \frac{31!}{6!25!}$$

3. **a)** 12 **b)** 17 **c)** 10
4. **a)** 5 **b)** 12 **c)** 7 **d)** 2 or 10
5. **a)** $x^{15} + 15x^{14}y + 105x^{13}y^2 + 455x^{12}y^3 + 1365x^{11}y^4 + 3003x^{10}y^5 + 5005x^9y^6 + 6435x^8y^7 + 6435x^7y^8 + 5005x^6y^9 + 3003x^5y^{10} + 1365x^4y^{11} + 455x^3y^{12} + 105x^2y^{13} + 15xy^{14} + y^{15}$
   b) Their sum is always 15. **c)** Their sum will always be n.
6. **a)** 715 **b)** 91 **c)** 480 700
7. **a)** 1 **b)** 3 **c)** after fourth person: 6 pairs; after fifth person: 10; after sixth person: 15; after seventh person: 21 **d)** These are the values in the third diagonal of Pascal's triangle.
8. **a)** $256x^8 + 1024x^7y + 1792x^6y^2 + 1792x^5y^3 + 1120x^4y^4 + 448x^3y^5 + 112x^2y^6 + 16xy^7 + y^8$
   b) $x^9 - 9x^8y + 36x^7y^2 - 84x^6y^3 + 126x^5y^4 - 126x^4y^5 + 84x^3y^6 - 36x^2y^7 + 9xy^8 - y^9$
   c) $1 + 24x + 240x^2 + 1280x^3 + 3840x^4 + 6144x^5 + 4096x^6$
   d) $243x^5 - 2025x^4 + 6750x^3 - 11250x^2 + 9375x - 3125$
   e) $256x^4 - 1792x^3y + 4704x^2y^2 - 5488xy^3 + 2401y^4$
   f) $19683x^{18} + 59049x^{16} + 78732x^{14} + 61236x^{12} + 30618x^{10} + 10206x^8 + 2268x^6 + 324x^4 + 27x^2 + 1$
9. **a)** 35
   b) i) 20 **ii)** 15 **iii)** 35
   c) This models Pascal's method, which gives ${}_7C_3 = {}_6C_2 + {}_6C_3$, or 15 + 20 = 35.
10. The rth term of the expansion of $(a + b)^n$ is given by ${}_nC_r a^{n-r}b^r$.
    a) In this case, $a = 7x$, $b = -5y$, and $n = 9$, so the 6th term is given by
    ${}_9C_6(7x)^{9-6}(-5y)^6 = 450\ 187\ 500x^3y^6$
    b) In this case, $a = 2x^3$, $b = 1$, and $n = 13$, so the 6th term is given by
    ${}_{13}C_6(2x^3)^{13-6}(1)^6 = 219\ 648x^{21}$
11. **a)** There are ${}_9C_1$ ways to start with $\angle$AOB and successively add adjacent angles. Similarly there are ${}_8C_1$ ways to start with $\angle$BOC and successively add adjacent angles in the clockwise direction. This continues with the rest of the angles to get a total of 45 angles.
    b) This is the second diagonal of Pascal's triangle. This problem can also be solved by selecting any two of the ten vertices to form an angle with vertex O. This can be done in ${}_{10}C_2$ or 45 ways.
12. **a)** 1 10 45 210 252 210 120 45 10 1
    b) 1 18 153 816 3060
    c) 15504 4845 1140 190
13. **a)** $(1.02)^{10} = (1 + 0.02)^{10}$;
    first four terms $\doteq$ 1.218 99
    b) $(0.999)^8 = (1 - 0.001)^8$;
    first four terms $\doteq$ 0.992 03
14. **a)** $(x + y)^5$ **b)** $(1 + x)^7$ **c)** $(2x - y)^4$ **d)** $(1 + 0.03)^6$
15. **a)**

Outcome	Number of Ways
5 correct, 0 wrong	1
4 correct, 1 wrong	5
3 correct, 2 wrong	10
2 correct, 3 wrong	10
1 correct, 4 wrong	5
0 correct, 5 wrong	1

    b) These are the values of the fifth row of Pascal's triangle.
16. **a)** ${}_{10}C_3(0.8)^7(0.2)^3 \doteq 0.201\ 33$
    b) ${}_6C_4(0.25)^2(0.75)^4 \doteq 0.296\ 63$
    c) ${}_{50}C_0(0.9)^{50} + {}_{50}C_1(0.9)^{49}(0.1)^1 + {}_{50}C_2(0.9)^{48}(0.1)^2 \doteq 0.111\ 73$
17. The general term is given by
    $${}_8C_k(3x^2)^{8-k}(-x^{-1})^k = {}_8C_k(3^{8-k})(-1)^{-k}(x^{16-2k})(x^{-k})$$
    $$= {}_8C_k(-1)^k(3^{8-k})(x^{16-3k})$$
    a) The term that is independent of x will have x^0. So, $16 - 3k = 0$ or $k = 5.3$. But k must be a natural number, so there is no term that is independent of x.
    b) The coefficient of each term is given by ${}_8C_k(-1)^k(3^{8-k})$. For this coefficient to equal $-13\ 608$, k must be an odd number. Check $k = 1, 3, 5, 7$. Successive trials yield $k = 3$, and so the term is $-13608(x^{16-3(3)})$ or $-13608x^7$.
18. 1.3048

Chapter 6

6.1 Basic Probability Concepts, pages 83–87

1. **a)** S = {each card in a standard deck of 52 cards}
   b) S = {1, 2, 3, 4, 5, 6, 7, 8, 9, 10}
   c) S = {any weight between 0 kg and 300 kg}
   d) S = {each name}
   e) S = {any time of day}
2. **a)** discrete **b)** discrete **c)** continuous
   d) discrete **e)** continuous

3. **a)** theoretical
b) subjective
c) experimental
d) theoretical
e) subjective

4. **a)** True. There is no 7 on a standard die.
b) False. There is an equal chance to get a tail.
c) False. Songs are selected randomly, and though the probability that a song will be repeated is low, it is not 0.
d) False. Two 6s have a sum of 12.

5. **a)** $\frac{26}{52} = \frac{1}{2}$; theoretical **b)** $\frac{2}{6} = \frac{1}{3}$; theoretical
c) $\frac{4}{20} = \frac{1}{5}$; theoretical **d)** $\frac{55}{80} = \frac{11}{16}$; experimental
e) $\frac{2}{10} = \frac{1}{5}$; experimental

6. **a)** $\frac{85}{861} = 0.099$ **b)** $\frac{98}{213} = 0.46$
c) if male: $\frac{422}{861} = 0.49$; if female: $\frac{439}{861} = 0.51$

7. **a)**

1 — 1: $1 + 1 = 2$; 2: $1 + 2 = 3$; 3: $1 + 3 = 4$; 4: $1 + 4 = 5$; 5: $1 + 5 = 6$; 6: $1 + 6 = 7$
2 — 1: $2 + 1 = 3$; 2: $2 + 2 = 4$; 3: $2 + 3 = 5$; 4: $2 + 4 = 6$; 5: $2 + 5 = 7$; 6: $2 + 6 = 8$
3 — 1: $3 + 1 = 4$; 2: $3 + 2 = 5$; 3: $3 + 3 = 6$; 4: $3 + 4 = 7$; 5: $3 + 5 = 8$; 6: $3 + 6 = 9$
4 — 1: $4 + 1 = 5$; 2: $4 + 2 = 6$; 3: $4 + 3 = 7$; 4: $4 + 4 = 8$; 5: $4 + 5 = 9$; 6: $4 + 6 = 10$
5 — 1: $5 + 1 = 6$; 2: $5 + 2 = 7$; 3: $5 + 3 = 8$; 4: $5 + 4 = 9$; 5: $5 + 5 = 10$; 6: $5 + 6 = 11$
6 — 1: $6 + 1 = 7$; 2: $6 + 2 = 8$; 3: $6 + 3 = 9$; 4: $6 + 4 = 10$; 5: $6 + 5 = 11$; 6: $6 + 6 = 12$

b) $S = \{2, 3, 4, 5, 6, 7, 8, 9, 10, 11, 12\}$ **c)** 11
d) $P(2) = \frac{1}{36}$, $P(3) = \frac{2}{36}$, $P(4) = \frac{3}{36}$, $P(5) = \frac{4}{36}$, $P(6) = \frac{5}{36}$, $P(7) = \frac{6}{36}$, $P(8) = \frac{5}{36}$, $P(9) = \frac{4}{36}$, $P(10) = \frac{3}{36}$, $P(11) = \frac{2}{36}$, $P(12) = \frac{1}{36}$
e) Sum $= \frac{1}{36} + \frac{2}{36} + \frac{3}{36} + \frac{4}{36} + \frac{5}{36} + \frac{6}{36} + \frac{5}{36} + \frac{4}{36} + \frac{3}{36} + \frac{2}{36} + \frac{1}{36} = 1$
f) $\frac{6}{36} \doteq 0.1667$

8. Sample answer: Historically, when there were conditions similar to those currently observed, it rained the next day 20% of the time. Value is based partly on experimental probability.

9. **a)** $\frac{14}{60} \doteq 0.233$ **b)** experimental **c)** 7

10. **a)**

Outcome	# of 6s	$P(6)$
1	0	$\frac{0}{1} = 0.00$
5	0	$\frac{0}{2} = 0.00$
2	0	$\frac{0}{3} = 0.00$
5	0	$\frac{0}{4} = 0.00$
3	0	$\frac{0}{5} = 0.00$
2	0	$\frac{0}{6} = 0.00$
4	0	$\frac{0}{7} = 0.00$
6	1	$\frac{1}{8} = 0.13$
4	1	$\frac{1}{9} = 0.11$
6	2	$\frac{2}{10} = 0.20$
2	2	$\frac{2}{11} = 0.18$
3	2	$\frac{2}{12} = 0.17$
1	2	$\frac{2}{13} = 0.15$
6	3	$\frac{3}{14} = 0.21$
2	3	$\frac{3}{15} = 0.20$
6	4	$\frac{4}{16} = 0.25$
1	4	$\frac{4}{17} = 0.24$
5	4	$\frac{4}{18} = 0.22$
5	4	$\frac{4}{19} = 0.21$
1	4	$\frac{4}{20} = 0.20$

b)

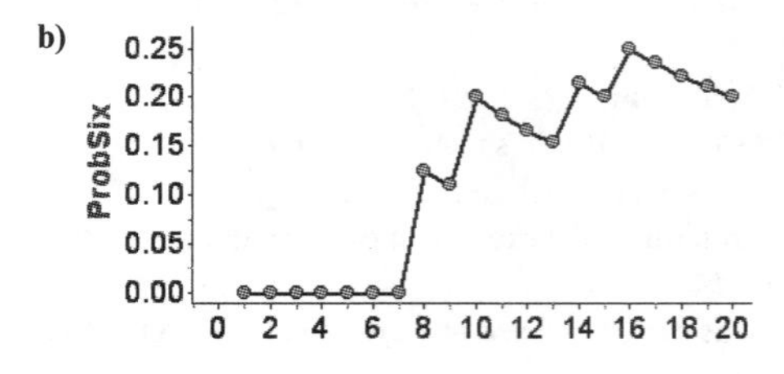

c) $P(6) = \frac{1}{6} = 0.17$ is near the experimental probability, but that probability is still too volatile to state that the values are close.

11. your sample space = {(1, 1), (1, 2), (1, 3), (2, 1), (2, 2), (2, 3), (2, 4), (3, 1), (3, 2), (3, 3), (3, 4), (3, 5), (4, 2), (4, 3), (4, 4), (4, 5), (4, 6), (5, 3), (5, 4), (5, 5), (5, 6), (6, 4), (6, 5), (6, 6)}, your friend's sample space = {(1, 4), (1, 5), (1, 6), (2, 5), (2, 6), (3, 6), (4, 1), (5, 1), (5, 2), (6, 1), (6, 2), (6, 3)}; You have the advantage: n(your sample space) = 24, n(your friend's sample space) = 12

12. **a)** S = {TTTT, TTTF, TTFT, TTFF, TFTT, TFTF, TFFT, TFFF, FTTT, FTTF, FTFT, FTFF, FFTT, FFTF, FFFT, FFFF}

b)

Situation	4F0T	3F1T	2F2T	1F3T	0F4T
Probability	$\frac{1}{16}$	$\frac{4}{16} = \frac{1}{4}$	$\frac{6}{16} = \frac{3}{8}$	$\frac{4}{16} = \frac{1}{4}$	$\frac{1}{16}$
Total	$\frac{1}{16} + \frac{4}{16} + \frac{6}{16} + \frac{4}{16} + \frac{1}{16} = \frac{16}{16} = 1$				

c) $P(\text{same}) = \frac{2}{16} = \frac{1}{8}$ **d)** $P(\text{at least 2T}) = \frac{11}{16}$

13 **a)**

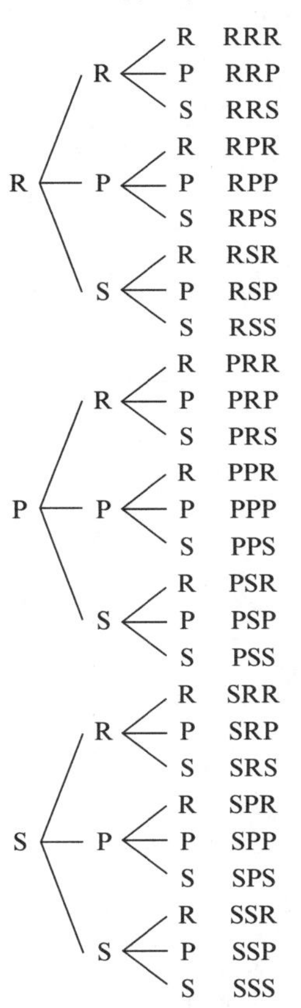

b) This rate is less than the expected probability: $P(\text{scissors}) = \frac{1}{3}$ or 33.3%;
$P(\text{scissors at least once in a game}) = \frac{19}{27}$ or 70.4%, which is more than twice the theoretical probability. **c)** Of nine cases where rock is thrown first, only four do not have repeated throws, so you must guess:
$P(\text{rookie sequence}) = \frac{1}{4}$ or 25%.

14. **a)**

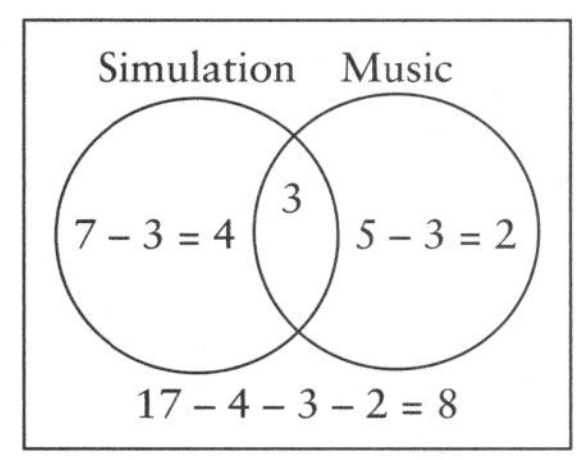

b) $P(\text{simulation not music}) = \frac{4}{17} = 0.235$ or 23.5%

c) $P(\text{not simulation or music}) = \frac{8}{17} = 0.47$ or 47%

15. **a)–b)**

	Probability	Average Number of Trials
i)	$\frac{1}{2} = 0.50$	{220, 300, 100, 200, 290} average = 222
ii)	$\frac{1}{6} \doteq 0.17$	{820, 1140, 1500, 680, 1800} average = 1188
iii)	$\frac{5}{36} \doteq 0.14$	{1100, 800, 900,1300, 1500} average = 1120
iv)	$\frac{1}{52} \doteq 0.02$	{3800, 1600, 2400, 6800, 3400} average = 3600

c) As the size of the sample space increases, the number of trials required before the experimental probability stabilizes around the theoretical probability also increases.

16.
$$P(\text{large prize}) = \frac{\text{area of inner circle}}{\text{total area}} = \frac{\pi r^2}{l^2} = \frac{3.14(1.5)^2}{10^2} = 0.071 \text{ or } 7.1\%$$

17. $P(\text{doubles or triples}) = \frac{96}{216}$ or 0.44

18. **a)** There are ${}_9C_3$ or 84 ways to choose three balls. Only 8 ways add to 15: {1 + 5 + 9, 1 + 6 + 8, 2 + 6 + 7, 2 + 4 + 9, 2 + 5 + 8, 3 + 5 + 7, 3 + 4 + 8, 4 + 5 + 6}. Therefore, $P(15) = \frac{8}{84}$ or 0.10.
b) 5-ball, since it appears more frequently (four times) than any other in the winning combinations
c) Sample answer: You can either choose from the numbers left to block your opponent or determine which of them have the highest probabilities of staying available.

6.2 Odds, pages 88–89

1. **a)** 1:3 **b)** 3:2 **c)** 3:2 **d)** 1:6 **e)** 7:13 **f)** 1:999

2. **a)** 1:49 **b)** 9:1 **c)** 21:5 **d)** 9:16 **e)** 5:1 **f)** 19:1

3. **a)** $\frac{3}{3+7} = 0.300$ **b)** $\frac{5}{5+35} = 0.125$ **c)** $\frac{5}{5+2} = 0.714$

4. **a)** 1:2.33 **b)** 1:7 **c)** 1:0.40

5. **a)**

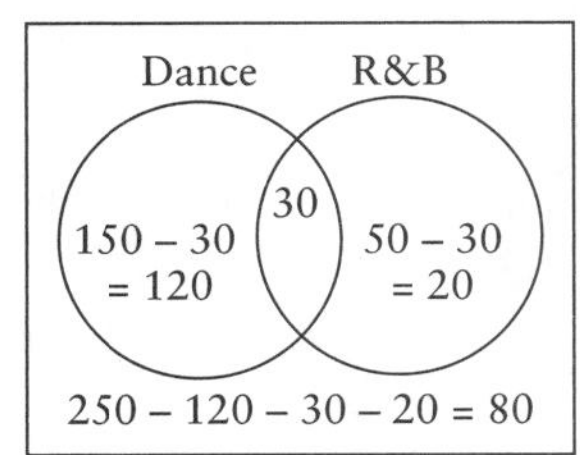

b) 12:13 **c)** 22:3

6. **a)** 1 : 12 **b)** 1 : 3 **c)** 2 : 11

7. **a)**

First	Second	Third	Outcome
W	W	W	WWW
		L	WWL
		T	WWT
	L	W	WLW
		L	WLL
		T	WLT
	T	W	WTW
		L	WTL
		T	WTT
L	W	W	LWW
		L	LWL
		T	LWT
	L	W	LLW
		L	LLL
		T	LLT
	T	W	LTW
		L	LTL
		T	LTT
T	W	W	TWW
		L	TWL
		T	TLT
	L	W	TLW
		L	TLL
		T	TLT
	T	W	TTW
		L	TTL
		T	TTT

b) 1 : 26 **c)** 7 : 20 **d)** 20 : 7

8. **a)** 9 : 10 **b)** Odds in favour of 9 : 10 means a probability of winning of $\frac{9}{19}$ or 47%, or 47 times out of 100. In 100 games you can expect to lose \$30 (i.e., the difference between 47 \$5 wins and 53 \$5 losses).

9. odds of 4 to 5 against = 5 to 4 in favour = $\frac{5}{9}$ or 56% chance of winning, as opposed to odds of 3 to 4 in favour = $\frac{3}{7}$ or 43% chance of winning

10. **a)** $P(\text{2 sides painted}) = \frac{36}{125}$, odds in favour are 36 : 89
b) $P(\text{3 sides painted}) = \frac{8}{125}$, odds against are 117 : 8
c) $P(\text{1 side painted}) = \frac{54}{125}$, odds in favour are 54 : 71 = 1 : 1.31; $P(\text{white}) = \frac{27}{125}$, odds in favour are 27 : 98 = 1 : 3.63; Pulling out a cube with one side painted has greater odds.

11. **a)** $P(\text{centre circle}) = \frac{28.26}{706.5} = \frac{1}{25}$, so the odds against are 24 : 1
b) $P(\text{shaded section}) = \frac{339.12}{706.5} = \frac{12}{25}$, so the odds in favour are 12 : 13

6.3 Probabilities Using Counting Techniques, pages 90–93

1. **a)** 6 **b)** $\frac{1}{6}$

2. **a)** 270 725 **b)** 0.055 **c)** 0.003

3. **a)** 20! or 2.43×10^{18} **b)** 0.05

4. **a)** 400 **b)** 0.025

5. 0.004

6. **a)** 40% **b)** 0.101

7. **a)** 0.053 **b)** 1 : 18

8. **a)** unrestricted case: ${}_{52}C_5$ = 2 598 960 different poker hands

${}_{13}C_1$	×	${}_{4}C_2$	×	${}_{12}C_3$	×	$({}_{4}C_1)^3$
First, choose one of the 13 ranks to be the rank of the pair.		Next, choose two of the four cards of the chosen rank.		Next, choose the ranks of the remaining three cards.		Finally, choose the suit for each of the three non-paired cards.

$= 13 \times 6 \times 220 \times 64$
$= 1\,098\,240$

Note that a common mistake is to use ${}_{48}C_3$ for the last three cards, since it appears they can be anything. But within that ${}_{48}C_3$ are threes-of-a-kind and other pairs that, if included, would change the hand having one pair to a hand with a full house (three of a kind plus a pair) or two pairs.

The probability of getting a pair is $\frac{1\,098\,240}{2\,598\,960}$ or 0.423.

b) My opponent needs a pair of 8s or higher to beat my pair of 7s, meaning my opponent can use only the JD or AC of the cards showing or form a pair of 8s, 9s, 10s, Qs, or Ks with his or her own cards. There are 45 cards left in the deck, so there are ${}_{45}C_2$ = 990 possible two-card sets. But because my opponent can make a pair only with jacks or aces, of which there are three of each in the deck, then the calculation below shows how to have one of the cards be a jack or ace.

${}_{2}C_1$	×	${}_{3}C_1$	×	${}_{39}C_1$
First, choose which rank will be the pair, jack or ace		Next, choose which of the three remaining cards of the chosen rank will complete the pair		Finally, choose the card that will not be used to make a pair. It cannot be the same rank as any of the five other cards already showing.

$= 2 \times 3 \times 39$
$= 234$

My opponent may also make a higher pair of 8s with his or her two cards in ${}_{3}C_2$ or 3 ways.

A higher pair of 9s, 10s, Qs, or Ks from those cards can be made in ${}_{4}C_1 \times {}_{4}C_2$ or 24 possible ways. In all, the probability is $\frac{234 + 3 + 24}{990}$, or 0.264, that my opponent has a higher pair.

9. **a)** 8.4×10^{-8}
b) 6.7×10^{-5}, which is about 800 times better

10. $P(\text{matching pair}) = 1 - \frac{{}_{n}P_k}{n^k}$

11. The total number of unrestricted groups of three is ${}_{10}C_3$ or 120.
a) $P(\text{all boys}) = \frac{{}_{3}C_3}{{}_{10}C_3} = \frac{1}{120}$ or 0.83%
b) Find the probability of 3 girls or 2 girls and 1 boy.
$P(\text{at least two girls}) = \frac{({}_{7}C_3) + ({}_{7}C_2)({}_{3}C_1)}{{}_{10}C_3} = 0.817$ or 81.7%.
c) Part a) becomes $P(\text{all boys}) = \frac{{}_{3}C_3}{{}_{9}C_3} = \frac{1}{84}$, or 1.19%, an increase of almost 43.37%. Part b) becomes

$P(\text{at least two girls}) = \dfrac{({}_6C_3) + ({}_6C_2)({}_3C_1)}{{}_{10}C_3} = 0.774,$
or 77.4%, a decrease of 5.26%.

12. **a)** $\frac{1}{2^{63}}$ or 1.1×10^{-19} **b)** 0.16 or 1.6%

13. **a)** 80.2% **b)** 43.6%
c) Because there are so many ways to pair songs, it is not surprising to hear an artist repeated.

14. **a)** ${}_5C_2 = 10$ **b)** 20%
c) No, since each choice has two ways to win.

15. **a)** The distribution of values is fairly random.

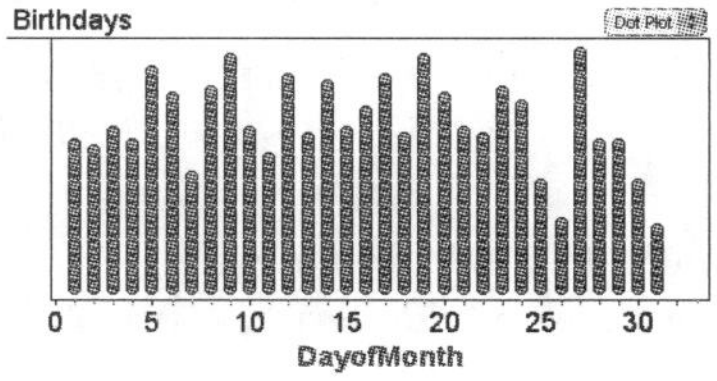

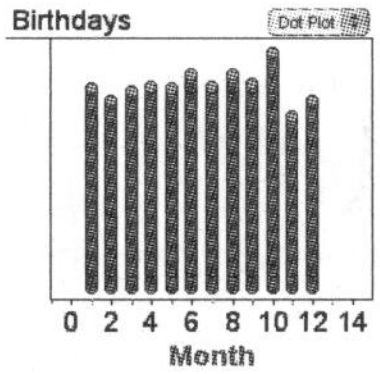

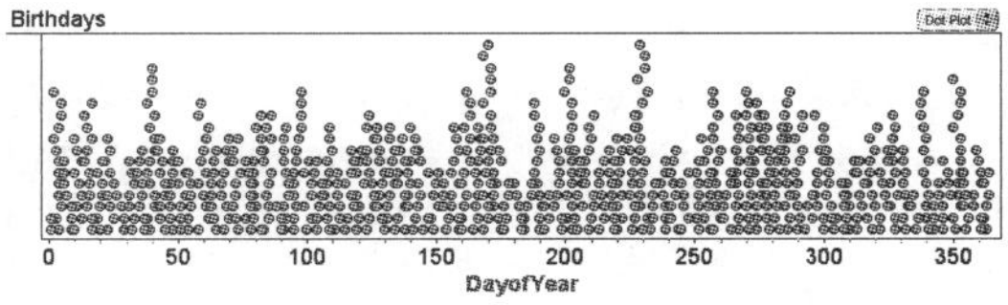

b) 155 matches after 305 trials, for an experimental probability of 51%

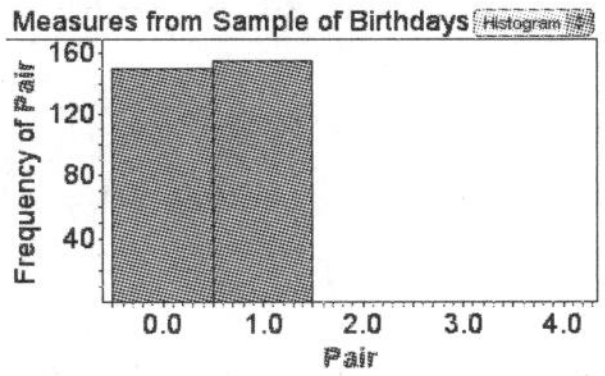

c) 347 cases with a pair after 505 trials, for an experimental probability of 68.7%

d) $P(\text{matching pair}) = 1 - \frac{{}_{365}P_{30}}{365^{30}}$, or approximately 70.6%. So, the answer in part c) is a bit less than the theoretical value.

16. 182 with pairs after 400 trials, for an experimental probability of 46%

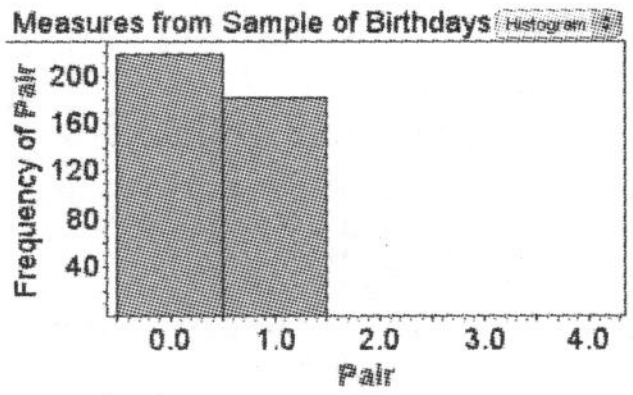

17. 30.9%

18. **a)** $P(\text{win}) = \frac{1}{{}_{49}C_6} = \frac{1}{13\,983\,816}$ or 7.2×10^{-8}

b) $P(\text{win})$ is 1000 times better but still very low:
$P(\text{win}) = \frac{1000}{{}_{49}C_6} = \frac{1000}{13\,983\,816}$ or 7.2×10^{-5}

c) There are ${}_6C_5$ ways to get five of the six numbers right, and there are 43 other wrong numbers that can be chosen for the sixth number.

$P(\text{5 numbers}) = \frac{{}_6C_5 \times {}_{43}C_1}{{}_{49}C_6} = \frac{6 \times 43}{{}_{49}C_6} = \frac{258}{13\,983\,816}$
or 1.8×10^{-5}

d) There are ${}_6C_3 = 20$ ways to have three of the six winning numbers. For the remaining three numbers, there are 43 to choose from, or ${}_{43}C_3 = 12\,341$ ways.

$P(\text{3 numbers}) = \frac{{}_6C_3 \times {}_{43}C_3}{{}_{49}C_6} = \frac{20 \times 12\,341}{13\,983\,816}$ or 1.8%

6.4 Dependent and Independent Events, pages 94–98

1. **a)** compound **b)** simple **c)** compound **d)** compound
e) compound **f)** compound

2. **a)** independent **b)** independent **c)** independent
d) dependent **e)** independent **f)** dependent

3. **a)**

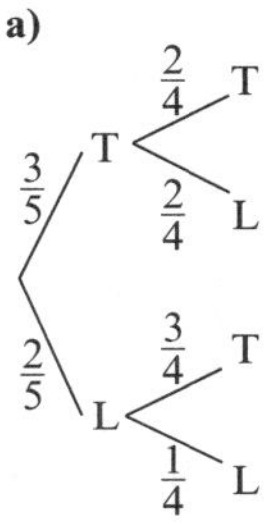

$P(\text{TT}) = P(\text{T}) \times P(\text{T}\,|\,\text{T}) = \left(\frac{3}{5}\right)\left(\frac{2}{4}\right) = 0.3$

$P(TL) = P(T) \times P(L\,|\,T) = \left(\frac{3}{5}\right)\left(\frac{2}{4}\right) = 0.3$

$P(\text{LT}) = P(\text{L}) \times P(\text{T}\,|\,\text{L}) = \left(\frac{2}{5}\right)\left(\frac{3}{4}\right) = 0.3$

$P(\text{LL}) = P(\text{L}) \times \text{P}(\text{L}\,|\,\text{L}) = \left(\frac{2}{5}\right)\left(\frac{1}{4}\right) = 0.1$

b) total of all probabilities = 0.3 + 0.3 + 0.3 + 0.1 = 1

4. The total number of ways to choose two of five coins is ${}_5C_2 = 10$.

Outcome	Combinations	Total	Probability
{T, T}	${}_3C_2$	3	$P(\text{TT}) = \frac{3}{10} = 0.3$
{T, L}	$({}_3C_1)({}_2C_1)$	6	$P(\text{TL or LT}) = \frac{6}{10} = 0.6$
{L, L}	${}_2C_2$	1	$P(\text{LL}) = \frac{1}{10} = 0.1$

5. **a)** $P(\text{2 spades}) = P(\text{spade}) \times P(\text{spade}\,|\,\text{spade})$
$= \left(\frac{13}{52}\right)\left(\frac{12}{51}\right)$ or $\frac{1}{17}$,
which is the same for two of any suit, so total probability is
$P(\text{2 same suit}) = 4\left(\frac{1}{17}\right)$ or $\frac{4}{17}$.

b) The number of ways to choose two cards is ${}_{52}C_2 = 1326$. The number of ways to choose two of the same suit is ${}_4C_1$ choices for the suit,times ${}_{13}C_2$ choices for the rank.

$P(\text{2 same suit}) = \frac{({}_4C_1)({}_{13}C_2)}{{}_{52}C_2} = \frac{312}{1326}$ or $\frac{4}{17}$

6. **a)** 7.1% **b)** 15.7%

7. **a)** 52.2% **b)** 56.2%

8. **a)** Since each card is either red or black, there is a 50% chance of guessing correctly. So, if the person only guessed, the person should get about half of them right.

b) $P(\text{25 wrong}) = \frac{1}{2^{25}} = \frac{1}{33\,554\,432} = 3.0 \times 10^{-8}$.

c) Sample answer: Guessing them all incorrectly is just as improbable as guessing them all correctly. If a person were to do this, it would strengthen their claim of being a psychic.

9. a)

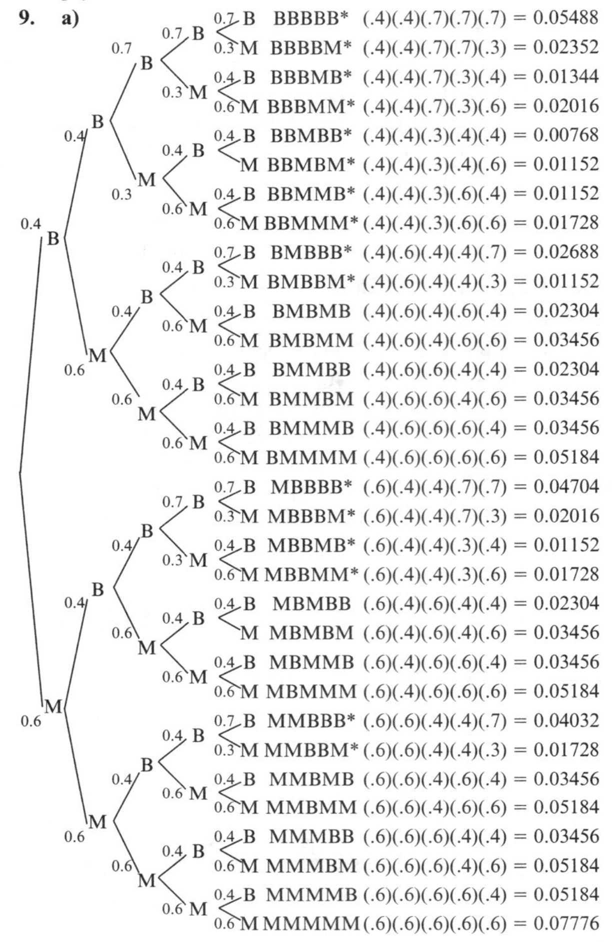

b) 16 (marked with an asterisk) **c)** Sample answer: Most studies have found there is no statistical evidence that the "hot hand" effect actually exists.

10. a) These are independent events, so **i)** 16.7% **ii)** 13.8% **iii)** 9.6% **iv)** 6.7%
b) Sample answer: Sample data for number of rolls in 20 experiments to get 1: {2, 16, 5, 6, 17, 1, 5, 1, 4, 2, 1, 3, 9, 11, 2, 1, 7, 9, 4, 35}. The mean number of rolls is 7.05 and the median is 4.5, so you could advise a player that he or she would likely roll a 1 after the fourth roll, so the player should stop at that point. **c)** After 300 trials, the mean number of rolls seemed to stay around 6.7 and the median around 5. The results appear to be similar.

11.

Choose 1 of 3 doors.		Probability of winning with a switch		
$\frac{1}{3} = 0.33$	Chooses the correct door	0	Switching will get the token	P(win) = (0.33)(0) = 0
		1	Staying gets the prize	P(win) = (0.33)(1) = 0.33
$\frac{2}{3} = 0.67$	Chooses one of the wrong doors	1	Switching gets the prize	P(win) = (0.67)(1) = 0.67
		0	Staying gets the token	P(win) = (0.67)(0) = 0

If you switch, you will win 67% of the time.

12. Regardless of when a person chooses a key, each player has an equal chance to win.

13. a) P(failing DM and E) = 7%
b) If the two events are independent, then
P(fail DM) × P(fail E) = P(fail DM and E)
P(fail DM) × P(fail E) = (0.10)(0.15) = 0.015 = 1.5%
The probabilities are not equal, so the events are not independent.

14. a)

First	Second	Third	Outcome	Probability
0.98 G	0.98 G	0.98 G	GGG	(0.98)(0.98)(0.98) = 0.941 192
		0.02 D	GGD*	(0.98)(0.98)(0.02) = 0.019 208
	0.02 D	0.98 G	GDG*	(0.98)(0.02)(0.98) = 0.019 208
		0.02 D	GDD	(0.98)(0.02)(0.02) = 0.000 392
0.02 D	0.98 G	0.98 G	DGG*	(0.02)(0.98)(0.98) = 0.019 208
		0.02 D	DGD	(0.02)(0.98)(0.02) = 0.000 392
	0.02 D	0.98 G	DDG	(0.02)(0.02)(0.98) = 0.000 392
		0.02 D	DDD	(0.02)(0.02)(0.02) = 0.000 008

b) 3(0.019 208) or 6%

6.5 Mutually Exclusive Events, pages 99–102

1. a) non-mutually exclusive
b) mutually exclusive
c) non-mutually exclusive
d) non-mutually exclusive
e) mutually exclusive
f) non-mutually exclusive

2. a) $\frac{7}{12}$ **b)** $\frac{9}{12}$ or $\frac{3}{4}$ **c)** $\frac{10}{12}$ or $\frac{5}{6}$

3. a) P(pair or sum of 8) = $\frac{10}{36}$ or $\frac{5}{18}$

DIE 2 \ DIE 1	1	2	3	4	5	6
1	2	3	4	5	6	7
2	3	4	5	6	7	8
3	4	5	6	7	8	9
4	5	6	7	8	9	10
5	6	7	8	9	10	11
6	7	8	9	10	11	12

b) $P(3 \text{ or sum of } 9) = \frac{13}{36}$

		DIE 1					
		1	2	3	4	5	6
D	1	2	3	4	5	6	7
I	2	3	4	5	6	7	8
E	3	4	5	6	7	8	9
	4	5	6	7	8	9	10
2	5	6	7	8	9	10	11
	6	7	8	9	10	11	12

c) $P(\text{sum of } 7 \text{ or sum of } 11) = \frac{8}{36}$ or $\frac{2}{9}$

		DIE 1					
		1	2	3	4	5	6
D	1	2	3	4	5	6	7
I	2	3	4	5	6	7	8
E	3	4	5	6	7	8	9
	4	5	6	7	8	9	10
2	5	6	7	8	9	10	11
	6	7	8	9	10	11	12

4. **a)** $\frac{4}{13}$ **b)** $\frac{4}{13}$ **c)** $\frac{1}{2}$
5. 3c), 4b), 4c)
6. Sample answers: **a)** rolling a 6 or a sum of 3
   b) rolling a 5 or a sum of 10
   c) rolling a sum of 5 or a sum of 6
   d) rolling a sum of 6 or a difference of 2
7. **a)** 58% **b)** 38% **c)** 42%
8. **a)** 43.2% **b)** 75%
9. **a)** 54% **b)** 64.2%
10. **a)** 51.7% **b)** 24.6%
    c) i) 96.0% **ii)** 88.8% **iii)** 43.6%
    d) Sample answer: Based on survival rates by class, it seems that "women and children first" was mostly adhered to for those not in third class.
11. Note that the five vacancies in the Senate should not count in any totals.
    a) 77.7% **b)** 33.6% **c)** 87.5% **d)** 22.3%
12. $P(A \text{ or } B \text{ or } C) = P(A) + P(B) + P(C) - P(A \text{ and } B) - P(A \text{ and } C) - P(B \text{ and } C) + P(A \text{ and } B \text{ and } C)$
13. **a)** 45% **b)** 23%

Chapter 7

7.1 Probability Distributions, pages 101–105

1. **a)** continuous **b)** discrete **c)** discrete **d)** continuous
   e) discrete
2. **a)** uniform; Each month is equally likely to be selected.
   b) not uniform; There are five weekdays and only two weekend days.
   c) not uniform; The chances of losing are much greater than the chances of winning.
   d) uniform; Each colour is equally likely to be selected.
   e) not uniform; The probability of getting three heads is less than the probability of getting a combination of heads and tails.
3. **a)** uniform: $P(x) = \frac{1}{12}$ **b)** Although the graphs do not match the shape of the theoretical uniform probability histogram exactly, they are close approximations.
   c) For the same set of data, the histograms will have the same shape.
4. **a)**

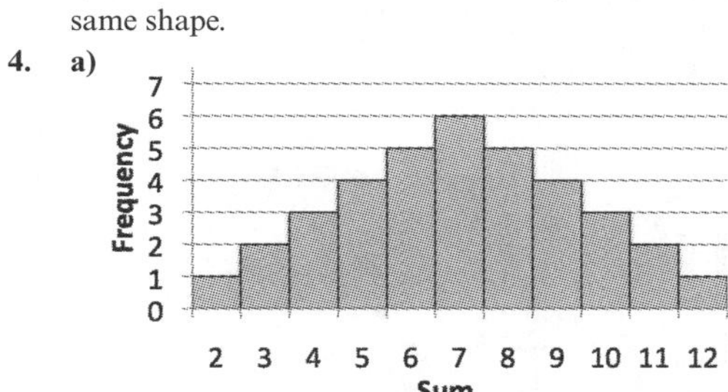

b)

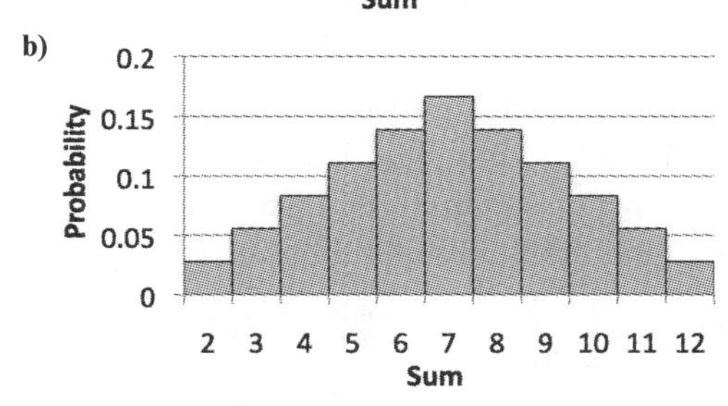

c) The graphs have the same shape.

5. **a)** 8

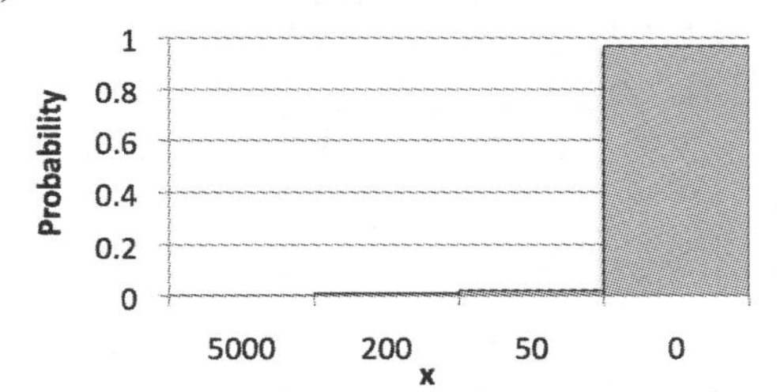

b) 5.7

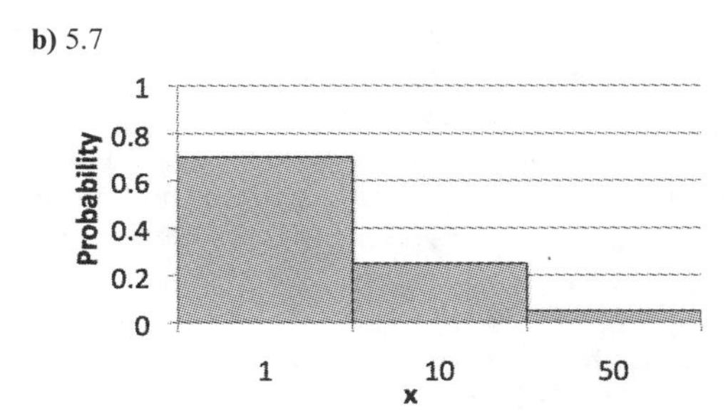

6. **a)**

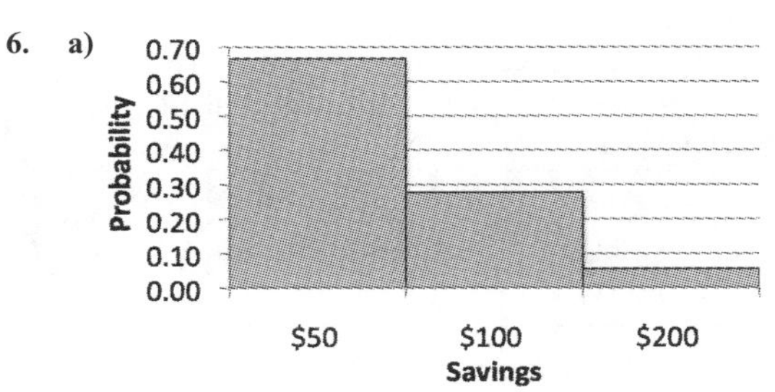

b) $72.22 **c)** Answers will vary.

7. **a)** your friend **b)** 0 **c)** The game is fair, since the expected value is 0.

8. **a)** 0.25 **b)** −$12.425 **c)** −$12.575 **d)** $24 850; The total of the prizes divided by the total number of tickets equals the expected payout. **e)** 994 **f)** Sample answer: A total of 500 prizes gives odds in favour of winning of 1 : 3, which makes the chances of winning appear to be reasonably good. It does not cost the school anything to offer prizes equal in value to the cost of tickets.

9. **a)**

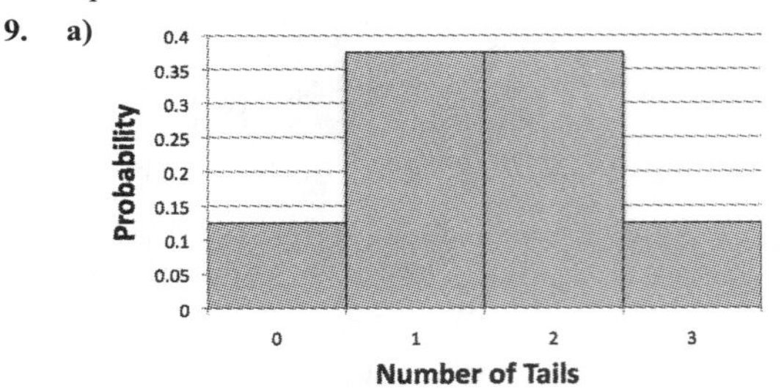

b) $1.75 **c)** Sample answer: The expected value is not 0, so this is not a fair game and is favourable to the player. If the cost to play is changed to $4, the expected value becomes −$0.25 in favour of the casino.

10. **a)**

# of Numbers Bet	Odds Against
1	37 to 1
2	18 to 1
3	11.7 to 1
4	8.5 to 1
5	6.6 to 1
6	5.3 to 1
12	2.2 to 1
18	1.1 to 1

b) In each case, the payout is less than the odds, but as the number of numbers bet on increases, the payout and the odds become closer. Betting on 18 numbers may be more advantageous to the player. Although the payout is less, the chance of winning is better.
c) $E(X) = -0.053$ in all cases except the five-number bet, where $E(X) = -0.079$. In the long run, no bet is more advantageous than any other and each slightly favours the casino.

11. **a)** uniform **b)** $131 477.54

12. Offering the option package is better, because the expected value of $13 300 is higher than the expected value of $12 000 without the option package.

13. Based on the expected number of points, it is more advantageous to try for a field goal inside the 20-yd line and a touchdown outside the 20-yard line.

	$E(X)$	
Field Position, x	**Touchdown**	**Field Goal**
5-yd line	4.2	2.97
10-yd line	3.5	2.94
20-yd line	2.8	2.91
30-yd line	1.75	2.7
50-yd line	0.7	1.8

14. Sample answers: **a)** Player 1 wins if the sum is 5, 6, 8, or 9, otherwise Player 2 wins.
b) Player 1 wins if the sum is 5, 6, 7, 8, or 9, otherwise Player 2 wins.
c) Player 1 wins if their die is larger than Player 2's, otherwise Player 2 wins.

15. **a)** $E(X)$ for the remaining cases is $69 387.50, a value that is twice the offer. By not making a deal, the player gets to open another case. At this point there is a 5 out of 6 chance it will not be the $400 000 case.
b) The new expected value is $82 265. This is still higher than the offer, but is close. Although the player still has a 4 out of 5 chance of not opening the $400 000 case, it might be a good time to make a deal.

7.2 Binominal Distributions, pages 106–109

1. **a)** binomial **b)** not binomial **c)** binomial **d)** not binomial **e)** not binomial

2. **a)**

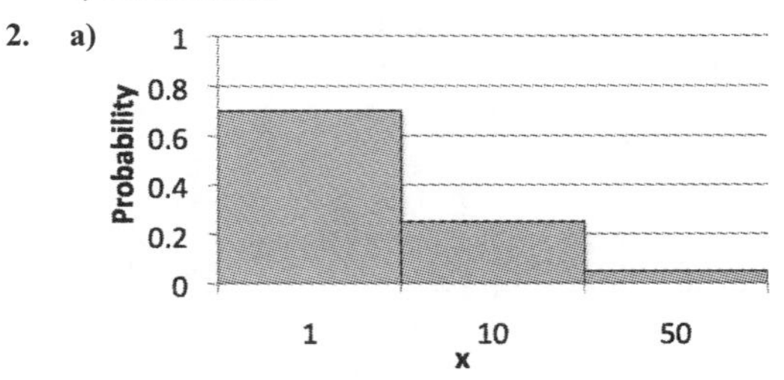

Number of Successes, x	$P(x)$	$xP(x)$
0	0.885 8	0
1	0.108 5	0.108 5
2	0.005 5	0.011 1
3	0.000 2	0.000 5
4	2.3×10^{-6}	9.2×10^{-6}
5	1.9×10^{-8}	9.4×10^{-8}
6	6.4×10^{-11}	3.8×10^{-10}
Totals	1.000 0	0.12

b)

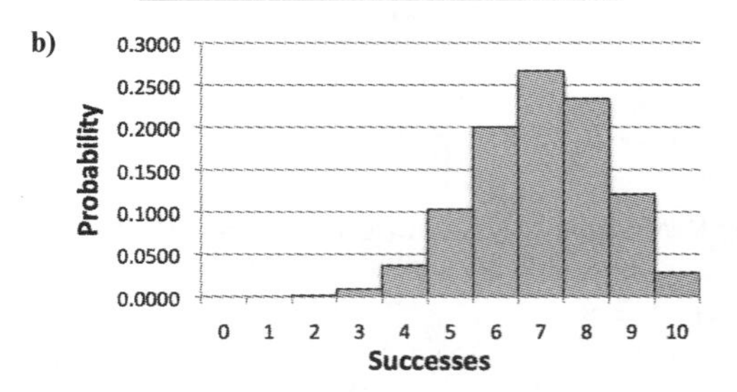

Number of Successes, x	$P(x)$	$xP(x)$
0	5.9×10^{-6}	0
1	0.000 1	0.000 1
2	0.001 4	0.002 9
3	0.009 0	0.027 0
4	0.036 8	0.147 0
5	0.102 9	0.514 6
6	0.200 1	1.200 7
7	0.266 8	1.867 8
8	0.233 5	1.867 8
9	0.121 1	1.089 5
10	0.028 2	0.282 5
Totals	1.000 0	7.0

3. a) In L2, enter **binompdf(9,.25,L1)**.

L1	L2	L3
0	.07508	------
1	.22525	
2	.30034	
3	.2336	
4	.1168	
5	.03893	
6	.00865	

L2 =binompdf(9,▮...

L1	L2	L3
4	.1168	
5	.03893	
6	.00865	
7	.00124	
8	1E-4	
9	3.8E-6	

L2(11) =

b)

	A	B	C
1	Success,x	P(x)	
2	0	=BINOMDIST(A2,9,0.25,0)	
3	1	0.22525406	
4	2	0.30033875	
5	3	0.2335968	
6	4	0.1167984	
7	5	0.0389328	
8	6	0.00865173	
9	7	0.00123596	
10	8	0.000103	
11	9	3.8147E-06	

c) Binomial

	x	Probability
=		binomialProbability (x, 9, 0.25)
1	0	0.0750847
2	1	0.225254
3	2	0.300339
4	3	0.233597
5	4	0.116798
6	5	0.0389328
7	6	0.00865173
8	7	0.00123596
9	8	0.000102997
10	9	3.8147e-06

4. a) probability of three successes with ten trials when $p = 0.91$
b) probability of five successes with six trials when $p = 0.15$
c) probability of fewer than three successes with 25 trials when $p = 0.1$
5. a) Evelyn: 0.276; Aislin: 0.346 b) Evelyn
6. a) 0.828 b) 0.234
7. a) 4.9×10^{-4}
b) Rolling the die is an independent event with only two possible outcomes, getting a 3 (success) or not getting a 3 (failure).
8. a) 0.143 b) 0.397
9. a) The lights are independent of each other and there are only two possibilities: not stopping (success = green) and stopping (failure = yellow or red).
b)

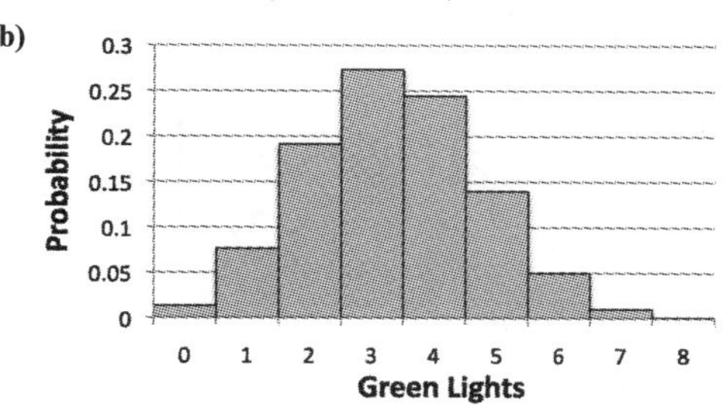

c) 3.33
10. Use a binomial distribution with $n = 3$ and $x \geq 1$. There is a 55.6% chance that at least one PlayBox will fail.
11. a) i)

Successes, x	$p = 0.2$ $P(x)$	$p = 0.4$ $P(x)$	$p = 0.6$ $P(x)$	$p = 0.8$ $P(x)$
0	0.107	0.006	0.000	0.000
1	0.268	0.040	0.002	0.000
2	0.302	0.121	0.011	0.000
3	0.201	0.215	0.042	0.001
4	0.088	0.251	0.111	0.006
5	0.026	0.201	0.201	0.026
6	0.006	0.111	0.251	0.088
7	0.001	0.042	0.215	0.201
8	0.000	0.011	0.121	0.302
9	0.000	0.002	0.040	0.268
10	0.000	0.000	0.006	0.107

ii)–iii)

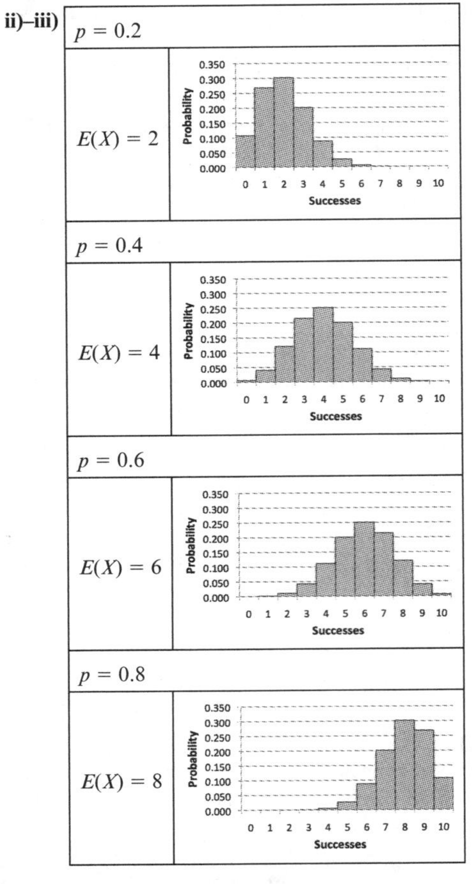

b) Sample answer: The distributions for $p = 0.2$ and $p = 0.8$ are reverses of each other, as are the distributions for $p = 0.4$ and $p = 0.6$. If there were 100 trials, the general shapes would be the same; however, the probabilities would be lower and the spread of the histograms would be greater.

12. The histograms have the same general shape, although the individual probabilities drop for each bar.

13. **a)** 2.1 **b)** 3.5 **c)** 13.3

14. **a)** 0.196 **b)** 1 return with a probability of 0.2
c) No, because the chance is 1 in 5.

15. **a)** The probability that one boy cycles to school (success) is $p = 0.033$. The probability that two or more boys cycle to school is 1 minus the probability that fewer than two boys cycle to school:
$P(x) = 1 - ({}_{15}C_0(0.033)^0(0.967)^{15} + {}_{15}C_1(0.033)^1(0.967)^{14})$
$= 0.086$ or 8.6%.
b) The probability of success (boy smokes cigarettes) is $p = 0.146$. The probability that exactly three boys smoke cigarettes is
$P(3) = {}_{15}C_3(0.146)^3(0.854)^{12}$
$= 0.213$ or 21.3%.
c) The probability of success (not right handed) is $p = 0.125$, so the expected value is $E(X) = 12(0.125) = 1.5$. The probability of exactly four girls not being right handed is
$P(4) = {}_{12}C_4(0.125)^4(0.875)^8$
$= 0.042$ or 4.2%,
which is low. The probability of having four girls who are not right handed is much higher than predicted.

16. 9

17. **a)** 1.7% **b)** 27 **c)** 20

18.

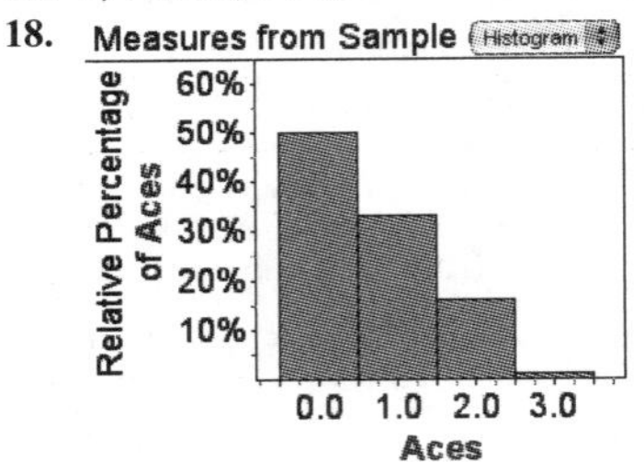

19. **a)** Assuming that men and women applied in equal numbers and that they were equally qualified, the probability of hiring a woman is $p = 0.5$, so $E(X) = 15(0.5) = 7.5$.
b) Assuming that the number of male and female applicants was equal, the probability that a female would be hired is $P(\text{female hired}) = P(\text{female}) \times P(\text{qualified female}) = (0.5)(0.3) = 0.15$. So, the expected number of female hires is $E(X) = 15(0.15) = 2.25$. Based on this, the fact that 3 women have been hired is in line with the company's claim.

7.3 Geometric Distributions, pages 110–112

1. **a)** yes **b)** yes **c)** yes
d) no, since the draws are not independent (no replacement)
e) no, since with each jump the ice will weaken more

2. **a)** $p = 0.5$, $q = 1 - 0.5 = 0.5$, waiting time: $E(X) = 1$

Number of Trials Before Success (Wait Time), x	$P(x) = q^x p$
0	$(0.5)^0(0.5) = 0.5$
1	$(0.5)^1(0.5) = 0.25$
2	$(0.5)^2(0.5) = 0.125$
3	$(0.5)^3(0.5) = 0.062\ 5$
4	$(0.5)^4(0.5) = 0.031\ 25$
5	$(0.5)^5(0.5) = 0.015\ 63$
6	$(0.5)^6(0.5) = 0.007\ 81$

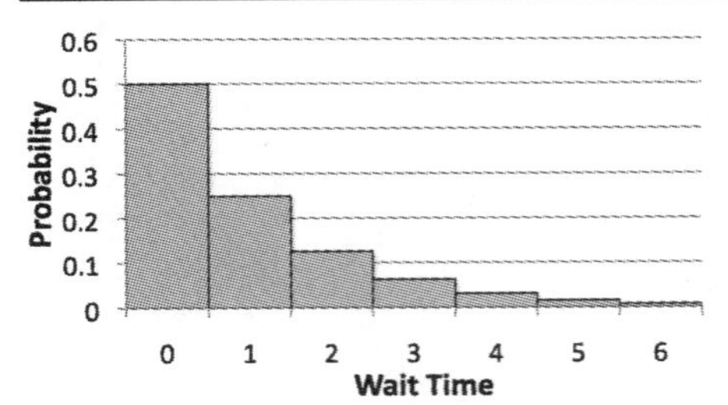

b) $p = 0.25$, $q = 1 - 0.25 = 0.75$, waiting time: $E(X) = 3$

Number of Trials Before Success (Waiting Time), x	$P(x) = q^x p$
0	$(0.75)^0(0.25) = 0.25$
1	$(0.75)^1(0.25) = 0.187\ 5$
2	$(0.75)^2(0.25) = 0.140\ 63$
3	$(0.75)^3(0.25) = 0.105\ 47$
4	$(0.75)^4(0.25) = 0.079\ 10$
5	$(0.75)^5(0.25) = 0.059\ 33$
6	$(0.75)^6(0.25) = 0.044\ 49$

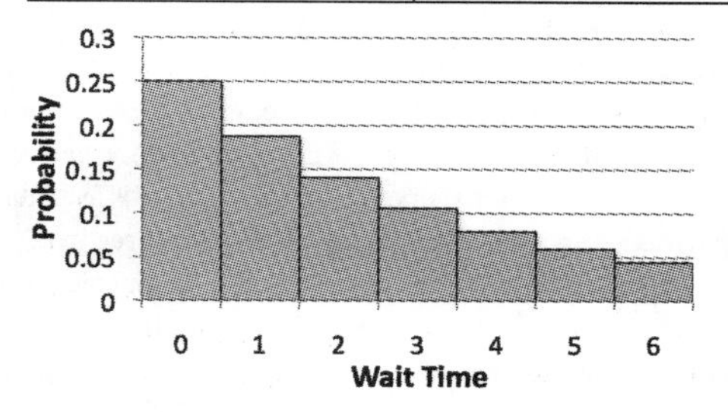

c) $p = 0.9$, $q = 1 - 0.9 = 0.1$, waiting time: $E(X) = 0.111$

Number of Trials Before Success (Waiting Time), x	$P(x) = q^x p$
0	$(0.1)^0(0.9) = 0.9$
1	$(0.1)^1(0.9) = 0.09$
2	$(0.1)^2(0.9) = 0.009$
3	$(0.1)^3(0.9) = 0.000\ 9$
4	$(0.1)^4(0.9) = 0.000\ 09$
5	$(0.1)^5(0.9) = 0.000\ 009$
6	$(0.1)^6(0.9) = 0.000\ 001$

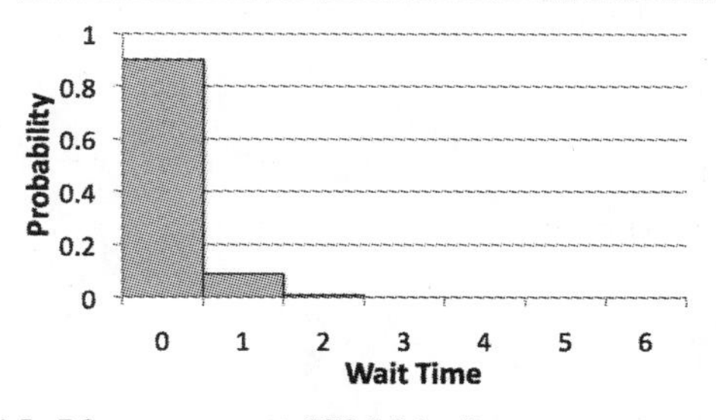

3. **a)** In **L2**, enter **geometpdf(0.5,L1+1)**.

L1	L2	L3 2
0	.5	------
1	.25	
2	.125	
3	.0625	
4	.03125	
5	.01563	
6	.00781	

L2 =...f(0.5,L1+1)

b)

	A	B	C	D
1	x	P(x)	p=	0.25
2	0	=(1-D$1)^A2*D$1		
3	1	0.1875		
4	2	0.140625		
5	3	0.10546875		
6	4	0.07910156		
7	5	0.05932617		
8	6	0.04449463		

c)

	WaitTime	Prob
=		geometricProbability (waittime, 0.9, 1, 1)
1	0	0
2	1	0.9
3	2	0.09
4	3	0.009
5	4	0.0009
6	5	9e-05
7	6	9e-06

4. Sample answer: Each histogram goes down at an exponential rate; however, as the probability of success becomes smaller, the rate of decrease from bar to bar becomes smaller and the expected waiting time increases.

5. **a)** $p = \frac{4}{52} = \frac{1}{13}$, $q = 1 - \frac{1}{13} = \frac{12}{13}$

Number of Trials Before an Ace, (Waiting Time), x	$P(x) = q^x p$
0	0.076 92
1	0.071 01
2	0.065 54
3	0.060 50
4	0.055 85
5	0.051 55
6	0.047 59

b) 12 draws

6. **a)** $p = 0.85$, $q = 1 - 0.85$, or 0.15

Number of Trials Before Success (Waiting Time), x	$P(x) = q^x p$
0	0.85
1	0.127 5
2	0.019 13
3	0.002 87
4	0.000 43
5	0.000 065
6	0.000 010

b) 5.67

7. **a)** 2.33 **b)** 6.41; about 4 more discs
8. **a)** about 5 **b)** 2
9. waiting time: 1.86, or approximately after every 2 people
10. The probability hardly changes: at start of year $p = \frac{1}{13\ 983\ 816}$, at end of year $p = \frac{1}{13\ 983\ 712}$.
11. **a)** 1% **b)** 9.6%

12. In this case, a failure of a parachute to open is considered the "success" outcome. Because the probability is low ($p = 0.001$), the waiting time will be long. Use a spreadsheet to calculate when the cumulative probability will exceed 50%. This occurs on the 693rd jump, after 692 "failures" of the parachute opening successfully.
b) At the higher failure rate, the cumulative probability will exceed 50% on the 69th jump, 624 fewer jumps than in part a).

13. The best strategy is to stop after 5 rolls, because the expected waiting time for rolling 1 is 5.

14. **a)** Since you will get any prize in the first box, it has a probability of $p = 1$. The waiting time is $E(X) = \frac{0}{1} = 0$.
b) The probability of getting any prize other than the first prize is $p = \frac{4}{5}$. The expected waiting time is $E(X) = \frac{\frac{1}{5}}{\frac{4}{5}} = \frac{1}{4} = 0.25$. You can expect to have two different prizes after 1.25 additional boxes, for a total of 2.25 boxes.
c) The probability of getting any prize other than the first two prizes is $p = \frac{3}{5}$. The expected waiting time is $E(X) = \frac{\frac{2}{5}}{\frac{3}{5}} = \frac{2}{3} = 0.67$. You can expect to have three different prizes after another 1.67 boxes, for a total of 3.92 boxes.
d) The expected waiting time for four different prizes is $E(X) = \frac{\frac{3}{5}}{\frac{2}{5}} = \frac{3}{2} = 1.5$, so you can expect to have four different prizes after 2.5 additional boxes, for a total of 6.42 boxes.
e) The expected waiting time to have all five prizes is $E(X) = \frac{\frac{4}{5}}{\frac{1}{5}} = \frac{4}{1} = 4$, so you can expect to have five different prizes after an additional 5 boxes, for a total of 11.42 boxes.

15. 18

16. The simulation gives a result close to the theoretical answer to question 14.

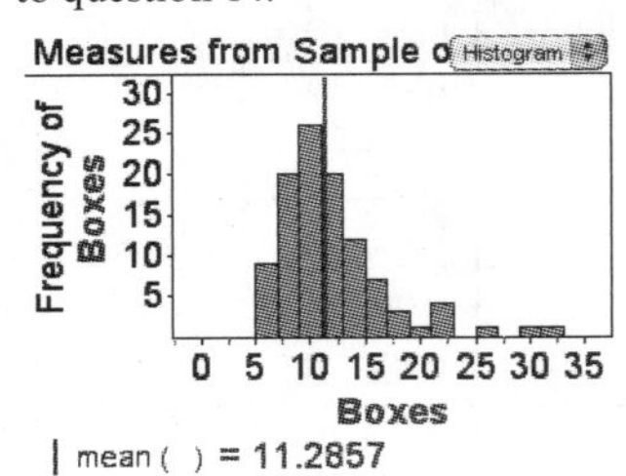

17. Determine the average waiting time. Then, use the expected waiting time formula to calculate the probability, p. Average waiting time is 70.5 and $p = 0.014$ or 1.4%.

7.4 Hypergeometric Distributions, pages 113–117

1. **a)** not hypergeometric; The trials are independent.
b) not hypergeometric; Only one part is being tested. Once the part is broken, it is not possible to have a successful trial after that.
c) hypergeometric; There two outcomes: 10 (success) or not 10 (failure), and they are dependent events, since there is no replacement.
d) hypergeometric; Each team member can be a girl (success) or not a girl (failure), and they are dependent events, since no girl can be chosen more than once.
e) hypergeometric; Any name can be drawn (success) or not drawn (failure), and since they are all drawn at once, this is like being drawn without replacement.
f) hypergeometric; There are two possibilities: pulling a red chip (success) or pulling a blue chip (failure), and since they are all drawn at once, this is like being drawn without replacement. (However, there is not enough information given to do any calculations, since only a percentage is available.)
g) hypergeometric; Even though there are three items, there are still only two outcomes: getting a blue marble (success) or not getting a blue marble (failure), and it is also done without replacement.

2. **a)**

x	$P(x)$	$xP(x)$
0	$\frac{({}_{10}C_0)({}_{20}C_6)}{({}_{30}C_6)} = 0.065\ 28$	0
1	$\frac{({}_{10}C_1)({}_{20}C_5)}{({}_{30}C_6)} = 0.261\ 11$	0.261 11
2	$\frac{({}_{10}C_2)({}_{20}C_4)}{({}_{30}C_6)} = 0.367\ 18$	0.734 37
3	$\frac{({}_{10}C_3)({}_{20}C_3)}{({}_{30}C_6)} = 0.230\ 39$	0.691 17
4	$\frac{({}_{10}C_4)({}_{20}C_2)}{({}_{30}C_6)} = 0.067\ 20$	0.268 79
5	$\frac{({}_{10}C_5)({}_{20}C_1)}{({}_{30}C_6)} = 0.008\ 49$	0.042 44
6	$\frac{({}_{10}C_6)({}_{20}C_0)}{({}_{30}C_6)} = 0.000\ 35$	0.002 12
	Sum	= 2

b) $E(X) = \frac{ra}{N} = \frac{(6)(10)}{(30)} = 2$. Thus, $E(X) = \sum_{i=0}^{r} x_i P(x_i) = \frac{ra}{N}$

3. **a)** $E(X) = 1.67$

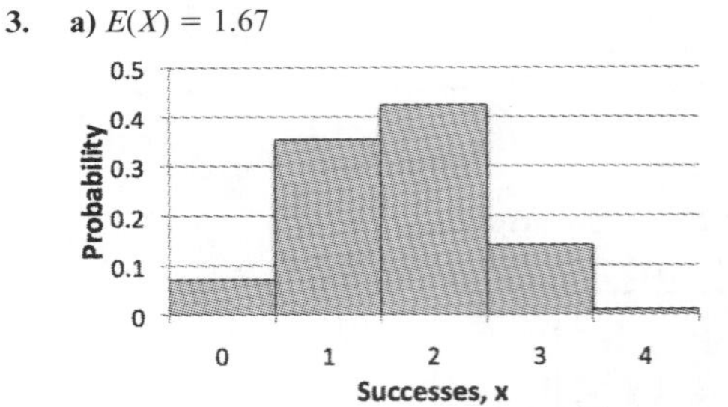

b) $E(X) = 1.25$

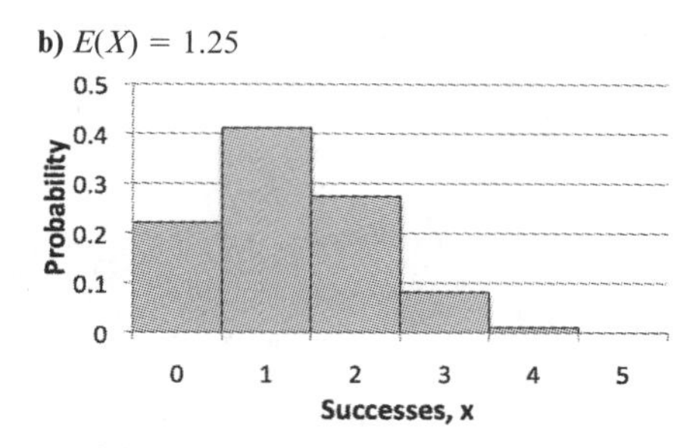

c) $E(X) = 1.73$

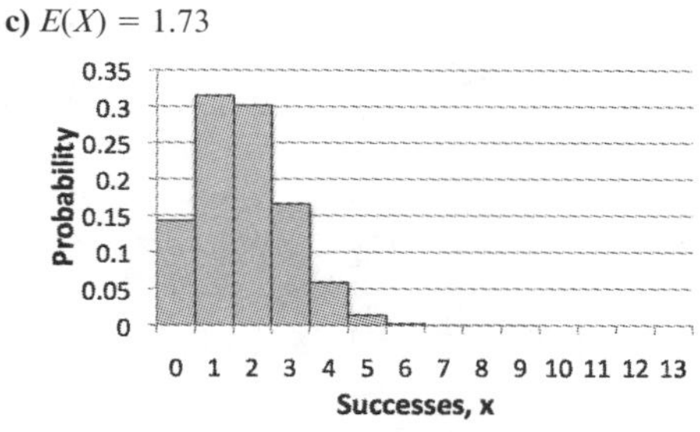

4. **a)** In L2, enter **(5 nCr L1*7 nCr (4-L1))/12 nCr 4**

L1	L2	L3
0	.07071	
1	.35354	
2	.42424	
3	.14141	
4	.0101	

L2 =(5 nCr L1*

b)

	A	B	C	D	E
1	x	P(x)			
2	0	=(COMBIN(13,A2)*COMBIN(39,5-A2))/COMBIN(52,5)			
3	1				
4	2	0.27427971			
5	3	0.08154262			
6	4	0.01072929			
7	5	0.0004952			

c)

	x	Prob
=		hyperGeomProbability (x, 150, 20, 13)
1	0	0.142924
2	1	0.314917
3	2	0.301685
4	3	0.165927
5	4	0.05828
6	5	0.0137579
7	6	0.00223706
8	7	0.000252571
9	8	1.97005e-05
10	9	1.04236e-06
11	10	3.61131e-08
12	11	7.69456e-10
13	12	8.94717e-12
14	13	4.23534e-14

5. **a)** The candies taken can be either green (success) or not green (failure), so there are only two possibilities. The candies are taken without replacement.

b) There are 80 candies in total, so $N = 80$. Based on the percent values, there are 20 each of orange and blue candies, and 10 each of brown, red, yellow, and green candies. Therefore, $a = 10$. Since 8 are taken, $r = 8$.

Number of Green, x	$P(x)$
0	$\frac{({}_{10}C_0)({}_{70}C_8)}{({}_{80}C_8)} = 0.325\ 67$
1	$\frac{({}_{10}C_1)({}_{70}C_7)}{({}_{80}C_8)} = 0.413\ 55$
2	$\frac{({}_{10}C_2)({}_{70}C_6)}{({}_{80}C_8)} = 0.203\ 54$
3	$\frac{({}_{10}C_3)({}_{70}C_5)}{({}_{80}C_8)} = 0.050\ 10$
4	$\frac{({}_{10}C_4)({}_{70}C_4)}{({}_{80}C_8)} = 0.006\ 64$
5	$\frac{({}_{10}C_5)({}_{70}C_3)}{({}_{80}C_8)} = 0.000\ 48$
6	$\frac{({}_{10}C_6)({}_{70}C_2)}{({}_{80}C_8)} = 1.75 \times 10^{-5}$
7	$\frac{({}_{10}C_7)({}_{70}C_1)}{({}_{80}C_8)} = 2.90 \times 10^{-7}$
8	$\frac{({}_{10}C_8)({}_{70}C_0)}{({}_{80}C_8)} = 1.55 \times 10^{-9}$

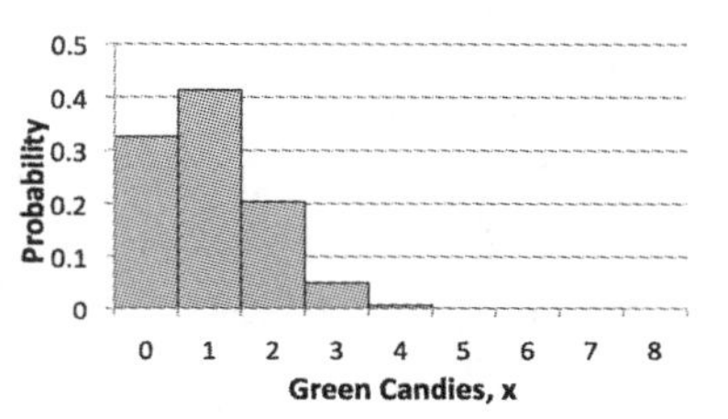

c) $P(4) = \frac{({}_{10}C_4)({}_{70}C_4)}{({}_{80}C_8)}$
$= 0.00664$ or 0.664%

d) $P(\text{at least } 4) = P(4) + P(5) + P(6) + P(7) + P(8)$
$= 0.00714$ or 0.7%

e) $E(X) = \frac{ra}{N} = \frac{(8)(10)}{(80)} = 1$, which make sense, since 12.5% of 8 is 1.

6. Sample answer:
${}_aC_x$: This represents the number of ways you can choose x desired things from the total number, a, of desired things.
${}_{N-a}C_{r-x}$: Since there are a ways to choose the desired thing, then there are $N - a$ ways to choose anything but the desired thing. Also, since x things have already been chosen from the desired things, the remaining number of trials must come from the non-desired things. So, this represents the number of ways you can choose the non-desired things to get the rest of the r items. The product of the two combinations in the numerator determines the total number of successes.
${}_NC_r$: This represents the total number of ways a sample of r items can be chosen from the population of N things.

7. **a)** Results of tree diagram:

Branch	Probability Distribution
SSSS	$\left(\frac{4}{15}\right)\left(\frac{3}{14}\right)\left(\frac{2}{13}\right)\left(\frac{1}{12}\right) = 0.000\ 73$
SSSF	$\left(\frac{4}{15}\right)\left(\frac{3}{14}\right)\left(\frac{2}{13}\right)\left(\frac{11}{12}\right) = 0.008\ 06$
SSFS	$\left(\frac{4}{15}\right)\left(\frac{3}{14}\right)\left(\frac{11}{13}\right)\left(\frac{2}{12}\right) = 0.008\ 06$
SSFF	$\left(\frac{4}{15}\right)\left(\frac{3}{14}\right)\left(\frac{11}{13}\right)\left(\frac{10}{12}\right) = 0.040\ 29$
SFSS	$\left(\frac{4}{15}\right)\left(\frac{11}{14}\right)\left(\frac{3}{13}\right)\left(\frac{2}{12}\right) = 0.008\ 06$
SFSF	$\left(\frac{4}{15}\right)\left(\frac{11}{14}\right)\left(\frac{3}{13}\right)\left(\frac{10}{12}\right) = 0.040\ 29$
SFFS	$\left(\frac{4}{15}\right)\left(\frac{11}{14}\right)\left(\frac{10}{13}\right)\left(\frac{3}{12}\right) = 0.040\ 29$
SFFF	$\left(\frac{4}{15}\right)\left(\frac{11}{14}\right)\left(\frac{10}{13}\right)\left(\frac{9}{12}\right) = 0.120\ 88$
FSSS	$\left(\frac{11}{15}\right)\left(\frac{4}{14}\right)\left(\frac{3}{13}\right)\left(\frac{2}{12}\right) = 0.008\ 06$
FSSF	$\left(\frac{11}{15}\right)\left(\frac{4}{14}\right)\left(\frac{3}{13}\right)\left(\frac{10}{12}\right) = 0.040\ 29$
FSFS	$\left(\frac{11}{15}\right)\left(\frac{4}{14}\right)\left(\frac{10}{13}\right)\left(\frac{3}{12}\right) = 0.040\ 29$
FSFF	$\left(\frac{11}{15}\right)\left(\frac{4}{14}\right)\left(\frac{10}{13}\right)\left(\frac{9}{12}\right) = 0.120\ 88$
FFSS	$\left(\frac{11}{15}\right)\left(\frac{10}{14}\right)\left(\frac{4}{13}\right)\left(\frac{3}{12}\right) = 0.040\ 29$
FFSF	$\left(\frac{11}{15}\right)\left(\frac{10}{14}\right)\left(\frac{4}{13}\right)\left(\frac{9}{12}\right) = 0.120\ 88$
FFFS	$\left(\frac{11}{15}\right)\left(\frac{10}{14}\right)\left(\frac{9}{13}\right)\left(\frac{4}{12}\right) = 0.120\ 88$
FFFF	$\left(\frac{11}{15}\right)\left(\frac{10}{14}\right)\left(\frac{9}{13}\right)\left(\frac{8}{12}\right) = 0.241\ 76$

b)

Number of Girls, x	$P(x)$
0	0.24176
1	0.48352
2	0.24176
3	0.03223
4	0.00073

c) Answers will vary.

8. **a)** Sample answer: In a hypergeometric distribution, which requires that there be no replacement, the probability changes after each item is chosen. Because the number of marbles is not known, there is not enough data to accurately determine a probability histogram. For a binomial distribution, it is necessary to know only the probability and that the trials are independent. Here, the probability is known, because the composition of the marble population is given. Although not told explicitly, you can then make the assumption to choose with or without replacement.

b)

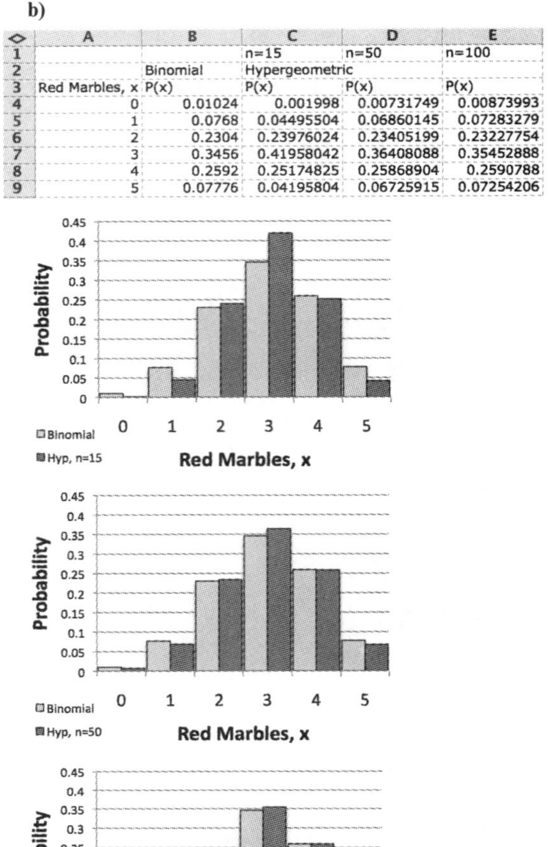

	A	B	C	D	E
1			n=15	n=50	n=100
2		Binomial	Hypergeometric		
3	Red Marbles, x	P(x)	P(x)	P(x)	P(x)
4	0	0.01024	0.001998	0.00731749	0.00873993
5	1	0.0768	0.04495504	0.06860145	0.07283279
6	2	0.2304	0.23976024	0.23405199	0.23227754
7	3	0.3456	0.41958042	0.36408088	0.35452888
8	4	0.2592	0.25174825	0.25868904	0.2590788
9	5	0.07776	0.04195804	0.06725915	0.07254206

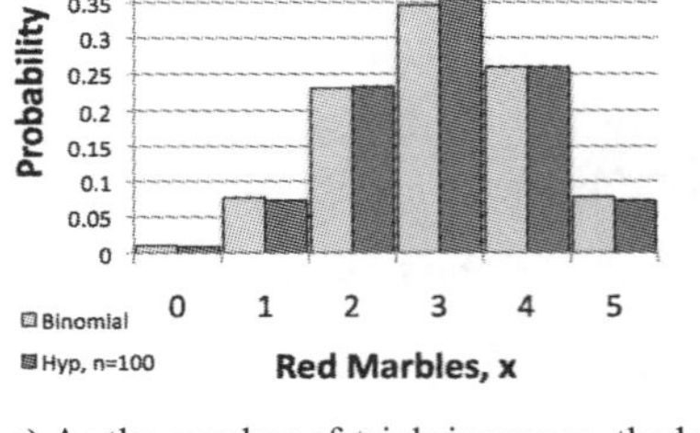

c) As the number of trials increases, the hypergeometric distribution seems to match the binomial distribution. So, if the number of trials was sufficiently large, the distributions would be almost the same.

9. **a)** If you try to calculate the probability when $x = 4$, you get $P(4) = \frac{({}_3C_4)({}_7C_0)}{({}_{10}C_4)}$, which is undefined, since you cannot choose four things from three. Though this probability is undefined mathematically, realistically it is zero, since you will never be able to have four red marbles.

b) The problem from part a) goes away since there are enough black marbles that ${}_4C_4$ is defined whereas ${}_3C_4$ is not.

10. **a)** 22.2% **b)** 0.24

11. a) 8

b)

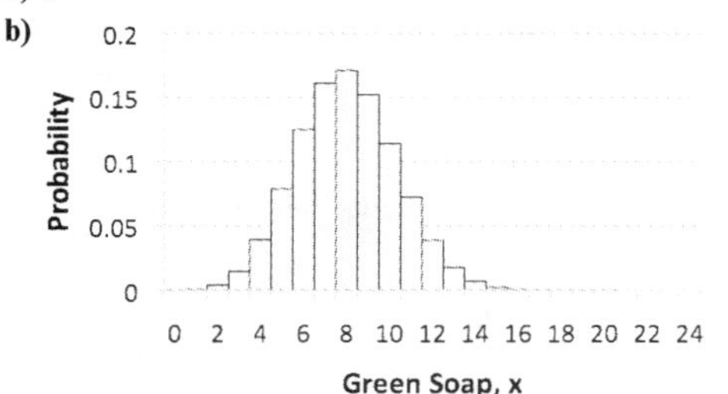

c) 6.7%

12. a) 39.8% **b)** 49.6% **c)** Sample answer: Getting two bad rings is not surprising, since the probability is 9.2%.

13. 25.1%

14. a) i) $E(X) = \frac{ra}{N} = \frac{(2)(3)}{(64)} = 0.094$

ii) $E(X) = \frac{ra}{N} = \frac{(5)(3)}{(64)} = 0.2234$

iii) $E(X) = \frac{ra}{N} = \frac{(10)(3)}{(64)} = 0.469$

b) The expected number of tainted cultures is less than 0.5 in each case. If one comes back tainted, that is double what is expected for 10 samples, four times what is expected for five samples, and ten times what is expected for two samples. In each case, the batch should be disposed of.

c) i) $P(0) = \frac{({}_3C_0)({}_{61}C_2)}{({}_{64}C_2)} = 0.91$ or 91%

ii) $P(0) = \frac{({}_3C_0)({}_{61}C_5)}{({}_{64}C_5)} = 0.78$ or 78%

iii) $P(0) = \frac{({}_3C_0)({}_{61}C_{10})}{({}_{64}C_{10})} = 0.60$ or 78%

It is a problem that no samples came back tainted, because that bit of information alone does not tell much. However, combine that data with the probability that none will come back tainted and the information is more helpful. For example, with only two samples it should not be a surprise if none come back tainted. But for ten samples, a 60% probability means that three out of five batches that have three bad cultures will come back with none. This means we can expect to catch about 40% of the bad batches.

d) By increasing the number of samples incrementally we can find out in what situation getting no tainted samples has less than a 50% chance. In this case, that occurs when you take 13 samples: $P(0) = \frac{({}_3C_0)({}_{61}C_{13})}{({}_{64}C_{13})} = 0.499\ 8$ or 50%. So, if the biologist samples 13 cultures, a bad batch can be expected to be caught half of the time.

15. a) 2500 **b)** $P(0) = 32.7\%$; $E(X) = \$33.65$

16. a) 65 **b)** 12

17. a) $E(X) = np = 20(0.5) = 10$

b) Sample answer: Knowing the result of the card affects the number of known red or black cards in the deck, so this then becomes a hypergeometric situation. By keeping track of the cards, you can increase the expected value.

c) Keep a running count of either the red or black cards. For the first card, you still have a 50% chance of guessing correctly. If the first card is a red one, regardless of whether you guessed it correctly, the chance the next one will be red is now $\frac{25}{51}$, or 0.490. Therefore, you have a slightly higher probability that the next card is black. So, guess black next. As the cards are drawn and revealed, add one if a red card is drawn and subtract one if a black card is drawn. If your total is positive, there are more black cards left, so, guess black. If the total is negative, there are more red cards left, so choose red. If the total is zero, randomly choose red or black.

18. Answers will vary.

Chapter 8

8.1 Continuous Probability Distributions, pages 118–123

1. a) continuous **b)** discrete **c)** discrete **d)** continuous **e)** discrete **f)** continuous

2. a) $\frac{1}{210-155} = 0.0\overline{18}$ **b)** $\frac{1}{0.57-0.52} = 20$ **c)** $\frac{1}{76-40} = 0.02\overline{7}$

d) $\frac{1}{13.25-13.1} = 6.\overline{6}$

3. a) Since the probability for any particular value is zero, it does not matter whether ≤ or < is used; the result will be the same. Area = (130 − 120)0.0625 or 0.625

b) (136 − 135)0.0625 or 0.0625

c) (128 − 122)0.0625 or 0.375

d) (130.3 − 129.5)0.0625 or 0.05

4. Since the area under a curve is always 1 and since, for a uniform distribution, the shape is always a rectangle, the area also is always a rectangle. So, if you do not already know the height, determine it by dividing 1 by the entire range and then multiplying that result by the range in question. Put simply, the area is the width of the interval times the height.

5. a) The area under the curve is 1.

b)

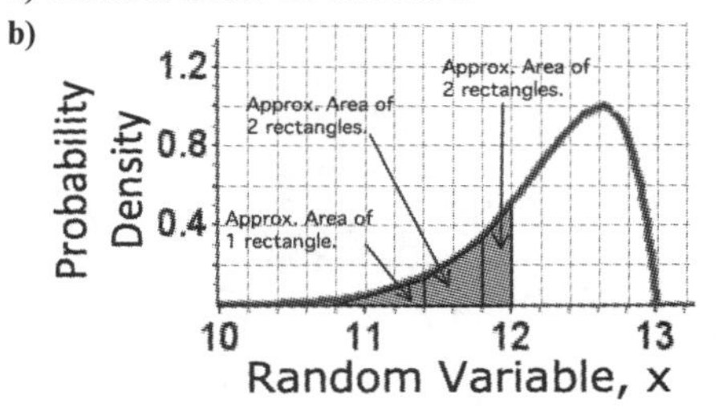

One rectangle of the background grid has an approximate area of (0.2)(0.2) or 0.04.

i) area of approximately 5 rectangles = 5(0.04) or 0.2

ii) $P(X > 11.4) = 1 - P(X < 11.4)$
$= 1 - \left(\frac{1}{2}\right)$ (area of rectangle) $= 1 - 0.02$
$= 0.98$

iii) area of approximately 7 rectangles = 7(0.04) or 0.28

6. **a)** First, the height of the distribution is needed: $h = \frac{1}{63-55}$ or 0.125. Then, for the probability, use the height to find the area of the rectangle.

	Range	Width, w	P = Area = hw
i)	$57 < X < 59$	2	0.25
ii)	$57.5 < X < 58.5$	1	0.125
iii)	$57.9 < X < 58.1$	0.2	0.025
iv)	$57.99 < X < 58.01$	0.02	0.0025
v)	$57.999 < X < 58.001$	0.002	0.00025

b) From part a), it can be seen that the probability gets smaller as the range gets smaller. In fact, the probability is always smaller than the area (in terms of magnitude). As the width gets smaller, the range gets closer and closer to the number 58 without actually touching it. So, the probability becomes closer and closer to $P(X = 58)$, which will be zero.

7. **a)**

Price Interval	Frequency	Relative Frequency	Density
\$1–\$2	5	0.020	0.020
\$2–\$3	8	0.032	0.032
\$3–\$4	69	0.276	0.276
\$4–\$5	88	0.352	0.352
\$5–\$6	51	0.204	0.204
\$6–\$7	16	0.064	0.064
\$7–\$8	8	0.032	0.032
\$8–\$9	5	0.020	0.020

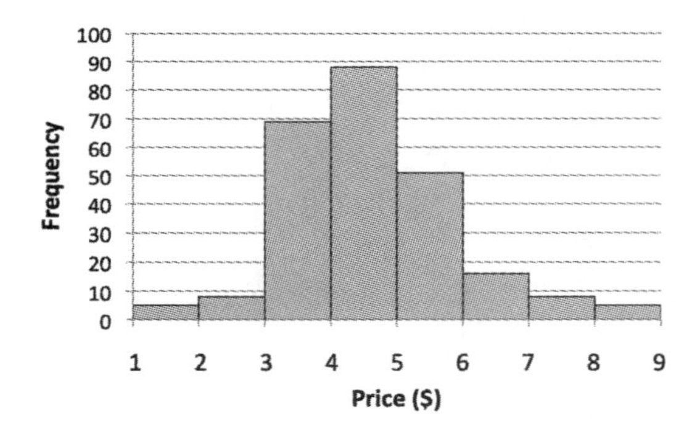

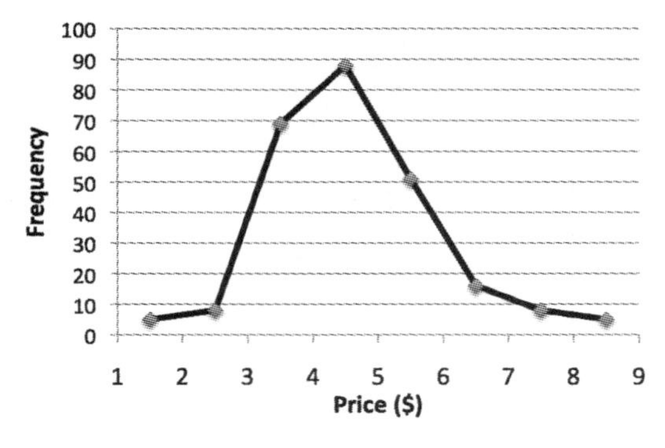

b) To make the bars, you need to know how tall the stack of bars would be if you totalled them up. This would produce one large rectangle with a total area of 1 and, in this case, a width of 1. Since there are 250 pieces of data, the height of each piece is $\frac{1}{250} = 0.004$ and the height of each bar is $0.004 \times$ the relative frequency (see table).

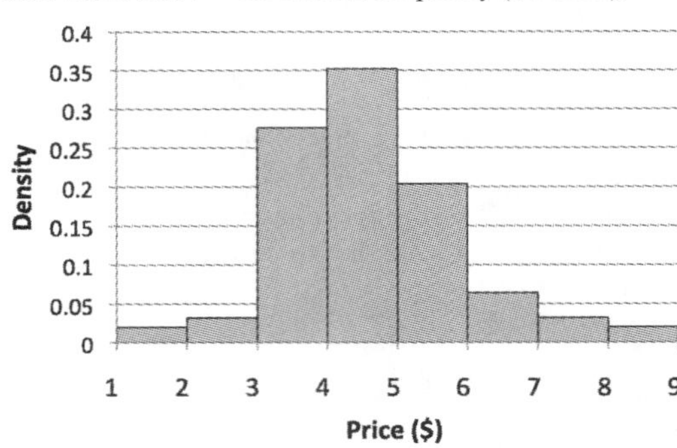

c) The area of each bar is the width of each bar times its height. Since the width in this case is 1, the actual value is equal to the density value. In fact, regardless of the width, the area of any bar is equal to the bar's relative frequency. This will be the case for any histogram. **d)** The area is the sum of the areas of the bars \$3–\$4 and \$4–\$5: $P(\$3–\$5) = 0.628$.

8. **a)** The height of the density is $\frac{1}{60}$ or $0.01\overline{6}$.

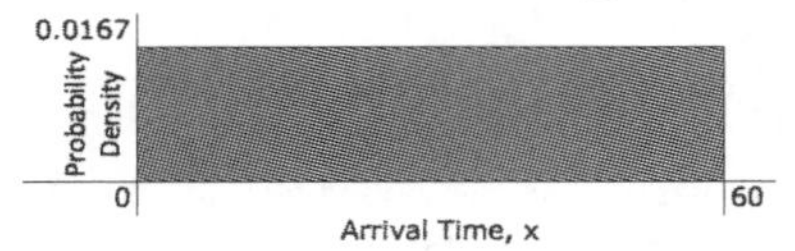

b) $P(0–5) = (5)(0.01\overline{6})$ or $0.08\overline{3}$
c) $P(15–25) = (10)(0.01\overline{6})$ or $0.1\overline{6}$
d) Go to *www.mcgrawhill.ca/links/MDM12*, select **Study Guide** and follow the links to obtain a copy of the file **WebserveSol.ftm** to see the solution.

9. **a)** The height is $\frac{1}{1.5}$ or $0.\overline{6}$.

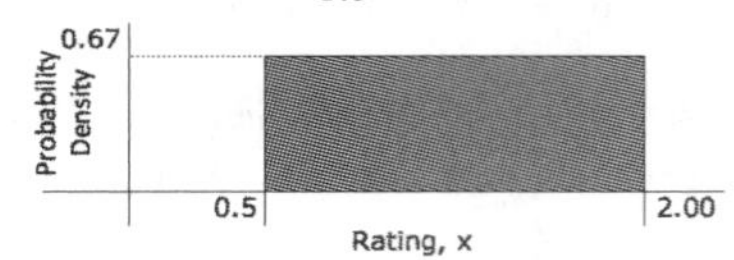

b) $P(1.4-2.0) = (0.6)(0.\overline{6})$ or 0.4
c) $P(1.0-1.1) = (0.1)(0.\overline{6})$ or $0.0\overline{6}$ **d)** The distribution is skewed right. The middle 50% of the data is between 0.846 and 0.971.

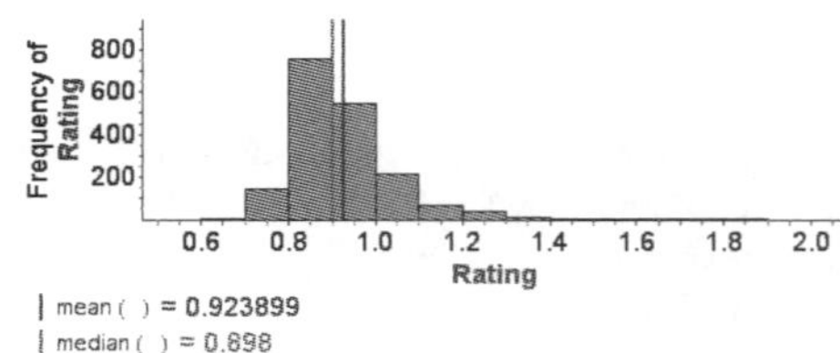

10. **a)** sample formula for the frequency:

Frequency	Rel. Freq.
3	0.006
=COUNTIF(A\$2:A\$501,">="&B3)-COUNTIF(A\$2:A\$501,">="&B4)	
COUNTIF(range, criteria)	0.084

Note that the bin mid-point is needed for the graphs.

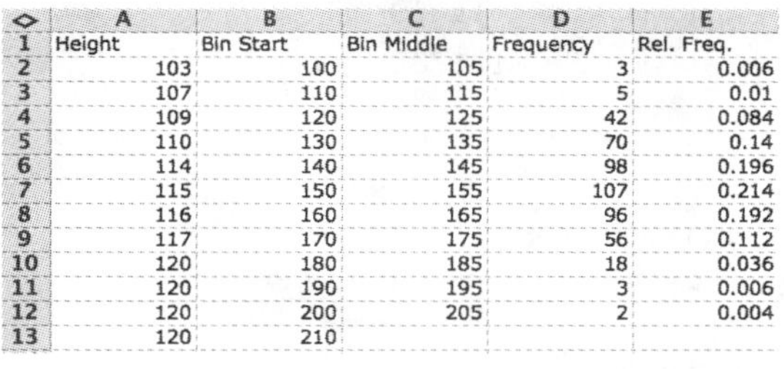

	A	B	C	D	E
1	Height	Bin Start	Bin Middle	Frequency	Rel. Freq.
2	103	100	105	3	0.006
3	107	110	115	5	0.01
4	109	120	125	42	0.084
5	110	130	135	70	0.14
6	114	140	145	98	0.196
7	115	150	155	107	0.214
8	116	160	165	96	0.192
9	117	170	175	56	0.112
10	120	180	185	18	0.036
11	120	190	195	3	0.006
12	120	200	205	2	0.004
13	120	210			

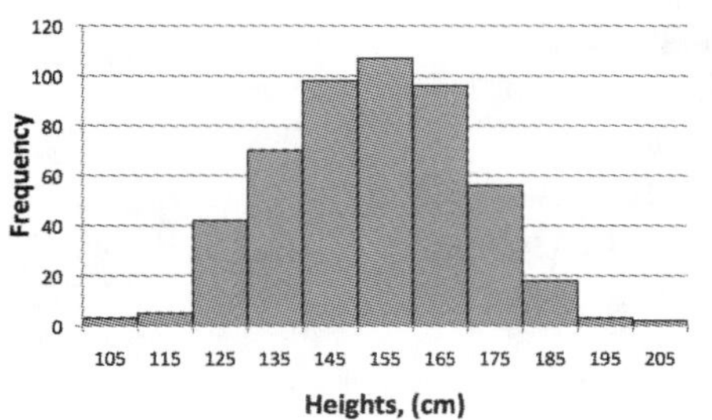

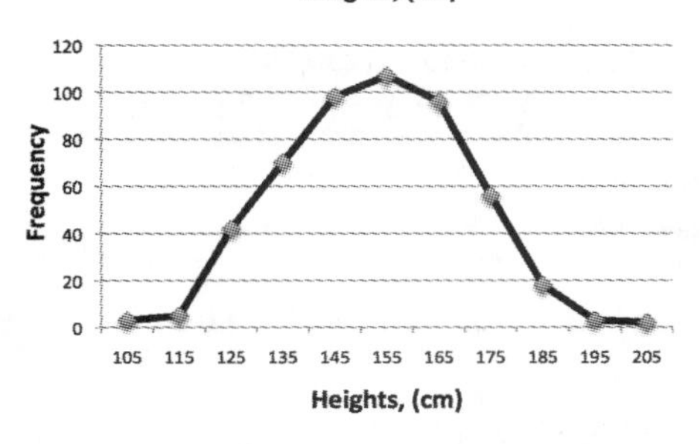

b) The approximate density value for each bar is the bar's relative frequency divided by 10, the interval (bin width).

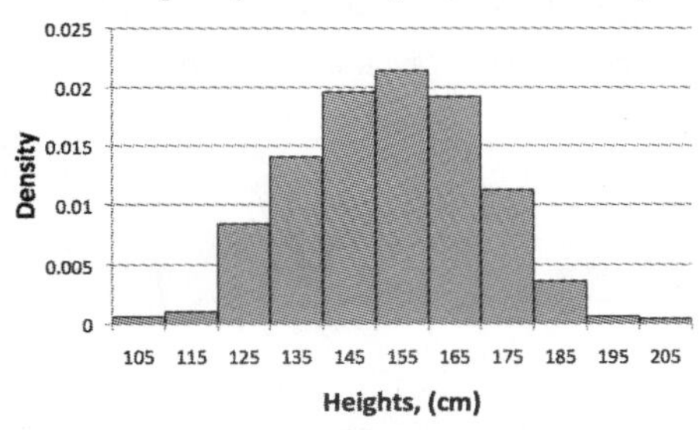

c) 0.842

d) between 140 and 170; $P(140 < X < 170) = 0.602$

11. **a)** 7898

b) skewed right (and also slightly bimodal), since data trail off to the right and the mean is to the right of the median

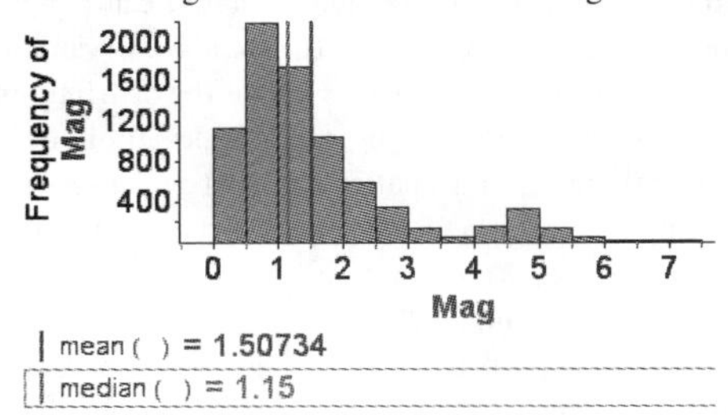

c) Sample answer: In general, the shape stays similar. However, as widths get smaller, gaps form (even though this is continuous data). So, if a frequency polygon was used, it would have a very jagged shape rather than one that follows the general shape of the data. Note also that the bimodal nature of the data becomes more evident as well.

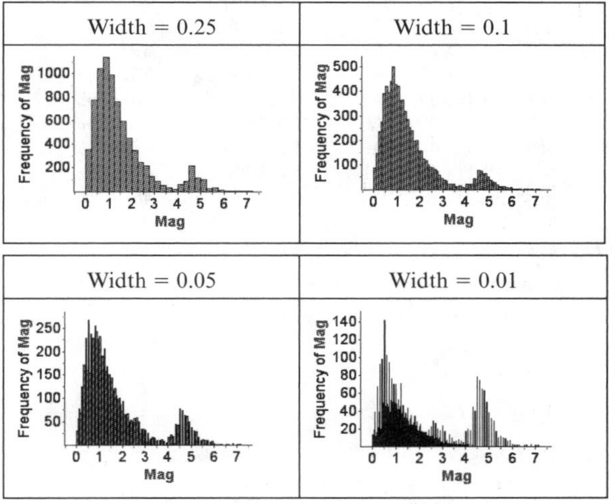

12. **a)** the number of people who started smoking when they were between 5 and 12 and are still smokers **b)** the total number of people who started smoking in that age interval

	A	B	C	D	E	F	G
1		Canadian			Ontario		
2		Canadian Smo	Canadian Quit	Totl Can Smok	Ontario Smok	Ontario Quitte	Tot Ont Smokers
3	5 TO 11 YEARS	145738	105612	251350	44258	29866	74124
4	12 TO 14 YEAI	1026210	899168	1925378	315047	279720	594767
5	15 TO 19 YEAI	3141313	3611030	6752343	1157858	1298666	2456524
6	20 TO 24 YEAI	766994	1152493	1919487	300816	441849	742665
7	25 TO 29 YEAI	208166	277474	485640	75318	108349	183667
8	30 TO 34 YEAI	90048	116200	206248	37900	38157	76057
9	35 TO 39 YEAI	39656	47413	87069	12885	18471	31356
10	40 TO 44 YEAI	26270	29579	55849	11671	9019	20690
11	45 TO 49 YEAI	13565	14507	28072	5783	5523	11306
12	50 YEARS OR	9094	11334	20428	3231	3030	6261

c) graph using the midpoint of each interval:

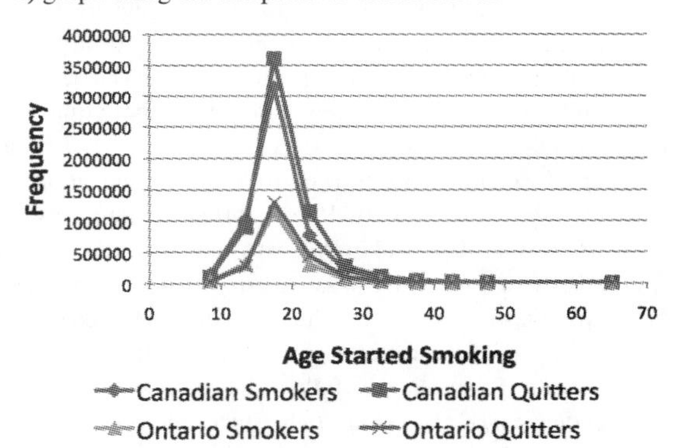

d) In most cases, the Canadian data are similar to the Ontario data. Most people seem to start smoking when they are younger than 20.

17		Canadian		Ontario	
18	Percentages	Canadian Smo	Canadian Quit	Ontario Smok	Ontario Quitters
19	5 TO 11 YEARS	2.67%	1.69%	2.25%	1.34%
20	12 TO 14 YEAI	18.77%	14.35%	16.03%	12.53%
21	15 TO 19 YEAI	57.46%	57.64%	58.93%	58.17%
22	20 TO 24 YEAI	14.03%	18.40%	15.31%	19.79%
23	25 TO 29 YEAI	3.81%	4.43%	3.83%	4.85%
24	30 TO 34 YEAI	1.65%	1.85%	1.93%	1.71%
25	35 TO 39 YEAI	0.73%	0.76%	0.66%	0.83%
26	40 TO 44 YEAI	0.48%	0.47%	0.59%	0.40%
27	45 TO 49 YEAI	0.25%	0.23%	0.29%	0.25%
28	50 YEARS OR	0.17%	0.18%	0.16%	0.14%
29	i)	78.90%		77.22%	
30	ii)		73.68%		72.03%
31	iii)	17.84%		19.14%	
32	iv)		22.83%		24.64%

e) You are least likely to quit if you started before the age of 15. You are most likely to quit if you start smoking in your 20s.

36	Rate of quittin	Canada	Ontario
37	5 TO 11 YEAR:	42.02%	40.29%
38	12 TO 14 YEAI	46.70%	47.03%
39	15 TO 19 YEAI	53.48%	52.87%
40	20 TO 24 YEAI	60.04%	59.50%
41	25 TO 29 YEAI	57.14%	58.99%
42	30 TO 34 YEAI	56.34%	50.17%
43	35 TO 39 YEAI	54.45%	58.91%
44	40 TO 44 YEAI	52.96%	43.59%
45	45 TO 49 YEAI	51.68%	48.85%
46	50 YEARS OR	55.48%	48.39%

13. a) $k = \frac{1}{\mu} = \frac{1}{18.3}$ and $P(X) = ke^{-kx}$

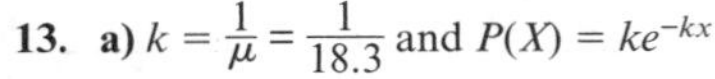

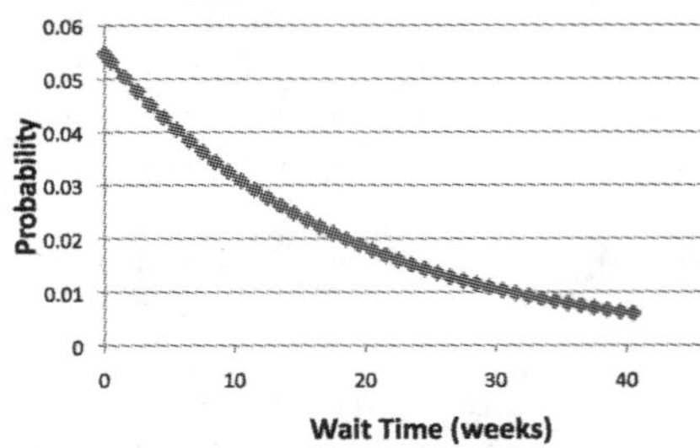

b) $P(X < 10) = 0.476$

c) Go to *www.mcgrawhill.ca/links/MDM12*, select **Study Guide**, and open the file **KneeWaitSol.ftm** to see the solution. Note that when using Fathom™, $P(X < 10) = 0.421$, which will be a more accurate answer.

14. $P(X > 4 \text{ min}) = 0.147$

15. In general, the patterns that existed for Canada and Ontario exist here. There are some anomalies. For example, the Territories consistently have the greatest percent of people who started smoking at a young age and also have the lowest percent of people who actually quit. BC, which has the lowest percent of both quitters and current smokers who started under the age of 20, also has the highest percent of quitters and current smokers who started in their 20s. If you started smoking between the ages of 40 and 44 in PEI, you have only a 17% chance of quitting, while if you started smoking after age 50 in Alberta, you have a 91% chance of quitting. However, in both of these last cases, there was a relatively small sample of data. For the full results, go to *www.mcgrawhill.ca/links/MDM12*, select **Study Guide**, and open the file **SmokersProvincesSol.xls**.

16. $P(0.75 < X < 1.00) = 0.789$; By making intervals with widths of 0.05, the relative frequency can be determined for each bin between 0.75 and 1.00.

8.2 Properties of the Normal Distribution, pages 124–129

1. a) −1.258 b) 0.841 c) −0.462

2. a) 0.2148 b) 0.9671 c) 0.9582

3. a) graphing calculator

b) spreadsheet

0.96709705 =NORMDIST(1981.4,1952.7,15.6,1)

NORMDIST(x, mean, standard_dev, cumulative)

c) Fathom™

Measure	Value	Formula
no_3c	0.958482	normalCumulative (13.742, 13.716, 0.015)

4. a) calculator

b) spreadsheet

0.18612201 =NORMDIST(230,245.8,22.9,1)-NORMDIST(210,245.8,22.9,1)

NORMDIST(x, mean, standard_dev, cumulative)

c) Fathom™

Measure	Value	Formula
no_4c	0.14151	normalCumulative (4.0, 3.56, 0.31) - normalCumulative (3.8, 3.56, 0.31)

5. a)

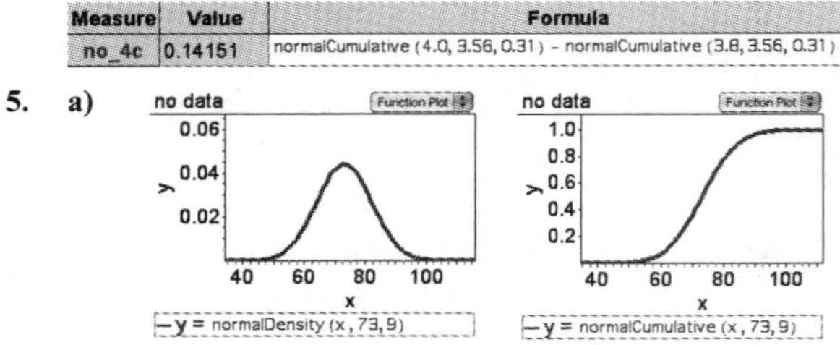

b) the total probability below any x-value

6. a)

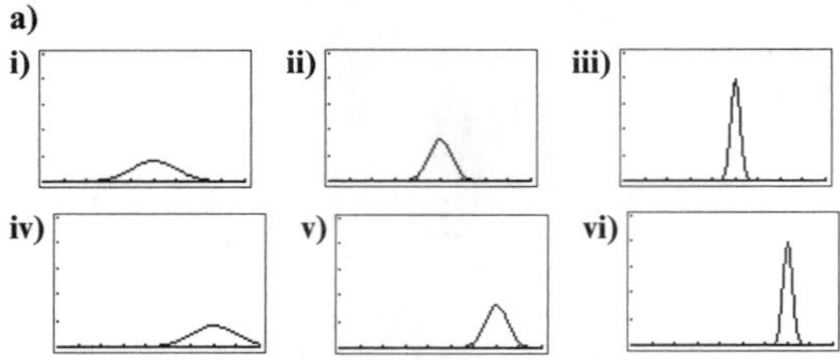

b) Moving the sliders should change the position and shape of the normal distribution.

c) Sample answer: Changing the mean while keeping the standard deviation constant preserves the shape of the distribution but shifts its position so that its centre is wherever the mean is. Changing the standard deviation while keeping the mean constant makes the distribution shorter and wider (for a larger standard deviation) or taller and thinner (for a smaller standard deviation) while preserving the position of the centre of the distribution.

7. a) $\mu = 92$, $\sigma = 4.2$ b) $\mu = 235$, $\sigma = 10.5$

8. a) i) 0.683 ii) 0.954 iii) 0.997

b) The ranges in parts i), ii), and iii) are one, two, and three standard deviations, respectively, on either side of the mean. Each has the expected probability.

9. **a) i)** 0.117 **ii)** 0.021 **iii)** 0.002
b) Sample answer: For a continuous distribution, the probability of any particular value is 0. In this case, the distribution is not quite continuous, so the probability of a particular mark is small but not equal to 0. The student is correct in the sense that the probability of achieving 80% is statistically improbable.
10. **a)** 57.9% **b)** 0.993 **c)** $46.75 < X < 62.65$
11. $\sigma_{\text{males}} = 11.2$, $\sigma_{\text{females}} = 12.22$
12. Sample answer: Although the Shift 1 worker has the least mean time, this worker has the greatest standard deviation and is the most inconsistent. The other two workers, even though they are slower, are more consistent and stay within the 54 s. The Shift 1 worker goes over the 54 s more often, so this worker should get some retraining.

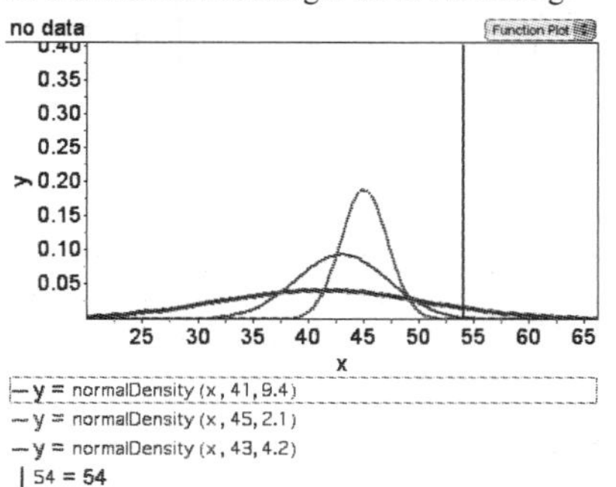

13. $X > 141.72$ or $X < 89.88$
14. **a)** By using the technology of your choice and some trial and error, you can get several scenarios. In this sample answer, Fathom™ is used with the normalCumulative() function used in the Measures section to see which has a probability of 0.95 of being smaller than $x = 140$. Go to *www.mcgrawhill.ca/links/MDM12*, select **Study Guide**, and open the file **TextMessageSol.ftm** to check this dynamically. Three solutions are shown: **i)** $\mu = 89$ and $\sigma = 31$ **ii)** $\mu = 96$ and $\sigma = 26.7$ **iii)** $\mu = 100$ and $\sigma = 24.3$

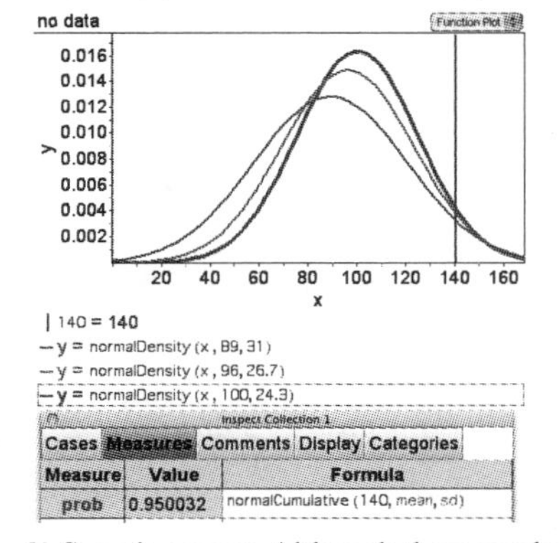

b) Sample answer: Although these numbers yield a probability of 0.95 for values less than 140, the left tail extends beyond zero into negative numbers. Although this is mathematically correct, realistically a message cannot have fewer than zero characters.

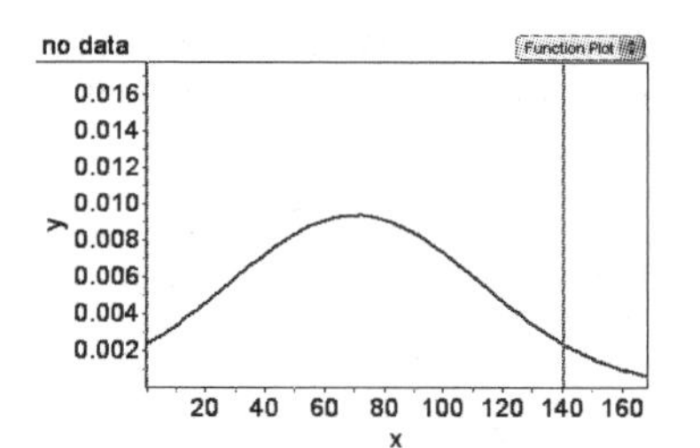

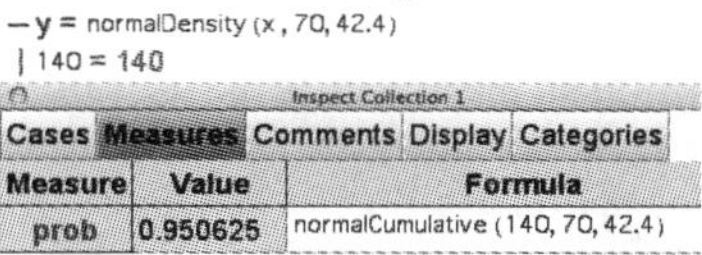

c) Sample answer: One reason is that the data are not continuous. What is more likely is that the shape is not normal, with many messages showing up as lower character counts. Thus, you would see a distribution that is skewed right. Part b) is another indication that a normal model may not be correct.
15. Sample answer: The class with the smaller standard deviation has the grades more closely spaced. A mark of 90% would be more unusual and less likely in this class, and therefore more meaningful
16. Sample answer: If data were normally distributed, the left tail would extend below 0 days. Since you cannot have a negative number of recovery days, this model does not work. This is more likely an exponential (waiting time) distribution or one that is skewed to the right.

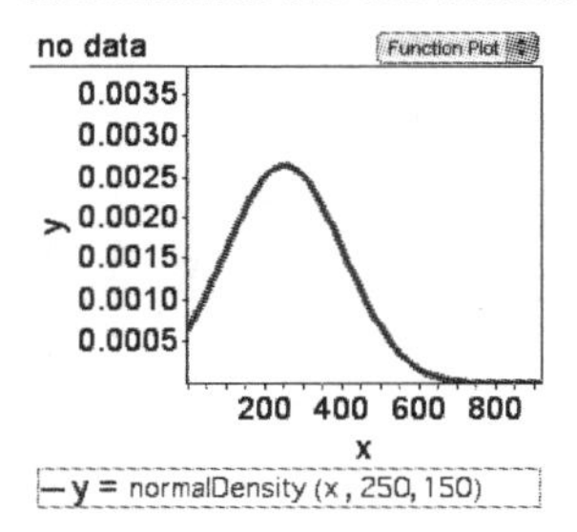

8.3 Normal Sampling and Modelling, pages 130–133

1.

	i) One Standard Deviation	ii) Two Standard Deviations	iii) Three Standard Deviations
a)	$37 < X < 53$	$29 < X < 61$	$21 < X < 69$
b)	$72.4 < X < 102.8$	$57.2 < X < 118$	$42 < X < 133.2$
c)	$13.762 < X < 13.808$	$13.739 < X < 13.831$	$12.716 < X < 13.854$

2. **a)** 83 **b)** 5.3 **c)** 0.401
3. **a)** Since the interval with the highest bar is centred within the data, assume that the midrange is close to the actual mean. Estimate 90 km/h as the mean. Since there are 70 pieces of data, there should be $(0.68)(70) = 47.6$ or about 48 data pieces within one standard deviation of the mean. Between 82.5 and 97.5 there are 46 pieces of

data. Approximate the standard deviation by dividing that interval in half to get $s = 7.5$.

b) Adjust the mean to 91:

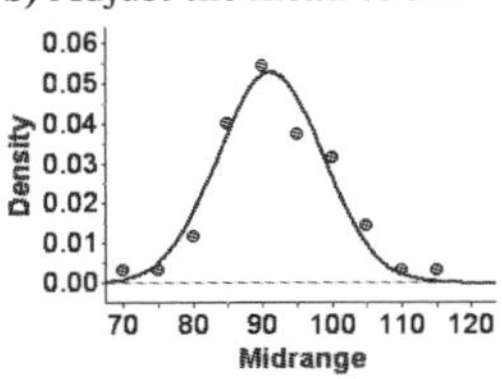

4. **a)** Sample answer: Count the data using the standard deviation to determine the ranges. Use half of the outermost interval counts to approximate 40 and 90 as boundaries. These percents closely match what is expected of a normally distributed set of data, though, as expected, the tails do not match as well as the areas closer to the mean.

Range	Percent of Data in Range
1 SD: $52.5 < X < 77.5$	65.7%
2 SD: $40 < X < 90$	95.5%
3 SD: $27.5 < X < 92.5$	97.9%

b) 0.655

5. **a)**

Interval	Frequency
120–130	2
130–140	7
140–150	11
150–160	13
160–170	8
170–180	5
180–190	3
190–200	1

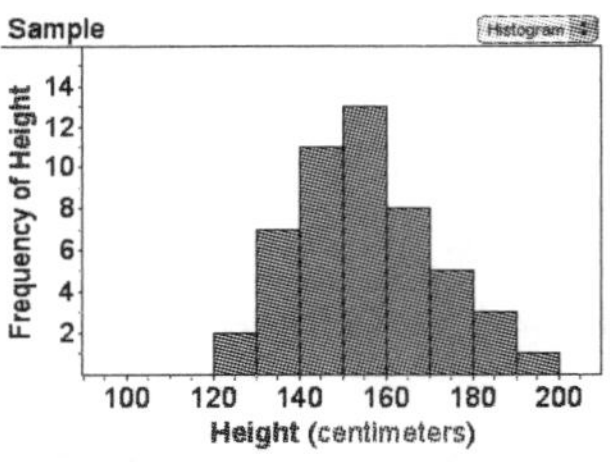

b) Sample answer: The distribution is slightly skewed right and is not quite normal, as about 72% of the data are within one standard deviation of the mean.

c) Sample answer: In general, the shapes are similar. However, the smaller the interval, the more gaps there are. You can see changes in the shape by shifting the starting position. This is mostly evident in the two sample graphs with intervals of 10.

Population:

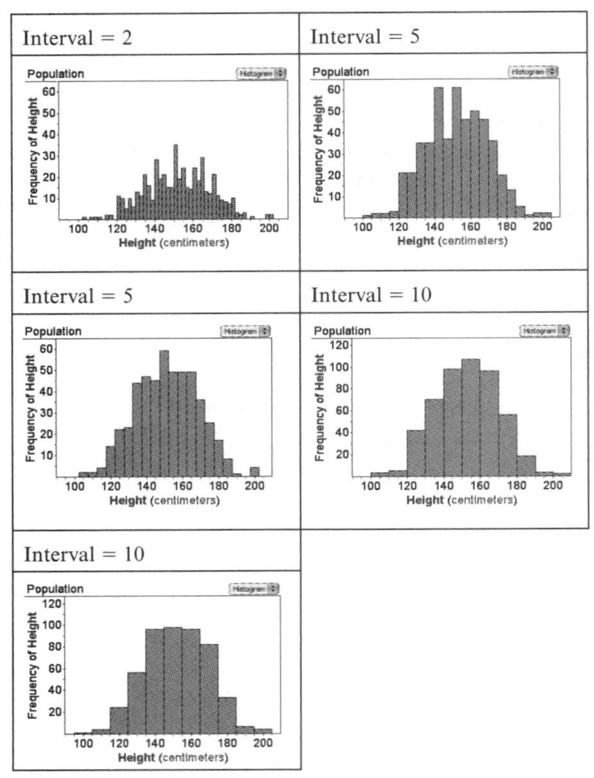

Sample:

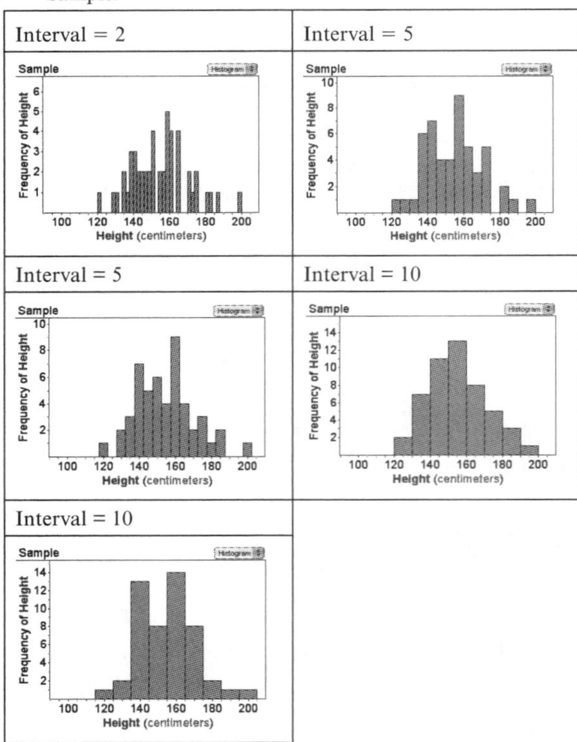

d) Sample answer: The shape is not quite the same and the mean and standard deviation are off a bit, but the sample still gives a reasonable idea of the population.

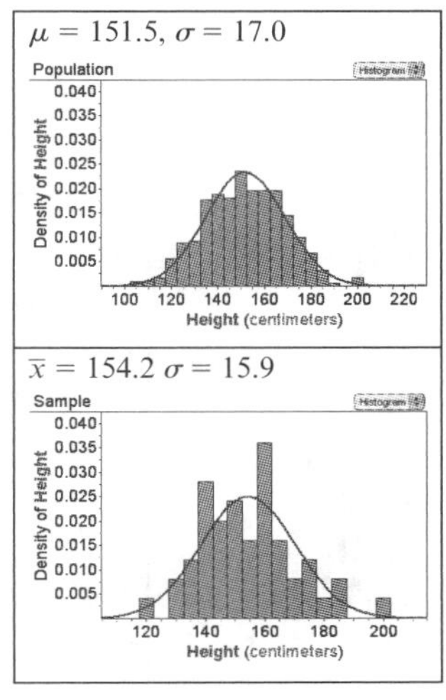

6. a) $\bar{x} = 12.4$, $s = 3.6$ **b)** 0.111 **c)** 0.512

7. **a)** Around the Bay - All

Sample answer: The distribution does not have a normal shape, because winning times include times for both men and women, who tend to run at different speeds, as well as times before 1982 when the course was slightly longer . Over time, winning times tend to get faster.

b) Sample answer: Neither fits a normal distribution very well. After 1982 the data seem to be less spread out.

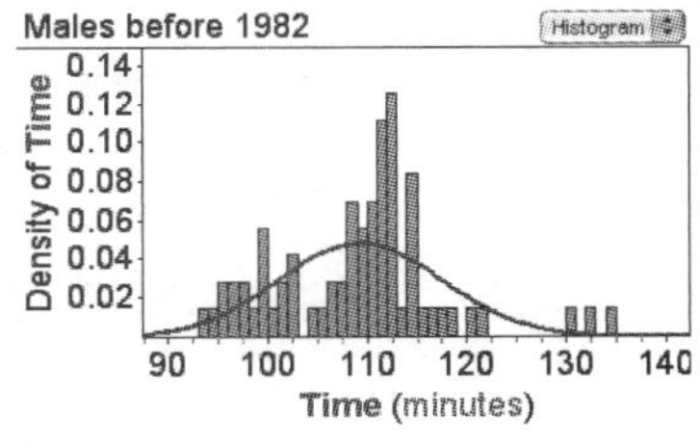

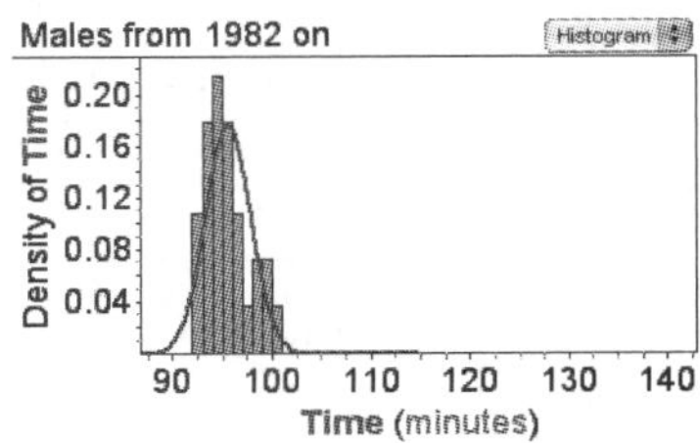

c) Before 1982, $P(X < 100) = 0.134$ and after 1982, $P(X < 100) = 0.975$. Sample answer: The values make sense, because after 1982 the course got shorter, so the times should be faster. In addition, the best times generally get faster over the years.

8. **a)** The distribution is bimodal.

b) older $\bar{x} = 2.51$, $s = 0.03$; newer $\bar{x} = 2.26$, $s = 0.04$

c) A coin weighing less than 2.3 g must be a new coin. First, determine the probability of getting a new coin. Then, use the probability of getting one less than 2.3 g:

$P(\text{new}) \times P(X < 2.3) = \left(\frac{1254}{1750}\right)(0.841)$ or 0.603

d) A coin that weighs more than 2.5 g is an old coin:

$P(\text{old}) \times P(X > 2.5) = \left(\frac{496}{1750}\right)(0.631)$ or 0.179

9. **a)** $P(X < 580.4) = 0.291$

b) $P(X < 600) = 0.679$. This is the probability of having a time below 10 h. There is a 32.1% chance of qualifying if you take 10 h to finish.

8.4 Normal Approximation to the Binomial Distribution, pages 134–138

1. **a)** $81.5 < X < 82.5$ **b)** $X < 3.5$

c) $X > 1186.5$ **d)** $X < 78.5$

2. **a)** $\mu = 90$, $\sigma = 7.04$

b) $\mu = 400$, $\sigma = 8.94$

c) $\mu = 9$, $\sigma = 2.85$

3. **a)** $\mu = 22.5$, $\sigma = 3.34$

b) $\mu = 11.25$, $\sigma = 3.02$

c) $\mu = 160$, $\sigma = 5.51$

d) $\mu = 41.1$, $\sigma = 5.27$

4. The mean and standard deviation are given: $\mu = 13.5$ and $\sigma = 3.4$. Since the data are discrete, in order to find the probability of "at least" 18 clips, it is necessary to use 18.5 in the calculation:

$P(X > 18.5) = P(Z > \frac{18.5 - 13.5}{3.4}) = P(Z > 1.47)$.

The solution may be found by determining the opposite, so $P(Z > 1.47) = 1 - P(Z < 1.47)$. For the value of $P(Z < 1.47)$, use the table of areas under a normal distribution curve on pages 606 and 607 of *Mathematics of Data Management* student edition. From the table, it can be seen that $P(Z < 1.47) = 0.9292$. So,

$P(Z > 1.47) = 1 - P(Z < 1.47) = 1 - 0.9292 = 0.0708$

5. **a) i)** 0.6172 **ii)** 0.6019

b) The answers are relatively close (to one decimal place).

c) i) 0.5535 **ii)** 0.5460

d) i) 0.5170 **ii)** 0.5146

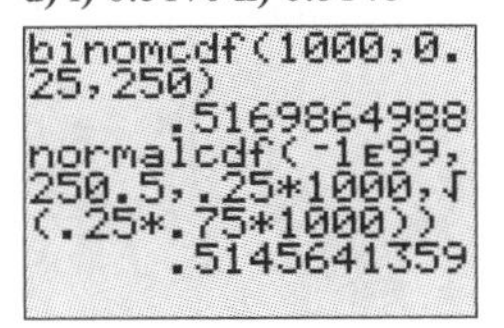

e) The approximation gets better as the number of samples increases.

6. **a) i)–v)**

p	Binomial	Normal	Difference
0.5	0.539 794 619	0.539 827 837	0.000 033 219
0.4	0.543 294 486	0.540 646 297	0.002 648 189
0.3	0.549 123 601	0.543 441 960	0.005 681 641
0.2	0.559 461 585	0.549 738 225	0.009 723 360
0.1	0.583 155 512	0.566 183 833	0.016 971 680

b) As the probability gets farther away from 0.5, the difference increases. The normal distribution is a better approximation when p is closer to 0.5.

7. **a)** $\bar{x} = 93.76$, $s = 6.33$

b) 0.0376

8. **a)** Both $np > 5$ and $nq > 5$, so a normal distribution is an appropriate approximation. There is a 100% chance that at least one quarter of accidents are related to cell phone use.

b) It is probably realistic.

9. Since $np = 10$ and $nq = 240$ are both greater than 5, a normal distribution can be used.

a) 99.9%

```
normalcdf(-1E99,
9.5,.04*250,√(.0
4*.96*250))
       .0010843798
1-Ans
       .9989156202
```

b) 12.8 %

10. **a) i)–iii)**

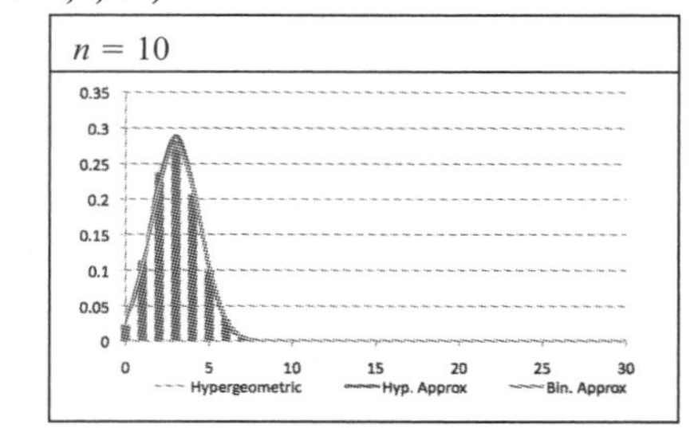

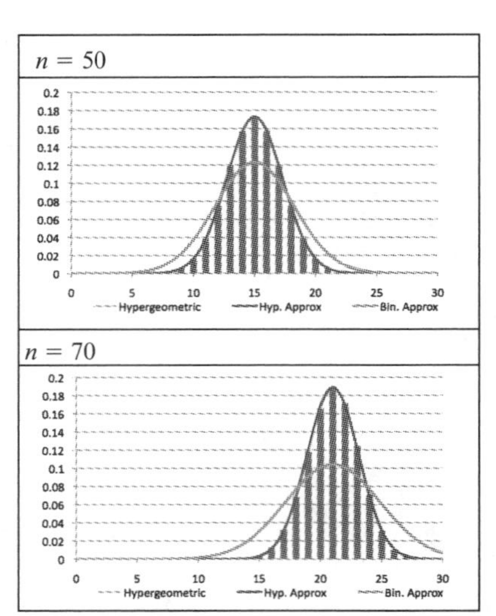

$n = 70$

b) As the proportion of the sample to the population increases, the normal approximation of the hypergeometric distribution improves.

11. **a)** Go to *www. mcgrawhill.ca/links/MDM12*, select **Study Guide**, and open the file **TownSurveySol.xls** to see the calculations.

					Hyp		Norm	
	N	n	p	a	x	$P(X \geq x)$	x	$P(X > x)$
Town A	550	60	0.16	88	9	0.649	8.5	0.659
Town B	1500	100	0.22	330	15	0.974	14.5	0.969
Town C	3500	200	0.18	630	30	0.894	29.5	0.891

b) The normal distribution is easier, because only one calculation is needed for each trial.

c) The hypergeometric distribution is more accurate, because the normal distribution is an approximation.

d) Town C

12. **a)** $P(X > 23.5) = 0.000\ 16$ **b)** $P(X > 129.5) = 0.025$

13. **a)** and **b)** Find the z-score of each age group and use the table to determine the percent. A continuity correction is needed, since this is a multiple-choice test.

Age Group	μ	σ	Z	$P(X < 32.5)$	IQ
Under 25	28	7.5	0.60	0.725 7	109
25–34	27	7.5	0.73	0.767 3	111
35–49	25	7.5	1.00	0.841 3	115
50–64	24	6	1.42	0.922 2	121
Over 64	20	5	2.50	0.994 6	138

8.6 Confidence Intervals, pages 139–142

1. **a)** $E = 2.5$, $212.5 < \mu < 217.5$

b) $E = 0.6$, $34.4 < \mu < 35.6$

c) $E = 0.20$, $9.45 < \mu < 9.85$

d) $E = 0.022$, $0.514 < \mu < 0.554$

2. **a)** 98 **b)** 38 **c)** 554 **d)** 300

3. **a)** $\hat{p} = \frac{30}{50} = 0.6$,

$$\hat{p} - z_{\frac{\alpha}{2}}\left(\sqrt{\frac{\hat{p}\hat{q}}{n}}\right) < p < \hat{p} + z_{\frac{\alpha}{2}}\left(\sqrt{\frac{\hat{p}\hat{q}}{n}}\right)$$
$$0.6 - 1.645\left(\sqrt{\frac{(0.6)(0.4)}{50}}\right) < p < 0.6 + 1.645\left(\sqrt{\frac{(0.6)(0.4)}{50}}\right)$$
$$0.6 - 0.114 < p < 0.6 + 0.114$$
$$0.486 < p < 0.714$$

b) $0.288 < p < 0.462$
c) $0.366 < p < 0.507$
d) $0.379 < p < 0.541$

4. **a)** 57 **b)** 222 **c)** 289 **d)** 14 440

5. Sample answer: As the number of flips increases, the confidence interval becomes smaller. That is, it becomes more likely that the average number of heads will be closer to the expected number of heads as the number of flips increases.

# of Flips	% Error	E	Confidence Interval
10	31	3.1	$1.9 < \mu < 8.1$
100	9.1	9.1	$40.9 < \mu < 59.1$
500	4.4	22	$228 < \mu < 272$
1000	3.1	31	$469 < \mu < 531$

6. Sample answer: Here are some sample data. For the most part, at least 19 of the 20 experiments had a percent of heads within the confidence interval.

# of Flips	Confidence Interval	Number Within Interval
10	$0.19 < \mu < 0.81$	20
100	$0.409 < \mu < 0.591$	19
500	$0.456 < \mu < 0.544$	18
1000	$0.469 < \mu < 0.531$	19

7. **a)** Sample answer: 62% in favour with a margin of error of 30%, 19 times out of 20 means that if the poll were conducted 20 times, the politician's approval rating would be $32\% < p < 92\%$ in 19 of those 20 polls.

b) $n = 4(0.6)(0.4)\left(\frac{1.96}{0.3}\right)^2 = 10.24$. Thus, approximately 11 people were polled.

c) $n = 4(0.6)(0.4)\left(\frac{1.96}{0.1}\right)^2 = 92.20$ or approximately 91 people

8. **a)** $\sigma = 0.3$, $\bar{x} = 18.8$, $n = 20$ gives $E = 0.13$ to make the interval $18.67 < \mu < 18.93$
b) In order to guarantee at least 18.9 L 95% of the time, they should adjust the machine so the average is 0.23 L higher, making the interval $18.9 < \mu < 19.16$.

9. **a)** $\mu = 1206$ mg **b)** $E = 8$ mg, $n = 7$, $z_{\frac{\alpha}{2}} = 2.576$, $\sigma = 8.2$ mg **c)** $n = 40$, $z_{\frac{\alpha}{2}} = 2.576$, $\sigma = 8.2$ mg, $\bar{x} = 1201$ mg, $E = 3.3$ mg, $1197.7 < \mu < 1204.3$
The low end of this interval is quite close to the low end of the original range, so in that respect it is not that different. However, this interval is much smaller, so there will be a larger proportion of capsules under 1200 mg than in the acceptable range.

10. **a)** Larry: $E = z_{\frac{\alpha}{2}}\left(\sqrt{\frac{\hat{p}\hat{q}}{n}}\right)$
$= 1.96\left(\sqrt{\frac{(0.48)(0.52)}{500}}\right)$
$= 0.044$
$43.6\% < p < 52.4\%$

Constance: $E = z_{\frac{\alpha}{2}}\left(\sqrt{\frac{\hat{p}\hat{q}}{n}}\right)$
$= 1.96\left(\sqrt{\frac{(0.46)(0.54)}{500}}\right)$
$= 0.044$
$41.6\% < p < 50.4\%$
Since the two intervals overlap, it is possible that Larry's actual percent is less than Constance's.

b) Larry: $E = z_{\frac{\alpha}{2}}\left(\sqrt{\frac{\hat{p}\hat{q}}{n}}\right)$
$= 1.96\left(\sqrt{\frac{(0.475)(0.525)}{1000}}\right)$
$= 0.031$
$44.4\% < p < 50.6\%$

Constance: $E = z_{\frac{\alpha}{2}}\left(\sqrt{\frac{\hat{p}\hat{q}}{n}}\right)$
$= 1.96\left(\sqrt{\frac{(0.465)(0.535)}{1000}}\right)$
$= 0.031$
$43.4\% < p < 49.6\%$

c) Sample answer: Neither is better than the other due to the margin of error. Both situations are such that either person's actual percent could be higher than that of the other candidate.

11. For the 95% confidence level, in the previous question where there was a sample population of $n = 500$, the margin of error would be $E = \frac{98\%}{\sqrt{500}} = 4.4\%$. Another example is question 3c), where the margin of error would be $E = \frac{98\%}{\sqrt{190}} = 7.1\%$. For the 99% confidence interval, consider question 3d): $E = \frac{129\%}{\sqrt{250}} = 8.2\%$.

12. 25: $E = 0.1$, $z_{\frac{\alpha}{2}} = 1.96$, $\sigma = 0.25$,
$$n = \left(\frac{z_{\frac{\alpha}{2}}\sigma}{E}\right)^2$$
$$= \left(\frac{1.96(0.25)}{0.1}\right)^2$$
$$= 24.01$$

13. Go to *www.mcgrawhill.ca/links/MDM12*, select **Study Guide**, and follow the links to obtain a copy of the file **BinomialFlips.xls**. In each case, when the probability of each possible outcome is summed within the range for a 95% confidence interval, each case comes out to at least 25%.

14. Looking at the confidence intervals, you can see that for Ken, the range goes from $52.6 < p < 59.4$ to $49.0 < p < 55.0$. Since there is overlap, it is just as likely that he had approval ratings in that overlap, so his rating has not really gone up or down.

	Sample Size	Probability	$z_{\frac{\alpha}{2}}$	E	Interval
Ken	$n = 812$	0.56	1.96	3.4%	$52.6 < p < 59.4$
Marge	$n = 812$	0.32	1.96	3.2%	$28.8 < p < 35.2$
Ken	$n = 1040$	0.52	1.96	3.0%	$49.0 < p < 55.0$
Marge	$n = 1040$	0.36	1.96	2.9%	$33.1 < p < 38.9$

Chapter 9

Case Studies, pages 143–148

Sample answers:

1. A familiar scenario can help the reader identify with the subject, and thus be interested in what the rest of the project might contain. The introduction could have included a description of what the primary data will be and where they will come from (this could also be included in the next section, where the data are first mentioned).
2. When discussing outliers, it is good to include more information about them. In this case, you could name the magazines and give specific numerical data.
3. Whenever graphs are being compared, it is essential that the scales be similar, so that readers can properly compare them visually. This way, any differences, if they exist, will be noticeable and real.
4. One thing that could have been considered is whether the independent and dependent variables are correct. That is, the number of pages allowable may be related to the number of advertisement pages sold, so the axes could be reversed.
5. Since the data can be separated by type, it would have been nice to see the same sort of analysis but with the magazines separated by type. This way, another story may appear that shows that there are different advertisement characteristics for different types of magazines.
6. The magazines at the lower end actually have a lower percent. For example, *Time* and *In Style* have 22% and 55% respectively, yet they are both near the line of best fit. This difference is due to the initial value of the relationship, which has a much larger effect on the proportion of advertisement pages as you move down the graph. By looking at this relationship, readers can get a deeper understanding of what is happening with the data.
7. When drawing any conclusion, you should always address any questions or hypotheses that were listed in the introduction (whether you found the answer you were expecting or not). This conclusion has not summed up answers to the initial questions fully.
8. It could be improved with something to capture readers' attention so that they can relate to the subject. Perhaps a history of gambling or casinos could have been used, or some statistics on how much revenue casinos generate.
9. Other bets that could be considered are other pairs of numbers (e.g. 2 and 3 or 8 and 9), sets of four (e.g. 2, 3, 11, 12) or a set of three numbers (e.g., 2, 7, 12).
10. Showing some sample calculations would make the results clearer. One possible variation could be to change the configuration of the numbers on the board or change the board entirely. This way, when combinations of numbers are chosen, the probabilities might have more variety. An alternative board is shown.

	2	12
6	3	8
4	7	10
9	11	5

11. The game and its results might have been compared to regular roulette. In that game, each individual number has the same probability, whereas here, each number has a different probability, so combining numbers is more complex.

Practice Exam

Practice Exam Answers, pages 149–154

1. **a)** Sample answer: With Excel, use **RAND()*(91–16)+16.**

◇	A	B
1	24	50.2
2	38	
3	88	
4	73	
5	67	
6	16	
7	26	
8	50	
9	68	
10	52	

b) Sample answer: With Fathom™, use **randomPick ("yellow","red","red","blue","blue","blue", "green","green", "green","green").**

Collection 1

	numbers
1	yellow
2	green
3	blue
4	green
5	green
6	green
7	green
8	green
9	green
10	blue

2. **a)** categorical, nominal; The most popular types of movies are adventure and comedy.
b) numerical, discrete; The most common family size is four.

3.

Height	Frequency	Relative Frequency	Cumulative Frequency
150–155	15	0.263	15
155–160	32	0.561	47
160–165	10	0.175	57

4. volunteer sampling, measurement bias
5. mean: 86.6, median: 84, mode: 61, standard deviation: 17.1, $Q1$: 74, $Q3$: 104, interquartile range: 30
6. **a)** accidental **b)** reverse cause-and-effect **c)** presumed
7. Although \$2.422 billion is a lot of money, this works out to only \$242.20 for each Haitian.
8. **a)**

DICE	IDCE	CDIE	EDIC
DIEC	IDEC	CDEI	EDCI
DCIE	ICDE	CIDE	EIDC
DCEI	ICED	CIED	EICD
DEIC	IEDC	CEDI	ECDI
DECI	IECD	CEID	ECID

b)

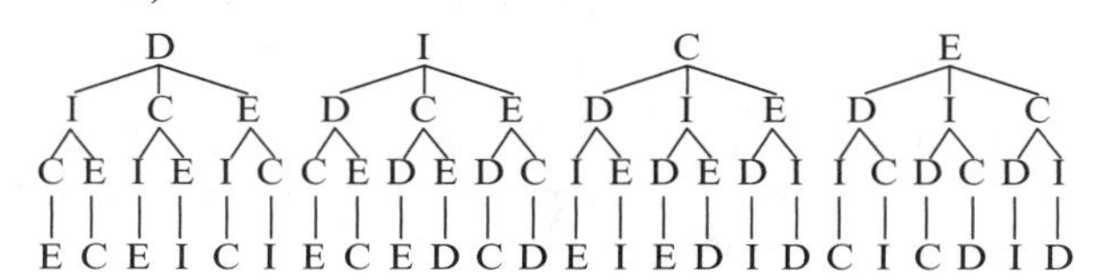

c) ${}_4P_4 = 24$

9. **a)** 212 520 **b)** 715 **c)** 44 102 888
10. **a)** SHI, SHF, SHT, SIF, SIT, SFT, HIF, HIT, HFT, IFT

b)

S — H — I: SHI
S — H — F: SHF
S — H — T: SHT
S — I — F: SIF
S — I — T: SIT
S — F — T: SFT
H — I — F: HIF
H — I — T: HIT
H — F — T: HFT
I — F — T: IFT

c) ${}_5C_3 = 10$

11. **a)** $\frac{15}{36}$ **b)** $\frac{5}{18}$
12. **a)** weight, amount of rain, temperature
b) age, number of children, number of people
13. 7.9
14. **a)** the number of correct answers on a multiple-choice test
b) drawing marbles from a bag with replacement until a red marble is drawn
c) drawing marbles from a bag without replacement until a red marble is drawn
d) measuring the width of pistons as they come off an assembly line
15. **a)** 67.5 **b)** $2.\overline{3}$ **c)** 5.625
16. **a)**

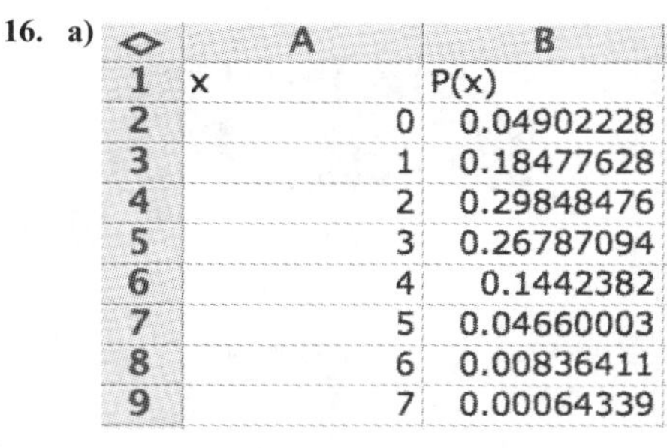

	A	B
1	x	P(x)
2	0	0.04902228
3	1	0.18477628
4	2	0.29848476
5	3	0.26787094
6	4	0.1442382
7	5	0.04660003
8	6	0.00836411
9	7	0.00064339

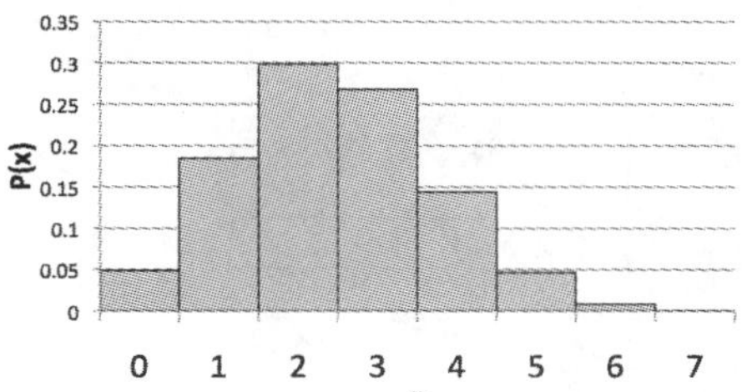

b)

	A	B
1	x	P(x)
2	0	0.4
3	1	0.24
4	2	0.144
5	3	0.0864
6	4	0.05184
7	5	0.031104
8	6	0.0186624
9	7	0.01119744

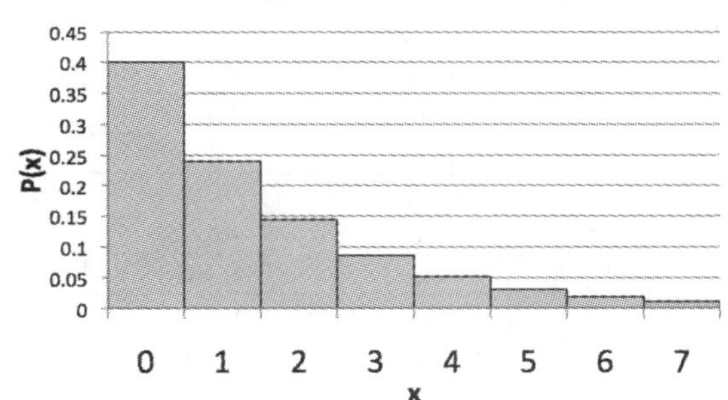

c)

	A	B
1	x	P(x)
2	0	0
3	1	0.000103
4	2	0.004334
5	3	0.047678
6	4	0.198658
7	5	0.357585
8	6	0.286068
9	7	0.095356

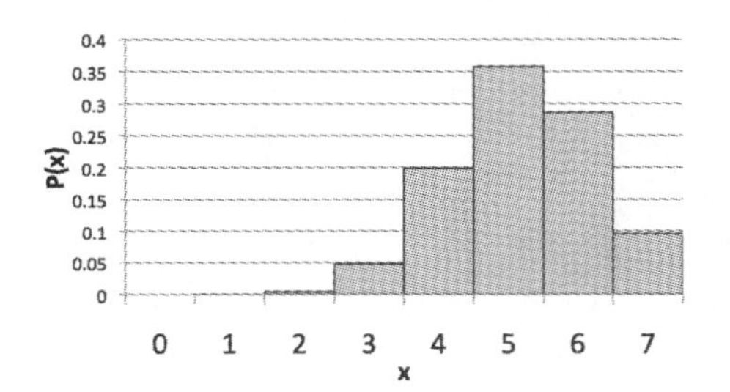

17. a)

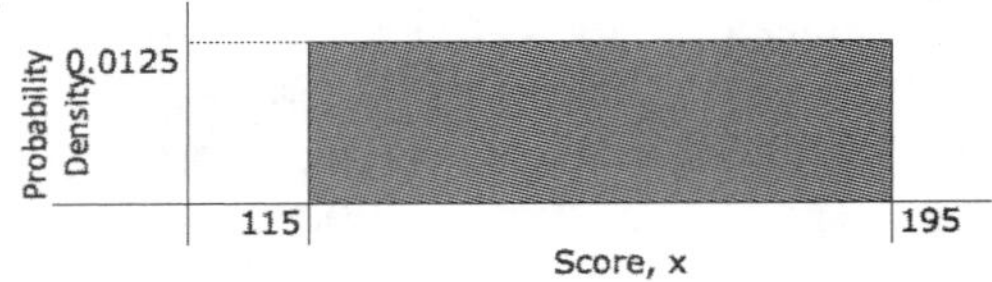

b) 0.625 **c)** 0

18. a) $0.536 < p < 0.714$

b) $0.433 < p < 620$

19. a) the total number of medals

b) Assign points: 3 for gold, 2 for silver, 1 for bronze to get the following results:

Rank	Country	Points
1	USA	70
2	Germany	63
3	Canada	61
4	Norway	49
5	Korea	32
6	Austria	30
7	Russia	26
8	China	23
8	Sweden	23
9	France	18

c) Using medals per million people:

Rank	Country	Medals per million
1	Norway	4.792
2	Austria	1.905
3	Sweden	1.118
4	Canada	0.765
5	Germany	0.366
6	Korea	0.281
7	France	0.168
8	USA	0.120
9	Russia	0.105
10	China	0.008

d)

Rank	Country	Value
1	Canada	\$9635.00
2	USA	\$9377.20
3	Germany	\$9293.80
4	Norway	\$7253.40
5	Korea	\$5028.80
6	Austria	\$3968.40
7	Sweden	\$3298.60
8	China	\$3298.60
9	Russia	\$3134.80
10	France	\$1994.40

20. a) and **b)** The data have a strong positive correlation, with $r = 0.917$.

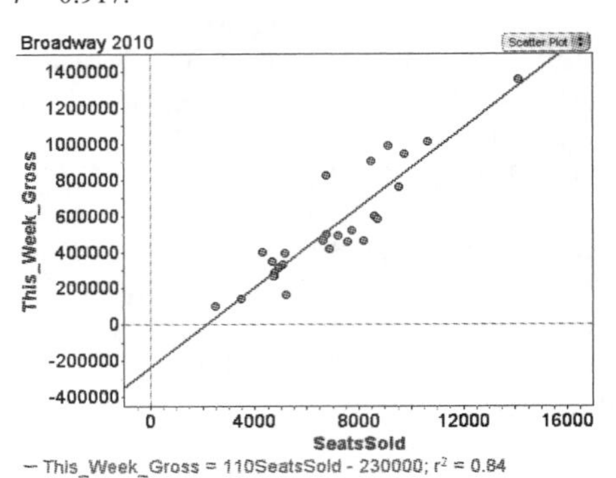

The equation of the line of best fit is $y = 110x - 230\,000$, where x represents the number of seats sold in a week and y represents the week's gross revenue.

c) The slope is the average price per seat sold.

d) The y-intercept could represent the amount invested in each production before it starts.

e) \$1 970 000. This may not be accurate due to the amount of extrapolation involved.

f) The data suggest that, once on Broadway, the play is probably deemed to be of high quality, so the only way for it to make more money is to use a bigger theatre.

21. a) Troy Matteson

b)

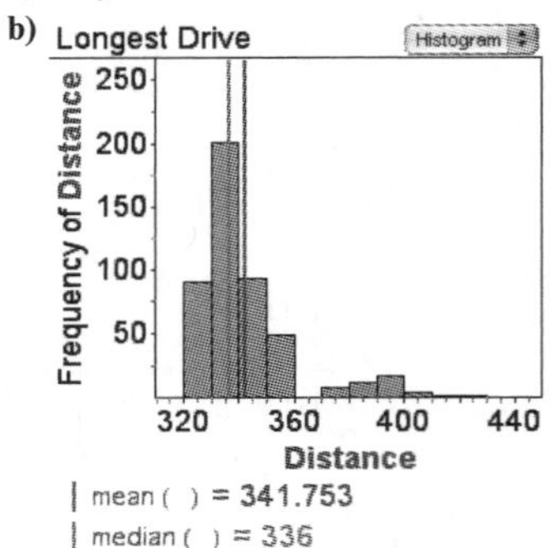

standard deviation 17.25

c) 41 outliers

d) This data is skewed to the right, since the data trails off to the right and the mean is larger than the median.

22. a) 308 900 776 **b)** 2.82×10^{12}

23. $729x^6 - 5832x^5 + 19\,440x^4 - 34\,560x^3 + 34\,560x^2 - 18\,432x + 4096$

24. a) 54 264 **b)** 18 480 **c)** 8640

25. $P(4 \text{ girls, } 2 \text{ boys}) = \dfrac{({}_{12}C_4)({}_9C_2)}{{}_{21}C_6} = 0.328$

26. **a)**

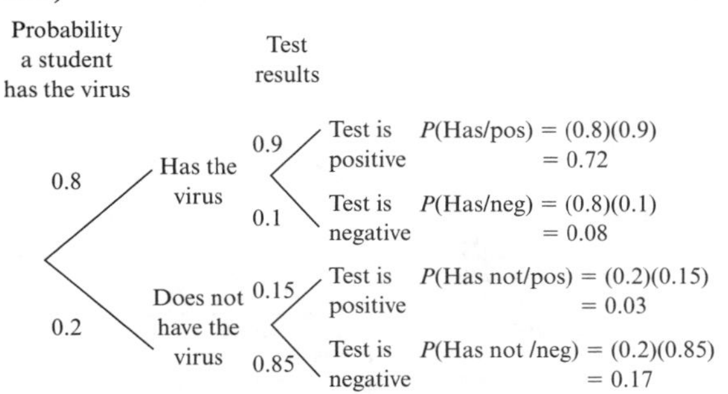

b) $P(\text{Has/pos}) = (0.8)(0.9) = 0.72$

c) $P(\text{negative correct}) = \dfrac{0.17}{0.17 + 0.08} = 0.68$

27. **a)** 10 **b)** $250; They need to charge more than $250 for collision insurance in order to make a profit.
28. **a)** 0.049 **b)** $E(X) = 0.05$
29. **a)** hypergeometric

b)

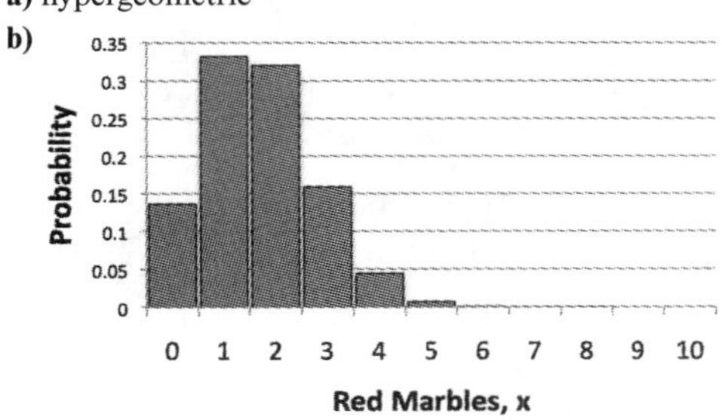

c) 0.211

30. **a)** 99 **b)** 1×10^{-12}
31. **a)** $P(\text{passing}) = 0.014$ or 1 in 71, which is better than Lloyd's chances.

b) $P(75\%) = 3.81 \times 10^{-6}$ or 1 in 262 467, which is better than Lloyd's chances.

32. **a)** 0.501 **b)** 0.001
33. **a)** $23.0 < \mu < 26.0$ **b)** According to part a), the longest mean expected ride time was 26 s. It is expected that 95% percent of the times should be within two standard deviations of the mean, so the longest a person would have to wait is 32.4 s, and so the gate should be set at that time.
34. **a)** Use any random number generator to generate numbers between two values; the smaller the interval is, the closer the data will be to uniform. Go to *www.mcgrawhill.ca/links/MDM12* and select **Study Guide**. Follow the links to obtain a copy of the file **Uniform.ftm** for an example.

b) In Fathom™, use the **RandomNormal(** function. Go to *www.mcgrawhill.ca/links/MDM12* and select **Study Guide.** Follow the links to obtain a copy of the file **Normal.ftm** for an example.

35. **a)** As x increases, you would expect Pi(x) to increase.

b)

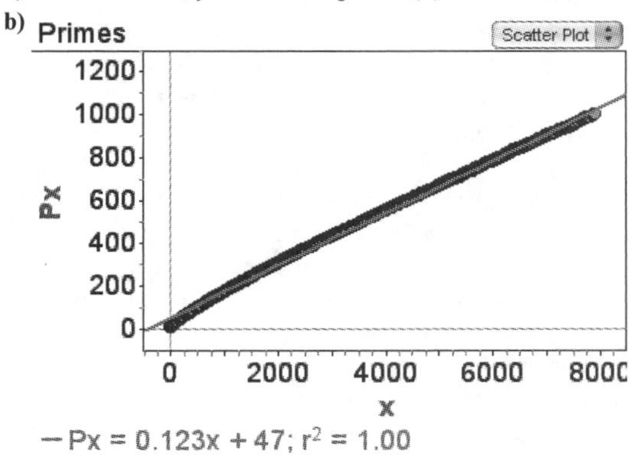

c) The data appear to be linear, with a strong positive correlation. Although $r^2 = 1.00$, it is clearly not perfect, since there seem to be more data below the line of best fit at either end.

d) 123 047

e) This relationship seems to have overestimated the number of primes. This may be due to the fact that the relationship is not really linear.

f) The fact that there is a pattern in the residuals indicates that the relationship is non-linear.

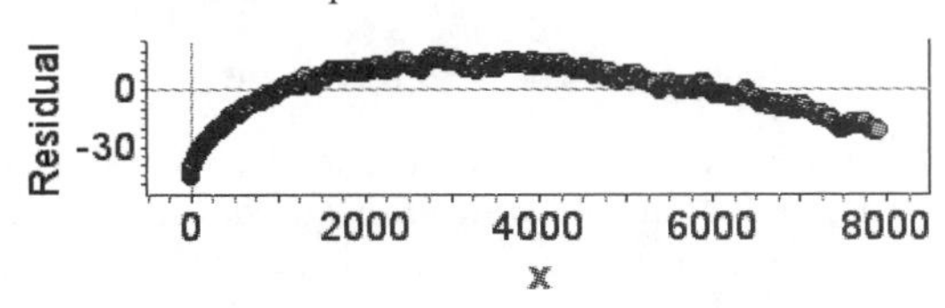

36. **a)** With 12 characters there are 5.6 billion names, but with 13 characters there are 11.5 billion names.

b) With 17 characters there are 9.9 billion names.

c) The actual number of characters, if letters can be repeated no more than twice, must be between 13 and 17.

37. **a)**

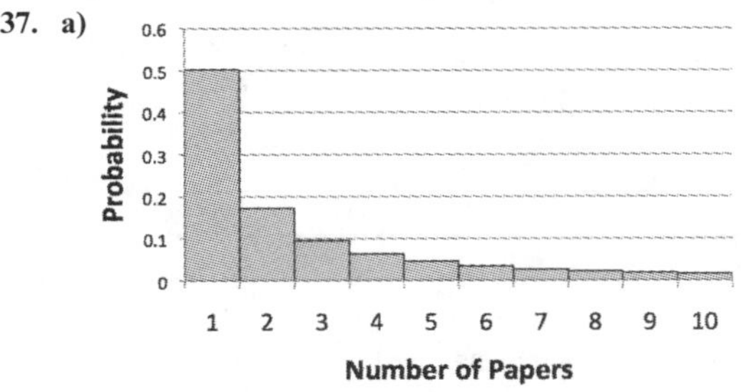

b) $E(X) = 2.5$ **c)** $E(X) = 8.4$

d) He was a Hungarian mathematician who published more papers than any other mathematician. An Erdös number indicates how many people away a person is from publishing a paper with Erdös. For example, if you published a paper with Erdös, your Erdös number is 1. If you published a paper with someone who had published a paper with Erdös, your Erdös number is 2, and so on. A Bacon number represents how many people you are away from the actor Kevin Bacon. If you acted in a movie with Bacon, your Bacon number is 1, if you acted in a movie with someone who acted in a movie with Bacon, your Bacon number is 2, and so on. A person's Erdös-Bacon number is the sum of the person's Erdös and Bacon numbers. One of the more famous people to have one is Natalie Portman, who starred in the more recent Star Wars films.

38. **a)** With a large list of numbers, the first digit should be approximately evenly distributed among all the digits.

b) The distribution is uneven; 1 appears as the first digit more often than any other digit.

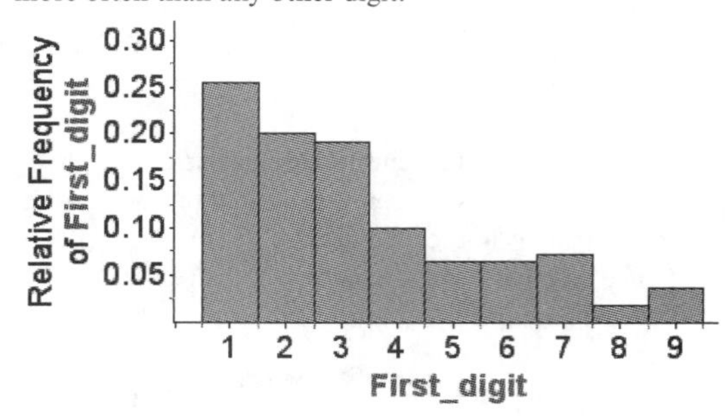

c) Although the export data do not match the frequency percents in Benford's law, in general the distribution starts with 1 as the most common digit, and the frequencies of the other digits decrease from there, roughly in order.

	Frequency (%)	
Digit	**Benford's Law**	**Car Exports**
1	30.1	25.4
2	17.6	20.0
3	12.5	19.1
4	9.7	5.0
5	7.9	6.4
6	6.7	6.4
7	5.8	7.3
8	5.1	1.8
9	4.6	3.6

d) Sample answer: One use is to detect accounting fraud. Often when fraud is suspected, the numbers in the accounting books are matched against Benford's law to see if a similar distribution occurs.

39.

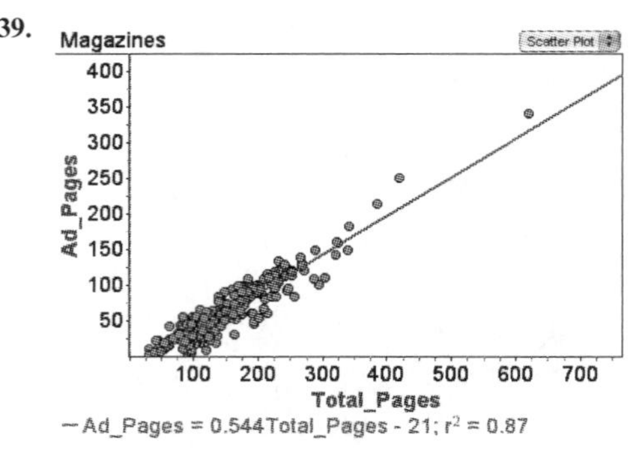

a) Residuals box plot with outliers:

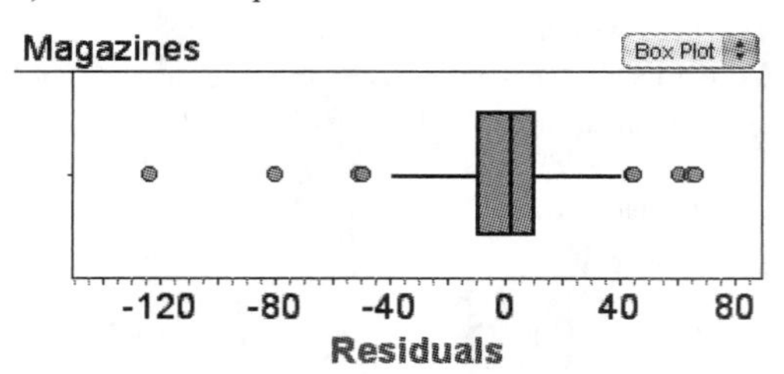

Scatter plot without outliers:

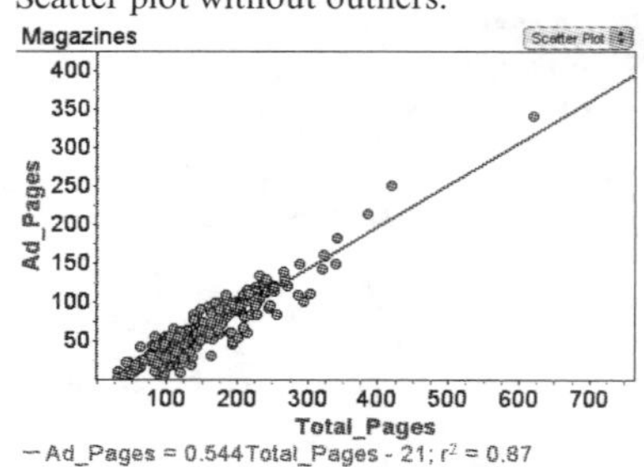

b)

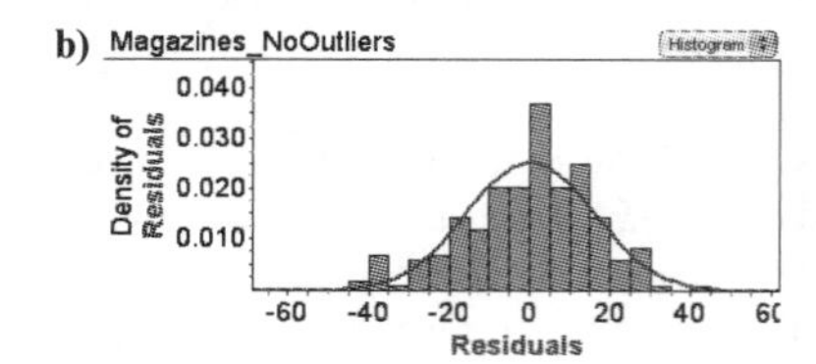